MOORES ROWLAND'S
TAXATION OF FARMERS
AND FARMING

AUSTRALIA AND NEW ZEALAND
The Law Book Company Ltd.
Sydney: Melbourne: Perth

CANADA AND U.S.A.
The Carswell Company Ltd.
Agincourt, Ontario

INDIA
N.M. Tripathi Private Ltd.
Bombay
and
Eastern Law House Private Ltd.
Calcutta and Delhi
M.P.P. House
Bangalore

ISRAEL
Steimatzky's Agency Ltd.
Jerusalem: Tel Aviv: Haifa

PAKISTAN
Pakistan Law House
Karachi

MOORES ROWLAND'S
TAXATION OF FARMERS AND FARMING

BY

NIGEL EASTAWAY, F.C.A., F.C.C.A., F.C.M.A., F.T.I.I., F.H.K.S.A., F.T.I.H.K., F.C.I.S.

AND

JOHN JEFFREY-COOK, F.C.A., F.C.I.S., F.T.I.I.

LONDON
SWEET & MAXWELL
1989

Published in 1989 by
Sweet & Maxwell Limited of
South Quay Plaza, 183 Marsh Wall, London E14 9FT
Computerset by Promenade Graphics Limited, Cheltenham
Printed in Great Britain by
Butler and Tanner Limited, Frome, Somerset

British Library Cataloguing in Publication Data
Eastaway, Nigel
Moores Rowland's taxation of farmers and farming.
1. Great Britain. Agricultural industries.
Farms. Taxation
I. Title II. Jeffrey-Cook, John
336.2'07

ISBN 0–421–41840–0

PREFACE

This book explains how the taxation rules apply specifically to farmers and landowners in sufficient detail to deal with most situations that arise in practice. Although it is likely to appeal to accountants, lawyers and taxation advisers who have farming clients, we have nevertheless avoided technical jargon so that those farmers who wish to know the taxation rules within which they have to work, knowingly or unknowingly, can follow our suggestions and the reasoning behind them. We believe that the taxpayer who has some basic understanding of the way in which the tax system affects him is likely to obtain much better value from his professional advisers than if his mind is closed to the rules under which the government takes a substantial proportion of his profits.

Farming is an area where it is particularly easy for the non-expert to be baffled by the terminology used so we are particularly pleased to have the Inland Revenue's permission to include the glossary of farming terms and conversion tables which inspectors of taxes use.

Being tax practitioners rather than farmers we are particularly indebted to John Padfield and Andrew Murray Threipland for their comments on farming aspects. We also thank Jennifer Ainsworth, Stephen Burwood, Chris Chadburn, Mike Gee, John Hunt and Jeremy Mitchell who gave us the benefit of their practical experience in dealing with tax matters for farming clients and Ronnie Ludwig, Edward Magrin and Graham Pringle who all provided useful information.

Finally we thank Ann Eastaway for her work as librarian, Celia Duncan for typing the manuscript and Richard Parry for helping to check the proofs.

The law and practice is stated as it stood at August 1, 1989, and includes changes made by the Finance Act 1989.

CONTENTS

CONTENTS

Contents

CONTENTS

TABLE OF CASES

xiii

TABLE OF STATUTES

TABLE OF STATUTORY INSTRUMENTS

LIST OF ABBREVIATIONS

A.H.A.	Agricultural Holdings Act
C.A.A.	Capital Allowances Act
C.G.T.	Capital Gains Tax
C.G.T.A.	Capital Gains Tax Act
F.A.	Finance Act
G.R.A.	General Rate Act
I.C.T.A.	Income and Corporation Taxes Act
I.H.T.	Inheritance Tax
I.H.T.A.	Inheritance Tax Act
L.G.F.A.	Local Government Finance Act
T.M.A.	Taxes Management Act
V.A.T.	Value Added Tax
V.A.T.A.	Value Added Tax Act

CHAPTER 1

Introduction

Farming

As this book is aimed at the accountant, solicitor or tax adviser as well as   **1–001**
at the farmer and landowner, a brief description of the main farming activities and the way in which they operate may be helpful.

The vast majority of farming in the U.K. is centred on cattle, sheep, pigs, poultry and crops (mostly cereals) and to a much lesser extent fruit and vegetables. Some farms engage in a variety of activities, others specialise in one or more particular areas.

Farming is big business with some three-quarters of the land area of England and Wales devoted to farming. About 30 million acres are under crops and 15 million are rough grazing, spread among some 240,000 farm holdings and 180,000 farmers employing more than half a million people.

Cattle

Cattle are kept either for their milk or to be slaughtered as beef.   **1–002**
Although a dairy cow at the end of her economic life is normally sold for meat, the prime purpose of rearing is for the milk production during her active life.

A cow is an animal, not simply a machine on four legs for turning grass into milk. She will, in the first instance, only produce milk after giving birth to a calf.

Newborn dairy calves are fed the mother's first milk or colostrum to build up immunity to disease but are weaned from the mother after 24 hours and weaned from milk powder by six to eight weeks. They are fed on milk powder followed by hay or silage and compounded calf foods known as concentrates.

Normally, within six to nine months of being born, the calf is turned out to grass, being housed again during the following winter.

Nowadays, it is increasingly common for a heifer to produce her first calf at about two years old and rarely after she is three years old. A calf may be conceived from either natural service by a bull or more commonly by artificial insemination or by embryo transfer; a heifer may not conceive first time.

At the end of the nine-month gestation period the heifer produces a calf.

1

After producing her second calf the heifer is then known as a cow. She will be inseminated again as close to two months after calving as possible. Then, at the end of her first lactation—normally of 10 months, with an additional two-month dry (rest) period—she produces her second calf, and so on.

Milk production

1–003   Modern feeding and breeding enable increasing yields of milk to be produced. A cow reaches peak production within three to four months of calving and then gradually her output declines. A good average yield for a cow is between 5,500 and 6,500 litres per lactation of 305 days. Normally, a cow produces milk for about five lactations—sometimes considerably more—after which she is sold for meat.

Milking cows are normally out to grass from mid-April to the end of September and then are housed indoors for the winter period. The winter ration is normally forage based, that is, hay, or more commonly, silage. Concentrate feeds, either straights or compounds, are fed alongside, the quantity depending on the yield required and how much milk is produced from the other feeds available. A relatively new approach is being considered with the introduction of B.S.T., a natural hormone the cow produces herself which can further stimulate milk production. Cows are always milked twice, and sometimes three times, a day.

Any surplus dairy calves and heifers are sold to other producers. In the main, the male Friesian and beef-cross calves are sold to specialist rearers for beef. Some producers breed all their cows with a beef bull, sell the beef-cross calves, and buy heifers in calf as replacements. Other farmers put the better half of the herd to a dairy bull and the remainder to a beef bull because only about 20 to 25 per cent. of the herd need be replaced each year. Of the dairy cows on a farm 95 per cent. may well have a calf each year.

The most popular dairy cow is the British Friesian, with a good milk yield and readily saleable surplus calves because the male calves make reasonable beef animals. Other breeds such as, for example, Jersey, have premium quality milk, but are less suitable for beef.

1–004   A dairy farmer wishing to sell milk has to register with the Milk Marketing Board (MMB) which is a farmers' co-operative, not a governmental organisation, but has a statutory monopoly on the trading of milk. Milk is sold to the MMB under a wholesaler's milk contract. The milk from the cows is stored under refrigerated conditions on the farm until collected by the MMB or its contractors. The amount paid for the milk depends on its protein and fat content as well as quantity. A farmer can obtain from the MMB a retail licence enabling him to sell direct to consumers and caterers and to sell farm-bottled milk to dairymen. Small scale producers may sell direct to registered producers or to consumers and caterers, but not to local dairies, under an exemption certificate.

1–005   The MMB as a statutory monopoly negotiates a price for milk with the Dairy Trade Federation which is made up of dairy retail companies. There is a differential price system, in that milk for liquid consumption is sold at a higher price than that used in manufacturing. The Board subtracts administration and haulage costs, and this determines the price that milk producers receive. The MMB pricing policies are strictly controlled by European Community regulations, which are designed to prevent the use of its monopoly position to disadvantage consumers and the dairy trade.

A registered producer-retailer has to keep detailed and complicated records to enable the MMB to run its differential price system.

The MMB also sells its milk testing services, farm management advice and consultancy services, as well as artificial insemination, mastitis control and a number of farm management schemes. It also acts as a monitoring agency for the Ministry of Agriculture under the Brucellosis Eradication Scheme.

Since March 1984, the MMB has operated milk quotas on behalf of the **1–006** Ministry of Agriculture, Fisheries and Food. These quotas are fixed by the EC under the Common Agricultural Policy, regulations 856 and 857 of 1984. They are applied to the U.K. under the Dairy Produce Quota Regulations 1984 (S.I. 1984 No. 1047). Under the milk quota scheme, farmers who produce more milk than their quota run the risk of paying a levy equivalent to 125 per cent. of the target price, so the consequences of overproduction are penal. However, the scheme allows shortfalls below quota to be allocated to those over quota, so producing a threshold percentage. Producers over quota pay levy only on the excess over this threshold. The threshold varies from year to year, but the more a farmer produces over quota the greater his risk of paying levy.

Milk quota can only be transferred with an interest in land, but the type of interest required may be no more than a grazing licence for six months. The amount of quota that can be transferred on each occasion can be up to 20,000 litres per hectare, with a minimum of five hectares per transaction. This in turn has meant that the Revenue regards the quota as effectively saleable as an asset on its own and would regard the sale of quota as giving rise to a chargeable gain on which roll-over relief is now available (see Chapter 15).

The introduction of milk quotas also gave rise to an outgoers' scheme whereby compensation is payable to dairy farmers who agree to discontinue milk production. The taxation consequences of the outgoers' scheme are considered in Chapter 8.

Beef production

Beef production starts with the purchase on birth of a calf which is then **1–007** reared, grown as a "store animal" and fattened for market. Farmers may deal with the entire range of production from breeding to final sale or may specialise in particular areas of production.

Beef cattle are either surplus calves from a dairy herd or bred specially for beef production. In hill farming areas, farmers tend to keep cattle purely for breeding for beef and a calf is reared naturally with its mother until it reaches the age of eight or nine months when it is sold to a specialist rearer; this is known as a single-suckler herd. In areas of better pasture the home-reared animals may be supplemented by week-old calves bought in so that each cow has two or three calves to feed; this is known as a multi-suckler herd.

In other cases, week-old calves are acquired from dairy herds and reared on substitute milk, fodder and concentrated food-stuffs until sold as store calves after three to six months, or as store cattle at nine to 12 months. Under a grass beef system animals are kept for 18 to 24 months and then sold for beef.

Store cattle at nine to 12 months are normally known as light stores and, if kept for a further period for fattening, are sold as medium or heavy stores or kept for final finishing.

1–008 Cereal beef is an intensive means of rearing cattle on energy concentrated food-stuffs indoors, under conditions which cause them to mature quickly so that they are sold as finished animals at 10 to 12 months. Such animals are particularly favoured by supermarkets and multiple butchers as the quality tends to be consistent.

The production of veal is a specialist occupation as the animals are kept indoors and intensively reared on a liquid diet for 12 to 18 weeks when they are sold for slaughter. In the U.K. this is normally carried out by grouping calves together and confining them to straw yards. To do this economically requires a lot of management, expensive food-stuffs and a large capital investment.

Cattle are normally sold in auction at a local market on a live weight basis as mature animals (fat cattle). Weight for weight, steers usually fetch rather more than heifers. Animals sold direct to a slaughterhouse or manufacturer are sold on the basis of the dressed carcass weight.

1–009 Although the EC runs an intervention policy for beef it is relatively unusual for beef to be sold into intervention directly by a farmer. Occasionally certain qualifying specialist dealers buy in bulk to put a floor to the market. The Ministry of Agriculture, Fisheries and Food, however, runs a special premium scheme under which a payment is made on a per head basis up to a maximum of 90 animals per producer per year and this is currently £29.19 per head. There is also the suckler cow premium scheme whereby payments of £33.40 per head are currently being made for all suckler cows that have been on a holding for a minimum of six months.

The Meat and Livestock Commission, financed by levies on the industry, promotes meat, collects data and engages in research and development.

Occasionally, sales of fat stock will be made to local butchers but this is relatively unusual because few butchers have their own licensed slaughterhouse or ready access to a public abattoir. Such sales are, however, not uncommon in certain areas.

Sheep

1–010 Sheep are normally kept for the production of lambs for slaughter and not milk, but wool is normally a useful by-product. In some countries, sheep are reared almost exclusively for the wool but this is unusual in the U.K.

Sheep mature relatively quickly and are normally used for breeding in their second year. Normally the ram runs with the flock, in the autumn, on the basis of one ram to 50 ewes. Artificial insemination is unusual. Ewes have a gestation period of about five months and frequently produce twins or triplets. Careful control of feeding and timing of mating can increase the proportion of twins in the flock. Male lambs are usually castrated to become wethers and are normally kept until finished for marketing, or sold as stores.

The sheep are often given additional food-stuff in the form of hay and finished for market on root crops.

Sheep are sometimes bred in intensive conditions, particularly by arable   **1–011** farmers, to produce early lambs for sale at about four to six months between March and May when prices are usually high.

Intensively reared sheep are kept indoors and fed on grain, roots and arable by-products. Although some farmers in favourable areas go for early lamb off the grass, it is more normal to arrange for lambing in February or March, or even April in the case of hill farms, and they are fattened off on the spring and summer pastures for sale between August and October. In some cases under-weight lambs are sold as stores in the spring after winter feeding or fattened with roots and food concentrates for sale in the autumn. Wintering lambs is often an off-season activity of arable farmers who can use otherwise under-occupied labour and arable by-products.

Hill sheep are bred in tough conditions and are sturdy enough to survive   **1–012** outside in all weathers. In many cases, however, hill farmers will not fatten lambs for market but sell them as stores for fattening in lowland areas; in the words of T. L. Peacock, "the mountain sheep are sweeter but the valley sheep are fatter."

Some farmers breed sheep as a specialist activity for sale to other farmers for rearing. Most farmers breed their own replacement ewes by holding back ewe lambs to replace elderly animals. If, as is often the case, the ram is of a different breed replacement, ewes will probably be purchased from specialist breeders rather than having a separate ram of the appropriate breed simply for the purpose of replacing stock animals.

First cross ewes, such as a cross between a hill ewe and a lowland ram, are much in demand as they tend to be prolific breeders of good quality lambs suitable for slaughter.

Sheep are usually sold either in the auction market as live animals or   **1–013** direct to large butchers and meat processors. Lambs sold direct to producers are usually calculated on the basis of a price based on the dressed carcass weight and, if sold in auction, on the live weight. As with beef, there is an EC support intervention scheme mainly for the trade. There is a sheep premium scheme and producers are paid at a current rate of £6.10 per head on all ewes that are in lamb and on the holding between January 1 and April 10 each year.

Wool production

Sheep are usually shorn before the summer period from April to July,   **1–014** and all wool is sold to a designated wool merchant allocated to the farmer by the Wool Marketing Board with whom he must register. The wool is paid for at a guaranteed price which is underwritten by the Government but not actively subsidised, because the Wool Marketing Board is expected to, and usually does, break even, with deficits cancelled by surpluses.

The quality of the wool can vary enormously and grading is a specialist art. The condition of the fleece is affected by the expertise of the farmer and is reflected in the price received.

Pigs

1–015   Pigs are usually intensively reared and there are two stages in production, breeding and fattening. Some farmers carry out both phases but others specialise. Pigs are transferred to a fattening unit as weaners normally after six to seven weeks but they may also be weaned earlier at about three weeks. A sow may normally be expected to produce on average about nine piglets at a time and will produce about two-and-one-quarter litters a year. A sow is normally slaughtered after producing four to eight litters. Most male pigs are castrated at birth.

Pigs are normally fattened for market and usually sold as porkers at about 17 to 18 weeks when they have a live weight of about 65 kilos, or as baconers at 23 weeks when they have a live weight of about 90 kilos. Pig farming used not to be particularly capital intensive and provided the classic illustration of a cyclical market, known to all students of economics as the hog/corn cycle. In this cycle good pork prices and cheap corn encourage marginal producers to come into the market; prices then fall, so marginal producers go out of business, then corn prices fall and pork prices rise, causing the cycle to start again.

Nowadays, however, pig farming has become a capital intensive business, although still cyclical.

Pigs are sold either direct to the slaughterer or manufacturer or through the livestock auction market. As with cattle and sheep, the price is normally quoted by reference to the live weight for auction sales and dressed carcass weight for sales to slaughterers and manufacturers.

Although there is an EC intervention scheme for pigs it is unusual to find a farmer selling directly into intervention.

The days of pigs roaming around the farm and being fed with swill are largely over so far as commercial production is concerned.

Poultry

Turkeys

1–016   Commercial turkey production is highly intensive and a very small number of firms account for over 85 per cent. of the market.

Some farmers, however, rear a number of turkeys for the Christmas trade. It takes about 12 weeks to breed a 16 lb. bird, with rather less for a light-weight bird and longer for a heavy one.

Ducks

1–017   As with turkeys, most commercially available ducks are intensively reared by specialist producers and sold direct to the food processors. Some farmers rear a few ducks on a casual basis but this is unusual.

Chickens

The main U.K. market for poultry is in chickens which are reared either **1–018** for eggs or for slaughter.

For egg production, chickens are usually put into a battery system at about 20 weeks and slaughtered for meat after 65 weeks laying, during which they would normally produce about 270 eggs. The most advanced units are highly capital intensive with food channelled in and eggs channelled out on conveyor belts. A rather less capital intensive method is the deep litter system under which the chickens are free to roam within a building but the egg production is likely to be little over 200 eggs per bird a year. This however is better than the 150 or so eggs a year likely to be obtained on a free-range basis, although feeding costs are lower and capital costs much lower for free-range chickens.

Most chickens produced for meat are intensively reared in a deep litter system in heated sheds and sold at a live weight of 4 lb. to 5 lb. after six weeks or so.

Chickens are generally reared for factory processors, or packers as they are sometimes known, under direct contract. The processors, who have a major investment in plant, deal with the chickens in a variety of ways depending on the state of the market.

Generally the chickens, bought live from the producers, are first killed, plucked and gutted. They are then onsold either fresh, frozen, or processed into prepared chicken joints or precooked birds.

Eggs have to be graded in conformity with the Eggs (Marketing **1–019** Standards) Regulations and about half the U.K. production is handled by specialist egg packers who collect and pack the eggs for distribution to wholesalers, retailers and processors.

About a quarter of the production is sold direct to retailers and the remaining quarter is sold direct by the producer who may well have his own packing and storage facilities as well as merely grading.

There are no EC intervention regulations dealing with eggs or poultry although for a time the U.K. did, unilaterally, ban the import of eggs.

Arable farming

The farm census return form divides crops into cereals for threshing: **1–020** wheat, winter barley, spring barley, oats, mixed corn and rye; maize for threshing or stock feeding; potatoes; sugar beet other than for stock feeding; hops and horticultural crops; crops for stock feeding—excluding maize—divided between beans, turnips and swedes, fodder beet and mangolds, rape or cole, kale, cabbage, savoy and kohlrabi and other crops, then on to rape grown for oil seed, lucerne, other crops, grass, clover and sainfoin and grass grown for seed.

The crops which the farmer chooses to grow depend very much on the part of the country where his farm is and the type and condition of the soil. Crop rotation is still common but large farms may use a combination of artificial fertilisers, pesticides and herbicides to avoid or reduce rotation. Traditional (Norfolk Course) rotation involving, for example, wheat, turnips, barley, grass and back to wheat or variations on such a theme is rarely practised, a more modern rotation being wheat, barley, rape. The arable farmer has to till the soil through ploughing, cultivating and harrowing but modern methods have considerably reduced the area of ploughed land. The crop is then sown, weeded, sprayed and fertilised as necessary and in due course harvested. The equipment used by a large modern arable farm is highly sophisticated and extremely expensive. A large farm with good equipment can be managed with only a handful of skilled employees.

Much of the production of an arable farm is used for animal food-stuffs and some such farms also have their own livestock.

Cereals

1–021   The Common Agricultural Policy seeks to manage the price of cereals both by surcharging imports and subsidising exports to and from the EC; it also buys cereals of acceptable quality into intervention stores. Normally the farmer sells to a grain merchant and the merchant in turn sells to the manufacturers or into intervention in accordance with the demands of the market. Cereal production is currently under review and constraints to avoid over-production may be imposed, as with milk and potatoes, and some arable land is likely to be taken out of production.

Some farmers sell the harvest as soon as it has been gathered in, others have facilities to store the crop for a period. Not unnaturally, grain prices tend to be lowest immediately after the harvest and increase until the next harvest is available. Less cereals are now sold on the spot market, direct to a local grain merchant. Forward sales can be made and there are still a number of corn exchanges of which the most important is in Mark Lane in London. Although the price available for a forward sale is likely to be higher than for a spot sale, the farmer has to take into account the costs of storage and interest on money tied up in stocks.

More commonly today farmers join farming co-operative groups for the marketing of their grain, with the benefit of bulk sales and greater influence on the market.

Some farmers sell locally to other farmers small quantities of grain as animal feeding stocks, but there are restrictions on what may be sold.

Potatoes

1–022   Potato production is controlled by the Potato Marketing Board and farmers who plant potatoes as a cash crop must register with the Board which controls production. The crop is sold to licensed merchants at a firm price and there is a guaranteed price under the Potato Marketing Scheme for purchases by the Potato Marketing Board. Potato quotas are valuable licences which may be transferred at a premium and qualify for roll-over relief for capital gains tax. About 40 per cent. of the potato crop is used for processing into crisps, pre-prepared chips, etc., and is usually sold direct to the manufacturer. Farm-gate sales of potatoes are not controlled but only one tonne of potatoes may be sold to local retailers direct without a licence. Early potatoes as opposed to main crop are not controlled and are sold direct to the merchants, retailers or public. The sale of seed potatoes is closely controlled to prevent disease.

Sugar beet

Sugar beet is sold in the U.K. to the British Sugar Corporation on terms **1–023**
agreed with the National Farmers Union. The price is based on a 16 per
cent. sugar content and is increased or reduced as appropriate for variations from the norm. The price paid depends on a number of factors,
some of which have to be estimated in advance. The price is therefore paid
on a estimated basis in the first instance and adjusted subsequently as
necessary. In addition a transport allowance is paid to growers. There are
also bonuses for early or late delivery to try to spread the harvest.

Fruit and vegetables

Some arable farmers grow vegetables on a large scale and may also have **1–024**
orchards for top fruit, such as apples and pears, or soft fruit such as rasp-
berries, strawberries, gooseberries, etc. Some farms have glass-house
crops such as tomatoes and cucumbers but these are normally confined to
specialist market garden operations. The growing of flowers and the operation of nurseries and garden centres is normally a specialist occupation
and is seldom carried on by a farmer.

Much horticultural produce is sold on commission to wholesalers rather
than at a fixed price. The wholesalers normally charge between 7.5 and 10
per cent. Some producers have set up wholesaling co-operatives to take
their produce and they may be substantial businesses in their own right or
merely joint marketing arrangements. Sales to manufacturers for processing are normally by direct contract. It is not uncommon for processors to
agree in advance with growers the area to be planted and the price at which
the product will be sold.

Sales of horticultural produce may also be made through farm shops and
a number of farms organise a pick-your-own arrangement whereby customers pick soft fruit and vegetables direct from the plants. Some farms set
up temporary stalls for the sale of produce if they are suitably located close
to a main traffic flow and, in some cases, sales are merely made on a casual
basis from the back door.

It should be appreciated that this brief summary merely sets out in very
general terms the main areas of farming activities in the U.K. and individual cases may differ substantially.

Constitution

Farming in the U.K. is carried out in a number of different ways, for **1–025**
example, by the farmer as landlord or tenant of the farmland trading personally in his own name as a sole trader. It is common practice to involve
other members of the farmer's family in running the farm and there are
many farming partnerships governed by the Partnership Act 1890. They
may be simple partnerships of the farmer and his wife or they may also
involve, as partners, other members of the family such as brothers or sis-
ters and their spouses, children, nephews, nieces, etc. It is, of course, possible to have a farming partnership between unrelated parties and this is

becoming more common in order to retain vacant possession and avoid a tenancy under the Agricultural Holdings Acts.

1–026 Farming may be carried on by a company, which may own the land, have a full lease of the land as an agricultural holding, or merely have a licence to occupy. It is unusual, but by no means unheard of, for farming to be carried on by a trust in which the trustees actually carry on the farming activities. The fear that the trustees may rightly have for their own position if the farming were to prove commercially unsuccessful could perhaps be overcome by setting up a trust of which the trustees were two limited companies with no other assets so that the farming liabilities were effectively limited to the farming assets while still preserving the taxation advantages of a trading trust.

It is not uncommon for trustees to own farmland under a will trust, for example where husband and wife farm in partnership and each leaves his or her share to the spouse for life. This can give rise to a problem in that roll-over relief for capital gains tax would not be available to the trustees if the farm were subsequently sold.

Farming co-operatives are not unknown although in most cases they cover specific activities related to the farming such as marketing the produce, bulk purchase of supplies, or the provision and use of farm equipment.

1–027 Farming lends itself to joint ventures of many descriptions which do not amount either to a tenancy of the farm or a partnership. Such arrangements vary from straightforward management agreements under which the farmer enters into an arrangement with an independent contractor to manage the farm, to sharecropping arrangements under which the landlord provides the land and raw materials and the contractor the labour, plant and equipment. Agisting and grazing agreements are relatively common, as are arrangements for seasonal cropping of, for example, vegetables. Each of these joint ventures needs to be considered carefully if the intended results are to arise, as it is otherwise very easy to end up with a partnership arrangement under which each partner would be jointly and severally liable for the partnership debts, or an arrangement giving the occupier of the land a right of occupation under the Agricultural Holdings Act 1986 or the Agriculture (Miscellaneous Provisions) Act 1976.

The advantages and disadvantages of these various arrangements from a taxation point of view are considered in Chapter 2.

Taxation

1–028 Farming is a trade and the farmer therefore has to keep books and records from which accounts may be prepared, forming the basis of the taxation computation from which the farmer's tax liability is calculated.

The profits shown by the accounts will normally require adjustment in accordance with the taxation rules in order to arrive at an adjusted profit for tax purposes. Items such as depreciation, entertaining, private consumption by the farmer or his family, capital expenditure and other expenses shown in the accounts are disallowed for tax purposes. Certain items of income such as rents, deposit interest or other investment income are excluded from the adjusted profit for tax purposes as these are taxed under different provisions of the tax legislation from those that apply to the farming profits. On the other hand, deductions for capital allowances are likely to be available.

When the adjusted profit has been arrived at the Revenue raises an assessment and the tax becomes due and payable on specified dates. If the Revenue has not agreed the accounts in time to raise the assessment on agreed figures it raises an estimated assessment against which the taxpayer must appeal within the appropriate time-limit.

Farmers carrying on business as sole traders or partnerships are taxed on **1–029** what is known as the preceding year basis, under Schedule D, Case I. Under these provisions the accounts for the accounting year ended in the preceding tax year form the basis of the assessment for the current year. The tax year is the year ended on April 5. This somewhat peculiar date arises from the changeover from the Julian to the Gregorian calendar in 1752 as a result of which 11 days were lost. The taxpayers not unnaturally objected to paying a year's tax for 354 days so these missing days were added to the old tax year end which was Lady Day, March 25, one of the four traditional quarter days in England.

If, for example, the farm accounts are made up to September 30, the accounts for the year ended September 30, 1989 would be for an accounting year ending in the tax year 1989–90 and would therefore form the basis of the assessment for the tax year 1990–91. Obviously the preceding year basis of assessment only works where there are accounts available for the preceding year and there are therefore special rules for the opening years under which the opening profits are assessed more than once and as a corollary, on cessation, certain profits fall out of assessment. As will be explained in Chapter 2 these rules can, on occasions, legitimately be used to the taxpayer's benefit.

Companies are not assessed on a preceding year basis but are liable to corporation tax on the adjusted profits of the accounting period. If accounts are made up for a period of more than 12 months this is split for tax purposes into two accounting periods: the first of 12 months and the second of the balance.

Most farmers need help to run their farms so they may well have **1–030** employees who are paid a wage or salary for their services. An employee enters into a contract of service with his employer and it is important to distinguish this from the self-employed contractor who enters into a contract for the provision of specific services. The farmer has no responsibility for the taxation liability of an independent contractor but in the case of an employee must withhold tax at source from any remuneration paid to the employee. This is done under the pay-as-you-earn system which must be operated for all full-time and part-time employees paid more than a very low basic amount.

The employer is also liable for the national insurance contributions of employees other than certain part-time employees. Part of this amount, the employee's contribution, is deducted from the wage or salary paid and the balance represents the employer's contribution which is an additional cost to the employer of having an employee.

The employer also has responsibilities under the statutory sick and maternity pay schemes and redundancy payment scheme to make various payments to employees, part of which may be recovered from the government. The farmer, as employer, may also set up or contribute to pension arrangements for the employees.

Value added tax

1–031   A large part of the sales of most farms is zero rated for V.A.T. purposes although items such as fees for the hire of equipment or contract services are standard rated. This means that a farm is often exempt from registering for V.A.T. purposes if its input tax would always exceed its output tax (C & E Leaflet 700/1/87, para. 7). However, many of the goods and services bought by a farm are subject to V.A.T. at the standard rate and in most cases it will be beneficial to register for value added tax purposes and to claim back the excess of V.A.T. input tax over output tax. Monthly V.A.T. repayments can be claimed on the grounds that input tax normally exceeds output tax, as explained in Chapter 19.

Capital gains tax

1–032   In the case of farming it is particularly common to have partial disposals of small parcels of land and to acquire additional areas of farm land, often on a piecemeal basis. In many areas farmers sell barns for conversion into houses. These barns realise a high price either sold as a building site or developed in conjunction with a developer. The tax liability arising on such disposals is dealt with in Chapters 15 and 16 and where land is sold for development the additional tax rules that have to be considered are dealt with in Chapter 17.

Clearly a disposal of the business outright will attract a potential C.G.T. liability but it is not always appreciated that a change in a partnership asset sharing ratio, which is normally the same as the sharing ratio for trading profits, can give rise to a capital gains tax charge on the partner whose share reduces and the acquisition of a chargeable asset by the partner whose share increases. This applies even though there is no disposal of the underlying farm assets and consequently no cash available to pay the tax.

Inheritance tax

1–033   One of the major problems of farming is the fact that the capital value of farmland is high in relation to its income earning capacity. This means that a business earning a modest income can produce a major I.H.T. charge on the death of the owner, and passing down the family farm can give rise to considerable problems in practice. Some of the ways in which inheritance tax can be alleviated, if not eliminated, are considered in Chapter 18.

Stamp duty

1–034   There is no stamp duty on gifts transferring farmland and other assets into trust or direct to members of the taxpayer's family during life. It remains a substantial cost on sale and it would appear that the more ingenious schemes to avoid stamp duty are unlikely to be effective.

In the case of a company, and in particular on a transfer of assets to a **1–035** company, it is essential to consider stamp duty. Although charged at only one per cent. it is, nonetheless, a material sum of money where the capital value of assets being transferred is substantial.

Rates

Although rates may justly be regarded as a form of local taxation it is **1–036** nonetheless a matter over which the taxpayer has little control except by way of challenge to a valuation for rating purposes. Domestic rates are being replaced by the community charge but business rates are explained in Chapter 20.

Tax avoidance

It is important to avoid, in the course of tax planning, any composite **1–037** transaction or series of transactions with an interposed step which is inserted purely for tax avoidance purposes. An early and famous attempt at the avoidance of taxation was that of the Duke of Westminster who decided to pay his gardener an annuity under a deed of covenant, which at that time was a deduction for tax purposes. This was held to be a perfectly reasonable form of tax mitigation in *I.R.C.* v. *Duke of Westminster* (1935) 19 T.C. 490. More recently however a series of cases, in particular *Ramsay* v. *I.R.C.* [1981] S.T.C. 174, *I.R.C.* v. *Burmah Oil Co. Ltd.* [1982] S.T.C. 30 and *Furniss* v. *Dawson* [1984] S.T.C. 153 have led to a radical change in the interpretation of tax legislation. This was spelt out by Lord Diplock in the *Burmah* case (at [1982] S.T.C. 32):

> "It would be disingenuous to suggest and dangerous on the part of those who advise on elaborate tax avoidance schemes to assume that Ramsay's case did not mark a significant change in the approach adopted by this House in its judicial role."

This was reinforced by Lord Scarman (at [1982] S.T.C. 39):

> "It is of the utmost importance that the business community (and others, including their advisers) should appreciate, as my noble and learned friend Lord Diplock emphasised, that Ramsay's case marks a significant change in the approach adopted by this House in its judicial role towards tax avoidance schemes."

This decision left the law uncertain and this uncertainty was emphasised **1–038** by Lord Scarman in *Furniss* v. *Dawson* (at [1984] S.T.C. 156):

> "The law in this area is in an early stage of development. Speeches in your Lordship's House . . . are concerned more to chart a way forward between principles accepted and not to be rejected than to attempt anything so ambitious as to determine finally the limit beyond which the safe channel of acceptable tax avoidance shelves into the dangerous shallows of unacceptable tax evasion. The law will develop from case to case . . . what has been established with certainty by the House in Ramsay's case is that the determination of what does and what does not constitute unacceptable tax evasion is a subject suited to development by judicial process."

The trouble however with judicial process is that it is a time consuming operation. In *Furniss* v. *Dawson* the final tax transaction took place on December 20, 1971 and the judgment of the House of Lords was on February 9, 1984. More recently the House of Lords in *Craven* v. *White, Bayliss* v. *Gregory* and *I.R.C.* v. *Bowater Property Developments Ltd.* [1988] S.T.C. 476 held that strategic tax planning under which transactions take place with a view to saving tax only if a further disposal or transaction take place is an acceptable form of tax avoidance. Tax planning is not acceptable where the ultimate disposal is already under negotiation, as this may mean that there is a preordained series of transactions or composite transaction.

CHAPTER 2

Choice of Trading Medium

Sole trader

Advantages

One of the major advantages of trading as a sole trader is the simplicity **2–001** compared with other forms of organisation, particularly compared with a company. Only the most elementary records are necessary to prepare a profit and loss account and a balance sheet and an audit of the accounts is not statutorily required. It should be emphasised, however, that the records should be sufficiently comprehensive to be able to identify receipts and payments and support them with invoices or other documentation, as a sole trader's accounts are much more likely to be subject to a detailed investigation by the Revenue than are those of a limited company. As the onus is on the taxpayer to prove that his accounts are correct and complete, it is important to ensure that the records are sufficient to enable this to be done to the satisfaction of the inspector of taxes and if necessary the General Commissioners. Although this may require little more than a cash book and files of paid and unpaid invoices, it is surprising how often even this limited information is not available.

Compared with a company, a sole trader or partner has the advantage that on a disposal of land or other chargeable assets there is only a single charge to C.G.T., whereas in the case of a company there is a charge on the gain made by the company and, to the extent that the net gain increases the value of the shares in the company, there is a further C.G.T. charge if the shares are disposed of. Retirement relief for C.G.T. purposes is also more likely to be available for a sole trader or partner than for a shareholder in a company, although these aspects are dealt with in more detail in Chapter 15.

Preceding year basis

When a trader first commences in business his profits are assessed for the **2–002** first tax year on the basis of the proportion, calculated on a time basis, of the adjusted profits from the date of commencement to the following April 5: I.C.T.A. 1988, s.61. The assessment for the second tax year is on the first 12 months from the date of commencement, again apportioning accounts profits, as adjusted for tax purposes, on a time basis where necessary.

Except in the unusual case where the business actually commenced on April 6, the accounts for a 12-month period ending in the preceding tax year (usually the first year's accounts) will be used as the basis of the assessment for the third tax year.

2–003 It will be appreciated that profits for the first 12 months may form the basis of assessment for the first three tax years and it is obviously desirable to keep the adjusted profits for this period as low as possible. It might, therefore, be sensible to keep a potential partner, such as a spouse, as an employee in the first year so that the remuneration paid is treated as a deduction for tax purposes and effectively is relieved two to three times, although the actual income paid, which is assessed as earnings under Schedule E, would only be assessed once on a current year basis. Similarly it could be beneficial to lease plant and machinery rather than buy it outright so that the rental charge reduces the profits. It might even be possible to rent premises for a period before outright purchase but this is unlikely to be workable in practice in the case of a farm, owing to the Agricultural Holdings Acts.

If the profits for the first period are reduced to the minimum in this way the profits assessed for the second and third years following commencement should be materially less than the actual profits.

2–004 EXAMPLE 1

Opening years—basis of assessment

Profits for the year ended April 30, 1988	£5,000
Profits for the year ended April 30, 1989	15,000
Profits for the year ended April 30, 1990	18,000
	£38,000

Assessments:

1987–88 Actual profits to April 5, 1988	
£5,000 × 11/12ths	£4,583
1988–89 Profits for first 12 months	5,000
1989–90 Profits in preceding year	5,000
Total profits assessed in the first 3 years	£14,583

If, for some reason, it is not possible to plan the profits in this way, and the actual profits apportioned on a time basis for the second and third tax years are less than the normal opening year assessments, the taxpayer can elect under I.C.T.A. 1988, s.62 for the assessments for both the second and third tax years to be adjusted to the actual profits arising.

EXAMPLE 2   **2–005**

Opening years—basis of assessment

Profits for the year ended April 30, 1988	£15,000
Profits for the year ended April 30, 1989	10,000
Profits for the year ended April 30, 1990	12,000
Total profits earned in first 3 years	£37,000

Assessments: Normal Basis

1987–88 Actual profits to April 5, 1988	
£15,000 × 11/12ths	£13,750
1988–89 Profits for first 12 months	15,000
1989–90 Profits in preceding year	15,000
Total profits assessed in the first 3 years	£43,750

Assessments: Election made under I.C.T.A. 1988, s.62

1987–88 Actual profits to April 5, 1988		
£15,000 × 11/12ths		£13,750
1988–89 Actual profits year to April 5, 1989		
£15,000 × 1/12th	1,250	
£10,000 × 11/12ths	9,167	10,417
1989–90 Actual profits year to April 5, 1990		
£10,000 × 1/12th	833	
£12,000 × 11/12ths	11,000	11,833
Total profits assessed in the first 3 years		£36,000

On cessation of trade, for example on the introduction of a partner into the business, there is an automatic cessation for tax purposes (subject to an election for continuation, see under partnerships below) which means that the profits in the final year from April 6 to the date of cessation form the basis of assessment for the final year. This means that the normal preceding year basis of assessment for that year is displaced and profits fall out of assessment. Which profits fall out of assessment will depend in the first instance on the date to which the trader makes up his accounts. If, for example, the accounting date is September 30 and the cessation takes place on say, June 30, 1989, the assessment for 1989–90 would be based on the profits from April 6 to June 30, 1989. The assessment for 1988–89 on the normal preceding year basis of assessment would be based on the accounts for the year ended in 1987–88, *i.e.* September 30, 1987. Therefore the profits from October 1, 1987 to April 5, 1989 drop out of assessment and are tax free.

2–006   EXAMPLE 3

Closing Years—Basis of assessment

1. Year end September 30; cessation June 30, 1989.

Profits year ended September 30, 1986	15,000
Profits year ended September 30, 1987	18,000
Profits year ended September 30, 1988	25,000
Profits 9 months ended June 30, 1989	12,000
	£70,000

1987–88 Profits to September 30, 1986	15,000
1988–89 Profits to September 30, 1987	18,000
1989–90 Profits April 6 to June 30, 1989	
£12,000 × say 3/9ths	4,000
	£37,000

Profits falling out of charge	£33,000

If, however, the business made up accounts to April 30 the 1988–89 assessment would be based on the accounts for the year ended April 30, 1987 ending in 1987–88. Therefore profits from May 1, 1987 to April 5, 1989 would drop out of assessment on cessation.

2–007   EXAMPLE 4

Closing Years—Basis of assessment

2. Year end April 30; cessation June 30, 1989.

Profits year ended April 30, 1986	15,000
Profits year ended April 30, 1987	18,000
Profits year ended April 30, 1988	25,000
Profits year ended April 30, 1989	12,000
Profits 2 months ended June 30, 1989	3,000
	£73,000

1987–88 Profits to April 30, 1986		15,000
1988–89 Profits to April 30, 1987		18,000
1989–90 Profits April 6 to June 30, 1989		
£12,000 × say 1/12th	1,000	
£3,000	3,000	4,000
		£37,000

Profits falling out of charge	£36,000

Conversely, if the accounts had been made up to March 31 each year the 1988–89 assessment would be based on the profits for the year ended March 31, 1988 and only the profits from April 1, 1988 to April 5, 1989 would drop out of assessment.

EXAMPLE 5   2–008

Closing Years—Basis of assessment

3. Year end March 31 cessation June 30, 1989.	
Profits year ended March 31, 1987	18,000
Profits year ended March 31, 1988	25,000
Profits year ended March 31, 1989	12,000
Profits 3 months ended June 30, 1989	4,000
	£59,000
1987–88 Profits to March 31, 1987	18,000
1988–89 Profits to March 31, 1988	25,000
1989–90 Profits April 6 to June 30, 1989 say	4,000
	£47,000
Profits falling out of charge	£12,000

It will be appreciated therefore that an accounting date early in the tax year gives the maximum delay between earning profits and paying tax on them, which can be important in times of high inflation, and also allows the longest period to drop out of assessment on a cessation. Nevertheless, in practice, most farms seem to make up accounts to March 31 or September 30.

In order to avoid the worst abuse of the cessation rules the Revenue has   2–009
the option to adjust the penultimate and antepenultimate years of assessment to an actual basis which mirrors the taxpayer's option on commencement. These provisions are contained in I.C.T.A. 1988, s.63 so in the case of an April 30 accounting date (example 4 above), the assessment for 1988–89 would be the actual profits apportioned on a time basis for the year ended April 5, 1989 and similarly for 1987–88 the actual profits for the year ended April 5, 1988 apportioned on a time basis. The 1986–87 assessment however remains unchanged and this would have been based on the accounts for the year ended April 30, 1985 and it is therefore the profits for the period from May 1, 1985 to April 5, 1987 which drop out of assessment and are tax free.

In order to take the maximum advantage of these rules it is obviously desirable to have a low profit followed by two years' high profits followed by a low profit and a cessation, often known as a low-high-high-low profit flow. Obviously it may not be possible to arrange for profits to arise in this manner, particularly in farming, but if the opportunity does arise a well chosen cessation can result in spectacular tax savings.

2–010   EXAMPLE 6
Commencement April 7, 1986, cessation April 8, 1990

	£
Profits for the year ended April 6, 1987	100
Profits for the year ended April 6, 1988	100,000
Profits for the year ended April 6, 1989	99,000
Profits for the year ended April 6, 1990	100
Profits for the two days to April 8, 1990	20
Total profits	£199,220

These are the profits as adjusted for tax purposes:

	£
The assessment for 1986–87—Actual profits 364/365ths of £100, say,	100
The assessment for 1987–88—first 12 months' profits	100
The assessment for 1988–89—preceding year's profits	100
The assessment for 1989–90—preceding year's profits	100,000
The assessment for 1990–91—actual profits 1/365th of £100 plus £20, say,	20
Total profits assessed	£100,320
Profits not assessed	£98,900

On these figures the penultimate and antepenultimate year adjustment is not available to the Revenue as the assessments for 1988–89 and 1989–90 on an actual basis (say £99,100) would be less than the £100,100 already assessed.

2–011   Farming does not normally lend itself to manipulation of profits to take advantage of the opening and closing years, although this may be feasible if there is a temporary expansion in farming activities, perhaps on some joint venture arrangement, or where there is a parallel company providing contracting services which can be carried out either by the sole trader or by the contracting company as required.

2–012   In some cases it is appropriate to have a company providing services to the farm, such as the employment of staff, although now that farming is so much less labour intensive than it once was, the use of a service company merely to employ staff on the farm is unlikely to be advantageous. It is important that any payment by the farmer to a contracting or service company can be justified on an arm's length pricing basis as otherwise the expense could be disallowed in the farm accounts but the income still be taxed in the company: *Stephenson* v. *Payne, Stone, Fraser & Co.* (1967) 44 T.C. 507.

Although an accounting date early in the financial year may be advantageous from a purely taxation point of view it is often found in practice that September 30 or October 31 are particularly popular accounting dates with farmers as by this time the crop has been harvested and the stock figure can be arrived at with the minimum of difficulty. As always with tax planning, the commercial advantage has to be compared carefully with the tax advantage to determine the most beneficial course of action.

Change of accounting date

A business can change its accounting date, which can be advantageous **2–013** but needs to be considered with care. The Revenue operates transitional averaging arrangements for the years of assessment affected by the change, the procedure for which is explained in Booklet IR26. This procedure was approved in *I.R.C.* v. *Helical Bar Ltd.* (1972) 48 T.C. 221.

A change in accounting date may cause the Revenue to look closely at tillages, the value of growing crops which are often ignored for accounting and tax purposes. With changing patterns over the years whereby autumn sown barley and wheat predominate, the value of the growing crops by the traditional year end of March 31 is considerable. A change in year end can be expensive in tax terms, if such values have to be included, notwithstanding the theoretical advantage of, say, an April 30 year end.

Other tax considerations

A further advantage of the unincorporated farm is the ability to average **2–014** fluctuating profits under I.C.T.A. 1988, s.96 which is dealt with in Chapter 3. Another is that farm losses of a sole trader or partner can be set against other income, whereas losses in a company can at best be set against other income of the company and cannot be set against the shareholders' personal income. Losses are dealt with in some detail in Chapter 4.

On occasions there may be advantages in transferring a farming business to a company, for example, to obtain the benefit of reduced valuations for minority shareholdings. C.G.T. reliefs are available to roll over the gain on a transfer of a business as a going concern to a company in exchange for shares under C.G.T.A. 1979, s.123, or, rather more unusually but perhaps advantageously for stamp duty purposes, as a gift of assets to a company where the gain is rolled over under C.G.T.A. 1979, s.126. See Chapter 15.

The disadvantages of an unincorporated trader, as opposed to a com- **2–015** pany, are that the rate of tax is higher in that a company currently suffers corporation tax at a maximum of 35 per cent. whereas an individual's higher rate is 40 per cent. In the recent past these rates of tax have been very much higher but the advantage of trading through a company and accumulating profits, subject only to corporation tax, used to be curtailed by means of a requirement to pay a dividend which was taxed on the recipient as investment income, or to suffer an apportionment of profits giving rise to a similar liability. Currently there is no requirement for the apportionment of trading profits.

The preceding year basis of assessment can sometimes work against the interests of the taxpayer, particularly giving rise to high taxable profits based on the results of a previous good year with the tax being payable in the year in which the results might not be nearly so good. The farm averaging provisions referred to above help to alleviate this problem. Similarly, on cessation, the Revenue power to adjust the penultimate and antepenultimate years to an actual basis of assessment might produce an unwelcome additional tax charge.

2–016   The main disadvantage of being a sole trader, particularly for the smaller farm, is, however, the proprietor's unlimited liability which can be particularly serious for a tenant farmer who has no substantial realisable asset to turn to in times of financial difficulty.

Partnerships

2–017   A partnership is defined by the Partnership Act 1890 as the relation which subsists between persons carrying on a business in common with a view to profit. A partnership is not a separate legal entity in England, although it is in Scotland, and it is not a taxable entity as such, even in Scotland. The profits of the partnership are divided among the partners in accordance with their profit sharing ratios which in turn are determined by the partnership agreement, the agreement of the partners amongst themselves, or the Partnership Act 1890.

    A tax liability of the partnership is merely the aggregate of the partners' individual liabilities and the partnership assessment is raised on the senior acting partner in the partnership name: I.C.T.A. 1988, s.118. Each partner is jointly and severally liable for the partnership debts which would include the remaining partners' tax liabilities. Many farming partnerships are family affairs and one advantage of such a partnership is that it is not necessary to justify each partner's share of profit by reference to the work actually done, whereas wages have to be reasonable for the services provided under *Copeman* v. *William Flood & Sons Ltd.* (1940) 24 T.C. 53; *Dollar and Dollar* v. *Lyon* [1981] S.T.C. 333.

2–018   Whether there is in fact a partnership is a matter of fact; attempts to push the partnership concept too far and, for example, involve minor children as partners, are likely to fail (*Alexander Bulloch & Co.* v. *I.R.C.* [1976] S.T.C. 514).

2–019   One of the advantages of including a spouse as a partner is the ability to split the income between husband and wife, both of whom have their own tax allowances and basic rate band. Until April 5, 1990 this may in worthwhile cases be achieved by making a wife's earnings election under I.C.T.A. 1988, ss.287, 288, whereby the wife's earnings are assessed on her as if she were a single person, leaving the husband to pay tax on his own earnings and his own and his wife's investment income, albeit with the benefit of only the single person's allowance instead of the married man's allowance.

2–020   If the wife does take part in the business, is a signatory to the bank account and does some of the secretarial activities, such as the book-keeping, letters, the ordering of supplies and the payment of suppliers, she should be accepted as a full partner without any need for her to become involved in the physical activities of the farm. Many farmers' wives, however, do become involved in all aspects of the farm, including physical work on the farm itself. The mere existence of a partnership agreement is not conclusive evidence that there is a partnership (*Hawker* v. *Compton* (1922) 8 A.T.C. 306, *Dickenson* v. *Gross* (1927) 11 T.C. 614, *Fenston* v. *Johnstone* (1940) 23 T.C. 29).

A partnership agreement can confirm the existence of a partnership **2–021** already in being but cannot retrospectively produce a partnership which did not exist at the time (*Ayrshire Pullman Motor Services and Ritchie* v. *I.R.C.* (1929) 14 T.C. 754, *Waddington* v. *O'Callaghan* (1931) 16 T.C. 187, *Taylor* v. *Chalklin* (1944) 26 T.C. 463) and more recently *Saywell & Others* v. *Pope* [1979] S.T.C. 824.

Just as a partnership agreement cannot produce a partnership where the facts are against it, the existence of an agreement denying a partnership and confirming a mere joint venture is not of itself proof that a partnership does not exist (*John Gardner & Bowring Hardy & Co. Ltd.* v. *I.R.C.* (1930) 15 T.C. 602). All the facts of the arrangement have to be taken into account in ascertaining whether a partnership exists (*I.R.C.* v. *Williamson* (1928) 14 T.C. 335, *Calder* v. *Allanson* (1935) 19 T.C. 293).

A change in the partners carrying on the partnership gives rise to an **2–022** automatic cessation for tax purposes under I.C.T.A. 1988, s.113(1). Such a cessation could arise on the death or retirement of a partner or on the introduction of a new partner. A change can be avoided by electing for continuation under I.C.T.A. 1988, s.113(2). Such a continuation election has to be made within two years of the partnership change and must be signed by all the partners before and after the change. In the case of a deceased partner the continuation election would be signed by his personal representatives. It is normally desirable to avoid a cessation on a partnership change and partnership agreements usually contain a clause requiring partners to join in a continuation election if required to do so by a majority of the remaining partners. Such a requirement would usually include an indemnity by the other partners indemnifying the partner to sign from any increase in his tax liability arising from a continuation rather than a cessation.

A partnership cessation gives rise not only to a possible penultimate and **2–023** ante-penultimate year adjustment by the Revenue on the old partnership but also to the modification of the commencement rules for the new partnership except where either a sole trader takes in a partner, or a partnership of two ceases and one of the partners continues the trade as a sole trader. Under I.C.T.A. 1988, s.62 the assessments for the first four tax years are adjusted to the actual profits apportioned on a time basis and only in the fifth year of assessment is the assessable income based on the adjusted profits for the accounting period ending in the preceding tax year. This means that it is not the first year's profits after the cessation which are assessed two to three times but those of the third and fourth years. The taxpayer has the option under I.C.T.A. 1988, s.62 to have the fifth and sixth years also assessed on an actual basis before going on to the preceding year basis of assessment. This merely changes the profits which are doubly assessed.

2–024   EXAMPLE

Farmco introduced a new partner into the business on January 1, 1986, on which date a partner also retired, and did not elect for continuation. The profits were as follows:

	£	£
Year ended December 31, 1982		48,000
Year ended December 31, 1983		50,000
Year ended December 31, 1984		60,000
Year ended December 31, 1985		40,000
Year ended December 31, 1986		70,000
Year ended December 31, 1987		80,000
Year ended December 31, 1988		90,000
Year ended December 31, 1989		65,000
Year ended December 31, 1990		75,000
Year ended December 31, 1991		55,000

The assessments were:

OLD BUSINESS	£
1983–84	
Year ended December 31, 1982	48,000
1984–85	
Year ended December 31, 1983	50,000
Penultimate and antepenultimate years, original	£98,000
1985–86	
9/12ths × year ended December 31, 1985	£30,000

Revenue option on cessation		
1983–84		
9/12ths × year ended December 31, 1983	37,500	
3/12ths × year ended December 31, 1984	15,000	52,500
1984–85		
9/12ths × year ended December 31, 1984	45,000	
3/12ths × year ended December 31, 1985	10,000	55,000
Penultimate and antepenultimate years—revised		£107,500

NEW BUSINESS	£
1985–86	
3/12ths × year ended December 31, 1986	17,500

		£
1986–87		
9/12ths × year ended December 31, 1986	52,500	
3/12ths × year ended December 31, 1987	20,000	£72,500
1987–88		
9/12ths × year ended December 31, 1987	60,000	
3/12ths × year ended December 31, 1988	22,500	£82,500
1988–89		
9/12ths × year ended December 31, 1988	67,500	
3/12ths × year ended December 31, 1989	16,250	£83,750
1989–90		
year ended December 31, 1988		90,000
1990–91		
year ended December 31, 1989		65,000
Fifth and sixth years—original		£155,000
1991–92		
year ended December 31, 1990		£75,000

Taxpayer's option

1989–90		
9/12ths × year ended December 31, 1989	48,750	
3/12ths × year ended December 31, 1990	18,750	67,500
1990–91		
9/12ths × year ended December 31, 1990	56,250	
3/12ths × year ended December 31, 1991	13,750	70,000
Fifth and sixth years—revised		£137,500

Assessments:

Old business 1983–84	52,500 }	107,500
Old business 1984–85	55,000 }	
Old business 1985–86	30,000	
New business 1985–86	17,500	
New business 1986–87	72,500	
New business 1987–88	82,500	
New business 1988–89	83,750	
New business 1989–90	67,500 }	137,500
New business 1990–91	70,000 }	
New business 1991–92	75,000	
	£606,250	

If an election for continuation had been made the assessments would have been:

Old business 1983–84 year ended December 31, 1982	48,000
Old business 1984–85 year ended December 31, 1983	50,000
Old business 1985–86 9/12ths year ended December 31, 1984	45,000
New business 1985–86 3/12ths year ended December 31, 1984	15,000
New business 1986–87 year ended December 31, 1985	40,000
New business 1987–88 year ended December 31, 1986	70,000
New business 1988–89 year ended December 31, 1987	80,000
New business 1989–90 year ended December 31, 1988	90,000
New business 1990–91 year ended December 31, 1989	65,000
New business 1991–92 year ended December 31, 1990	75,000
	£578,000

An election for continuation would have been beneficial in this case. However, the assessments on the old business were reduced by the cessation which clearly benefited the retiring partner. The election would have to have been made by December 31, 1987, long before the effect of an actual basis could be calculated under the recommencement rules.

Partnership losses are dealt with by the partners individually in the same way as losses from activities as a sole trader, see Chapter 4.

2–025   In many cases of partnership, partners are given salaries, interest on capital or amounts in lieu of rent as a prior charge against profits before dividing the remainder of profits in the profit-sharing ratio. These prior charges are a means of allocating the partnership profits and merely affect the way in which the partnership assessment is divided among the partners. They are, therefore, all part of the Schedule D, Case I assessment and interest on capital is not, for example, assessed under Schedule D, Case III, salaries under Schedule E or rent under Schedule A or Schedule D, Case VI.

An employee may be called a partner and even held out as a partner to other creditors, but for tax purposes such a salaried partner is taxed on his salary under Schedule E in the same way as any other employee and this would be an expense in arriving at the partnership profits before those profits are divided among the equity partners.

A farming partnership is limited to a maximum of 20 partners; this is rarely a problem and most farming partnerships have far fewer partners.

Partnerships with companies

A company can be a partner in which case the company pays corporation **2–026** tax on its share of the partnership profits on a current year basis, whereas the remaining partners divide the remainder of the profits and are taxed as if there were no corporate partner. The advantage of a corporate partner is that in poor years the company's share of profits could be reduced in order to give the individuals who are partners the income they require to live on, and in very good years the corporate partner could be given a much higher share of profit which would suffer corporation tax at 35 per cent. instead of income tax at 40 per cent.

Corporate partners are covered by I.C.T.A. 1988, ss.114, 115 and a change in the corporate partner does not give rise to a cessation, nor does it require a continuation election to preserve the continuing basis of assessment for the partners who are individuals: I.C.T.A. 1988, s.114(3)(b). A corporate partner can therefore give a number of the tax advantages of incorporation without losing the flexibility of the partnership for the individual partners.

The use of a corporate partner artificially to manipulate tax losses is largely prevented by I.C.T.A. 1988, s.116.

The disadvantages of a partnership, apart from the commercial one of an **2–027** unlimited liability for the partnership debts, is that profits ploughed back into the business are nonetheless taxed as the partners' income in the same way as drawings and therefore the reinvestment is limited by the tax at the higher rate on the undrawn profits. Although it is arguable that it is possible to put profits to reserve in a partnership, subject only to tax at the basic rate, following *Franklin v. I.R.C., Swaythling v. I.R.C.* (1930) 15 T.C. 464, the Revenue is unlikely to accept in practice that any transfer to reserves for the normal commercial operation of the farm could be covered by the sort of contingency arrangement envisaged by the *Franklin* case.

Limited partnerships

It is possible to trade as a limited partnership under the Limited Partnership Act 1907. Under these provisions there has to be at least one partner **2–028** with unlimited liability, called the general partner, who takes part in the management of the business, and other partners with limited liability who do not take part in the management of the business but receive a share of the profits and are liable for a share of their losses, up to the amount of their partnership capital but no further. A limited partner is taxed in the same way as an unlimited partner except that his tax losses are limited to his partnership capital under I.C.T.A. 1988, ss.117, 118.

Because a limited partner is unable to participate in the management of the business it is unusual to find a limited partnership involved in farming. A limited partner could be involved in a subordinate activity on the farm and it could perhaps be appropriate for someone putting money into a farm but not getting involved in the management activities to be a limited partner in view of the commercial risks of unlimited liability.

When considering a partnership involved in farming it is important to **2–029** ascertain how the land is held. It could be owned personally by one or more of the partners and let or licensed to the partnership, or it could be held by the partnership as partnership property. In the latter case any change in the partnership asset-sharing ratio could crystallise a capital gain but retirement relief would be available; this might not be the case where the land was owned by the individual partner and let to the partnership. This problem is considered in more detail in Chapter 16.

Joint ventures

2–030   It is important to distinguish a partnership from a joint venture under which each of the joint venturers is liable only for his own tax and other liabilities in accordance with the joint venture contract, unlike a partnership. Often one party to such a venture will own the farm land and the other party will have farming expertise. It may well be, however, that what the parties regard as a joint venture is in fact a partnership. For example, in *George Hall & Son* v. *Platt* (1954) 35 T.C. 440 a firm of merchants entered into an agreement with a farming partnership to grow carrots on land occupied by the farmers. The expenses contributed by each party were directly reimbursed out of the proceeds of the sale of the crop and the balance was divided equally between them. The result was a partnership, not a joint venture. On the other hand an agreement to share receipts and expenses with the former proprietor who continued to work for a period after the sale of his business did not amount to a partnership in *Pratt* v. *Strick* (1932) 17 T.C. 459.

Farming joint ventures include sharecropping and share flocking arrangements, seasonal renting of vegetable land and grazing and agisting agreements. In every case the precise contractual arrangements need to be looked at carefully.

2–031   It is normally desirable to avoid a joint bank account and for the land-owner joint venturer to ensure that he retains ownership of the crops in the ground, subject to the contractual rights of the joint venturer, and that he participates in the management decisions and does not merely receive rent calculated on the basis of the crop value: *Gittos* v. *Barclay* [1982] S.T.C. 390, *Fry* v. *Salisbury House Estate Ltd.* (1930) 15 T.C. 266. The path between a partnership on the one hand, and a tenancy arrangement on the other, is a narrow one. In a valid share farming joint venture each party is carrying on farming in his own right and is taxed accordingly.

Trading trusts

2–032   The possibility of having a trading trust to carry on a farm or to be a partner in a farming partnership now has little advantage compared with the more usual case where the trustees hold shares in a farming company. A trust is liable to tax at 25 per cent. (for 1989–90) in the case of distributed profits and 35 per cent. in the case of undistributed profits, which would effectively leave 65 per cent. to be accumulated in the trust. If the trustees were to receive a comparable dividend the amount accumulated would be after corporation tax paid by the company which would reduce the amount available unless the small companies rate applied. The advantage of a trading trust was considerably higher when the rate of corporation tax was 52 per cent.

Where payments are made out to beneficiaries, credit for the tax suffered by the trust is available.

EXAMPLE

2–033

	Trust £	Company standard rate £	Small companies rate £
Farming profits	1,000	1,000	1,000
Corporation tax at 35% (25%)	—	(350)	(250)
Available for dividend	—	650	750
Add tax credit	—	217	250
Gross trust income	1,000	867	1,000
Income tax at basic rate (25%)	(250)	(217)	(250)
Income tax at additional rate (10%)	(100)	(87)	(100)
Available for distribution or accumulation (65%)	650	563	650
Tax payable by beneficiary at 40% less 35% (5%)	(50)	(43)	(50)
Net income of beneficiary	£600	£520	£600

C.G.T. relief is available for transfers of business assets into trust and again out of trust in favour of a beneficiary under C.G.T.A. 1979, ss.126, 147A.

Companies

A company will usually have limited liability which has an obvious com- 2–034
mercial advantage, although it would be possible to trade as an unlimited company without affecting the taxation position in any way. It is normally easier to arrange outside capital for borrowing by a company as it is very easy to arrange a floating charge over the company's assets to give the lender security. Outside capital can be obtained by issuing shares, although farming is not a qualifying activity for the business expansion scheme: I.C.T.A. 1988, s.297(2)(j).

Where farming is carried on by a company it can issue shares to members of the family not connected directly with the farming and issue different classes of shares with different rights. The possibilities of reducing the capital value of the farm and simplifying the problem of transferring a farm to succeeding generations is considered in Chapter 18.

Shares can be given to key employees such as the farm manager, perhaps 2–035
under an approved share option scheme within I.C.T.A. 1988, s.185 and

Sched. 9 or under some other arrangement, but anti-avoidance provisions tax gains under an unapproved share option or share incentive scheme as income instead of as capital gains: see I.C.T.A. 1988, ss.135–140.

2–036   The rate of corporation tax applicable to a company making profits of less than £150,000 in the year ending March 31, 1990 is 25 per cent. This means that some 75 per cent. of gross profit can be reinvested in the company after corporation tax compared with only 60 per cent. of the surplus in the case of a sole trader or partnership paying at the higher rate of income tax. A company making profits of over £150,000 is subject to corporation tax at 35 per cent. with marginal relief up to £750,000. These figures are reduced if there are associated companies.

If a company does declare a dividend to its shareholders the dividend is made out of profits after corporation tax. However, the net dividend has attached to it a tax credit of one-third which is paid over to the Revenue by the company as advance corporation tax. Advance corporation tax may be set against the mainstream corporation tax liability subject to leaving in charge a minimum mainstream liability. The tax credit so imputed to the dividend is available to set against the shareholders' tax charge on the total of the dividend and tax credit, which means that a basic rate taxpayer has no further tax to pay. A trust accumulating the income or a higher rate taxpayer would pay only the difference between its or his actual liability and the basic rate deemed to have been suffered already.

2–037   One of the disadvantages of a company is the requirement for an audit, and the legal formalities necessary for a company, which include the filing of accounts for public inspection with the Registrar of Companies. These requirements obviously increase the cost of running a business as a company compared with a sole trade or partnership. The theory of limited liability is often severely dented by a requirement which a lending banker may make for a guarantee by the shareholders and directors in respect of loans made to the company. Obviously the giving of such guarantees will be avoided where possible but if bank facilities are required it might be unavoidable.

Although a partner can draw funds from the business without any formalities, where a company is involved the directors are officers or employees of the company and pay tax personally under Schedule E on their earnings. Amounts drawn from the company are therefore subject to pay as you earn (P.A.Y.E.).

The requirement to deduct P.A.Y.E. also exists if amounts are credited to a director's account, even though these amounts may not be drawn until much later.

2–038   There is also an obligation to deduct national insurance Class 1 contributions from amounts paid, or credited to directors, and other employees. The employer is also liable to national insurance contributions at a rate of 10.45 per cent. on employees' and directors' earnings, although lower rates are payable if earnings fall below £165 per week (1989–90).

The combined employer's and employee's national insurance contributions are considerably higher in the case of a company employee or director than for a partnership or sole trader.

There are a number of rules relating to company employees on benefits in kind, in particular in respect of loans from the company and the provision of company cars and fuel, which give rise to taxable income. A partner or sole trader, on the other hand, would have part of the cost of such expenses disallowed to cover any private usage.

Loans to company directors are prohibited by the Companies Act 1985 **2–039** except in certain limited cases. As well as giving rise to a benefit in kind, unless interest is charged at a full rate, such loans also give rise to an additional tax charge on the company under I.C.T.A. 1988, ss.419, 420 if the borrower is a shareholder or associate of a shareholder in the company as would normally be the case in a family farming company. This means that the loan is grossed up at the appropriate rate of advance corporation tax and the grossing addition, being an amount equivalent to advance corporation tax, is paid over to the Revenue as an interest-free loan which is only released when the loan to the participator is repaid. The Revenue has been known to treat drawings by directors on which P.A.Y.E. has not been properly applied as loans to which these provisions apply.

The main disadvantage of a company, however, is the difficulty and **2–040** expense of taking assets out of the company other than by putting it into liquidation. Even in a liquidation, if the company owns assets on which there is a material capital gain there is a double charge, first on the realisation by the company and secondly by the shareholder when the proceeds in liquidation are paid out to him producing a gain on the disposal of his shares. It is easy to put assets into a company without any taxation penalty but extremely difficult to get them out again later should it be desirable to do so. Although there are provisions for demergers under I.C.T.A. 1988, ss.213–218 the rules are highly restrictive. The Revenue in July 1987 published a discussion paper on disincorporation; its proposals would alleviate this problem in due course, but not eliminate it.

CHAPTER 3

Computation of Profits

Farming

3–001   Farm land is defined in I.C.T.A. 1988, s.832(1) as "land in the United Kingdom wholly or mainly occupied for the purposes of husbandry, but excluding any dwelling or domestic offices, and excluding market garden land," and farming is to be construed accordingly.

The same section defines market garden land as "land in the United Kingdom occupied as a nursery or garden for the sale of the produce (other than land used for the growth of hops)" and "market gardening" is to be construed accordingly.

For the purposes of agricultural buildings allowances the definition of husbandry is in C.A.A. 1968, s.69 as extended by F.A. 1986, s.56(4)(c) which states that husbandry "includes any method of intensive rearing of livestock or fish on a commercial basis for the production of food for human consumption."

This implies that for purposes other than agricultural buildings allowances husbandry does not include such intensive rearing activities; although these amount to carrying on a trade it is not the trade of farming, and therefore it does not qualify for those allowances which apply only to farming.

The definition of husbandry was considered in a number of early cases when farming was dealt with under Schedule D. It was held that land occupied for husbandry was not confined to cultivating the soil and included sheep grazing (*Kier* v. *Gillespie* (1919) 7 T.C. 473) but did not extend to dealing with agricultural produce, such as a creamery (*I.R.C.* v. *Cavan Central Co-op* (1917) 12 T.C. 1). In *Lean and Dickson* v. *Ball* (1925) 10 T.C. 341 poultry farming was held to be husbandry, as it was in *Jones* v. *Nuttall* (1926) 10 T.C. 346.

In *I.R.C* v. *William Ransom* (1918) 12 T.C. 21 husbandry included the occupation of land for growing herbs for medicinal purposes. In *Long* v. *Belfield Products* and *Thornber* v. *MacInnes* (1937) 21 T.C. 221 poultry farming which was husbandry was separated from custom hatching and egg and chick dealing which was not. In *Lord Glanely* v. *Wightman* (1933) 17 T.C. 634 stallion fees were not separable from the husbandry activities of the stud, and in *Peter Reid* v. *I.R.C.* (1947) 28 T.C. 451 a poultry farm amounted to husbandry although the chickens were in part intensively reared.

A potato merchant who effectively rented land for part of the year from farmers and provided the seed, fertiliser and manual labour was held to be occupying land for the purpose of husbandry in *Back* v. *Daniels* (1924) 9 T.C. 183 but this decision was criticised in *Dawson* v. *Counsell* (1937) 22 T.C. 149.

One trade

It is provided that all farming and market gardening in the U.K. shall be **3-002** treated as the carrying on of a trade or part of a trade and is taxed under Schedule D, Case I: I.C.T.A. 1988, s.53(1). It is also provided that all farming carried on by any particular person or partnership or body of persons shall be treated as one trade: I.C.T.A. 1988, s.53(2).

The occupation of land in the U.K. for purposes other than farming or market gardening is treated as a trade if the land is managed on a commercial basis for example, as a caravan park, I.C.T.A. 1988, s.53(3) although there are special rules for woodlands carried on on a commercial basis, I.C.T.A. 1988, ss.53(4), 54, (see Chapter 12).

The fact that all farming in the U.K. is a single trade is extremely helpful as not only does it mean that the profits and losses of different farms, even in widely separated areas of the country, are aggregated for tax purposes; it also means that if a loss-making farm is sold, but farming continues in other areas, the losses are still available against the profits of the remaining farming (*Bispham* v. *Eardiston Farming Co. (1919) Ltd.* (1962) 40 T.C. 322)). It also means that if a farmer of a loss-making farm acquires a profitable farm the losses should be available against the entire activities and not only against the profits arising from the farm in which the losses were made. It also enables a farmer with a single loss-making farm to sell this and acquire a profitable farm and carry the losses forward. Similarly if a loss-making farming company is acquired by the proprietor of a profitable farm, that farm could be transferred into the loss-making company. It would also be open to a loss-making company to acquire a profitable farm on the open market. In either event the accumulated losses should be available for relief against the future profits without any requirement to segregate the results.

In the case of companies, however, it is necessary to bear in mind the provisions of I.C.T.A. 1988, s.768, which disallow losses if within any period of three years there is both a change in the ownership of the company and a major change in the nature or conduct of a trade carried on by the company, or there is a considerable revival of the trade where the scale of activities has previously become small or negligible (see Chapter 4).

Trading receipts

Precisely what profits should be included as trading receipts from farm- **3-003** ing has been a matter of dispute in some cases. It has been held that profits from the sale of trees on farmland, which did not amount to a separate trade of running a commercial woodland, should be included in the farming profits, both where the trees had originally been planted by the farmer and where they were the proceeds of long-standing trees: (*Elmes* v. *Trembath* (1934) 19 T.C. 72). Similarly proceeds from the sale of turf by a farmer were included as farming profits although in other circumstances such sales could amount to income from the land assessable under Schedule A within I.C.T.A. 1988, s.15(1)(c): (*Lowe* v. *J.W. Ashmore Ltd.* (1970) 46 T.C. 597).

Grassland let for seasonal grazing was treated as occupied for husbandry by the landlord in *I.R.C.* v. *Forsyth Grant* (1943) 25 T.C. 369 and by the lessee in *Drummond* v. *I.R.C.* (1951) 32 T.C. 263 and *Mitchell* v. *I.R.C.* (1943) 25 T.C. 380.

3–004   Difficulties can arise on a change in ownership, although in the case of growing crops a payment specifically so made for these is a trading expense as they are not part of the land but the result of working the land, *fructus industriales*, and the amount received is a trading receipt (*Gunn* v. *I.R.C.* (1955) 36 T.C. 93). This can be contrasted with the case of a fruit orchard in which it was held that the trees formed part of the land, as *fructus naturales*; and although the proceeds of the sale of the crop were taxable on the purchaser as a trading receipt he could not claim a deduction for the value of the growing fruit at the time of acquisition, which conversely would mean that the sale proceeds would be treated as a capital receipt of the vendor (*I.R.C.* v. *Pilcher* (1949) 31 T.C. 314).

Income from grazing whether casual or under a grazing agreement would be a normal trading receipt as would keep fees, agisting fees or other income from allowing other farmers' animals to occupy farmland on a temporary basis. The use of land as a stud farm would be a separate trade, not farming, *Dawson* v. *Counsell* (1937) 22 T.C. 149.

Futures trading

3–005   Farmers often trade in the commodities markets, particularly in potatoes and wheat, as a hedge against adverse price movements. Such activities in commodities which are actually farmed will be treated as part of the farm trading profits or losses. Dealing in commodities not farmed may be an investment activity qualifying for capital gains treatment as explained in Chapter 15.

Income from letting race horses may be charged to tax under Schedule D, Case VI (*Norman* v. *Evans* (1964) 42 T.C. 188).

Income from land, including royalties for sand and gravel, can be assessed under Case VI (*T & E Homes Ltd.* v. *Robinson* [1979] S.T.C. 351) as can payments to a retired partner under a consultancy agreement where the services rendered are negligible (*Hale* v. *Shea* (1964) 42 T.C. 260). Casual commissions where a trade is not being carried on are assessed under Case VI (*Ryall* v. *Hoare* (1925) 8 T.C. 521, *Sherwin* v. *Barnes* (1931) 16 T.C. 278, *Lyons* v. *Cowcher* (1926) 10 T.C. 438, *Grey* v. *Tiley* (1932) 16 T.C. 414) and this includes insurance commissions other than to a broker or professional agent (*Way* v. *Underdown* (Nos. 1 and 2) (1974) 49 T.C. 215, 648, *Hugh* v. *Rogers* (1958) 38 T.C. 270).

Other income

Grants, subsidies and compensation receivable are dealt with in Chapter 8, the taxation of rents is covered in Chapter 9, sporting rights in Chapter 10, mineral rights in Chapter 11, woodlands and forestry in Chapter 12, receipts from the sale of land in Chapters 15 and 16. **3–006**

Miscellaneous receipts

Annual profits and gains, including one-off receipts, not falling under any other Case of Schedule D are assessed under Schedule D, Case VI: I.C.T.A. 1988, s.18(3) on the income arising in the tax year, that is on a current year basis: I.C.T.A. 1988, s.69. **3–007**

Wayleaves

A wayleave payment for electricity pylons or telegraph poles comprises a rent element and can also include compensation. **3–008**

The rent element might be 50p per pole and if the total rent under one agreement exceeds £2.50, tax is deducted at source: I.C.T.A. 1988, s.119(2). As many telegraph lines have been put in bit by bit many agreements are for less than £2.50 and no tax is deducted. An individual farmer might receive wayleaves under numerous agreements. There are no provisions for consolidating agreements.

A much larger payment is for compensation where, *e.g.* a pylon or telegraph pole is in the middle of a field. This can amount to £6 per post and, in the case of compensation element of the wayleave, tax is not deductible. **3–009**

Where wayleaves are received under deduction of tax in accordance with I.C.T.A. 1988, s.348(2)(b) a certificate of deduction of tax should be received in accordance with I.C.T.A. 1988, s.352. The gross income however is normally included as part of the farm trading income in accordance with I.C.T.A. 1988, s.119 and credit is claimed against the income tax or corporation tax payable on the farm profits for any tax deducted at source. If the tax so deducted is in excess of the liability it is reclaimable from the Revenue.

Expenses

As with other trading activities the rules for the deduction of expenses allowable for tax for farming are those expenses wholly and exclusively laid out or expended for the purposes of the trade: I.C.T.A. 1988, s.74(a) other than those specifically disallowed. **3–010**

The legislation operates in the negative by disallowing expenses not wholly and exclusively incurred for business purposes, and not by specifically allowing those that are. This means that farm accounts are prepared on normal accounting principles, *Odeon Associated Theatres Ltd.* v. *Jones* (1971) 48 T.C. 257, although ultimately it is a matter of law whether an expense is deductible for tax purposes or disallowed as capital, *Heather* v. *P-E Consulting Group Ltd.* (1972) 48 T.C. 293.

Normal accounting principles require the matching, as far as practicable, of income and related expenditure. Accounts are made up on an accruals or earnings basis, *D & GR Rankine* v. *I.R.C.* (1952) 32 T.C. 520, *Wetton, Page & Co.* v. *Attwooll* (1962) 40 T.C. 619 which take into account stock, work-in-progress, debtors and creditors.

The Revenue will not allow a cash basis for a trade such as farming, SP/A27, November 20, 1969.

3–011   Nearly all expenses in running a farm will be allowable for tax purposes, such as purchases of livestock, food, seed, sprays and fertiliser, wages, rent and rates, water, electricity, fuel, heating, insurance, telephone, training, interest, veterinary fees, repairs to premises and equipment, contractors charges, audit and accountancy fees and sundry expenses. This list is by no means exhaustive.

One expense shown in the accounts for which no allowance is available is depreciation of plant and premises. However, the cost of capital expenditure on such items is, in many cases, allowed for tax purposes through the capital allowances system, see Chapter 5.

There are a number of points on particular types of expenditure which are considered below.

Private expenditure

3–012   There is a specific disallowance for private expenditure, including expenses for the maintenance of the farmer and his family: I.C.T.A. 1988, s.74(b). One-eighth of the subscription to the National Farmers' Union is normally disallowed as providing private benefits not relating to the trade.

Accommodation

3–013   Accommodation costs, including rent, for the farmhouse are allowable as a business expense to the extent that the property is used for the purpose of the trade. Only exceptionally may more than two-thirds of any rent paid be allowed: I.C.T.A. 1988, s.74(c). In practice the normal allowance is unlikely to exceed one-third of the total costs, but this proportion has no statutory authority and should be such proportion as is appropriate having regard to the size of the farmhouse in relation to the farm and the extent to which it is used for business purposes. A limit of one-third applies to capital expenditure on the farmhouse and in this case the limit has statutory authority under F.A. 1986, Sched. 15, para. 2(2).

Own consumption

3–014   Where a farmer personally consumes produce, sold commercially, the market value of the amount consumed must be brought into the credit of the trading results, *Sharkey* v. *Wernher* (1955) 36 T.C. 275. If the produce is not commercially sold it may only be necessary to disallow part of the cost. No adjustment is necessary for the farmer's private activities such as vegetables grown for his own consumption. Where the farm is run as a company the only adjustment necessary is a benefit-in-kind charge on the cost to the company of produce consumed by employees earning £8,500 or more and directors.

Repairs

The whole of the expenditure on maintaining farm cottages and other **3–015** farm buildings and works is normally allowable as an expense but a provision for future repairs is not deductible until expended: I.C.T.A. 1988, s.74(d), *Clayton* v. *Newcastle-under-Lyme Corporation* (1888) 2 T.C. 416 and *Peter Merchant Ltd.* v. *Stedeford* (1948) 30 T.C. 496). Any improvement element is capital and as such is specifically disallowed: I.C.T.A. 1988, s.74(g), *Wynne-Jones* v. *Bedale Auction Ltd.* [1977] S.T.C. 50, *Highland Railway Co.* v. *Balderston* (1889) 2 T.C. 485 and *Vale* v. *Martin Mahony & Brothers Ltd.* [1947] I.R. 30, 41).

The prohibition against a deduction for capital expenditure includes any allowance for depreciation. Allowances for capital expenditure are given by way of agricultural buildings allowances or capital allowances for plant as described in Chapter 5. In appropriate cases, expenditure on improvements can be taken into account for capital gains, see Chapter 15.

The borderline between repair expenditure and expenditure on improvements is ill–defined.

One of the main sources of argument often concerns the case where a **3–016** building is replaced in whole or part because it is uneconomic to repair it. If the building is a separate entity the cost of replacement is capital: (*O'Grady* v. *Bullcroft Main Collieries Ltd.* (1932) 17 T.C. 93, *Phillips* v. *Whieldon Sanitary Potteries Ltd.* (1952) 33 T.C. 213, *Brown* v. *Burnley Football and Athletic Co. Ltd.* [1980] S.T.C. 424). However, the replacement of part of a single building can qualify as a repair and the cost can be written off as a revenue expense (*Samuel Jones & Co. (Devondale) Ltd.* v. *I.R.C.* (1951) 32 T.C. 513).

The mere fact that a replacement building is a separate entity is not however conclusive, because if it forms an integral part of a larger unit the replacement could still be regarded as revenue expenditure (*Hodgkins* v. *Plunder & Pollak (Ireland) Ltd.* [1957] I.R. 58). For example, the replacement of a building that houses the sterilisation unit for a dairy could, even though a separate building, be regarded as part of a larger entity. The replacement of a Dutch barn by another similar unit, even if on the same site, would however be the replacement of an entire unit and therefore the cost would be capital. It would not be possible to argue that the whole farm represents an entirety so that the replacement of any building on the farm would be allowable as a repair.

A substantial element of improvement in the replacement would tend to indicate capital expenditure rather than a repair and it is not possible to argue that a deduction should be available for the notional cost of repairing the existing building where a replacement is constructed: *William P. Lawrie* v. *I.R.C.* (1952) 34 T.C. 20, *Thomas Wilson (Keighley) Ltd.* v. *Emmerson* (1960) 39 T.C. 360, *Curtin* v. *M Ltd.* [1960] I.R. 97 and *Margrett* v. *The Lowestoft Water & Gas Company* (1935) 19 T.C. 481.

Where expenditure can be argued to be on repairs or allowable replacements there is no requirement that the building be repaired to its former state and the inclusion of structural alterations would not necessarily lead to disallowance as capital expenditure, (*Conn* v. *Robins Brothers Ltd.* (1966) 43 T.C. 266).

Expenditure on converting a building to a different use would normally be capital expenditure (*Lothian Chemical Co. Ltd.* v. *Rogers* (1926) 11 T.C. 508).

3–017   Arguments often arise over the cost of repairing an asset acquired in a dilapidated condition. If the asset was nevertheless usable for the purpose for which it was acquired the costs of repair should be deductible as a revenue expense (*Odeon Associated Theatres Ltd.* v. *Jones* (1971) 48 T.C. 257). If, however, the repairs were essential before the asset could be brought into use, the cost would be disallowed as capital expenditure (*Law Shipping Co. Ltd.* v. *I.R.C.* (1923) 12 T.C. 621, *Jackson* v. *Laskers Home Furnishers Ltd.* (1956) 37 T.C. 69). If the asset has been acquired just before commencing to trade, expenditure on improvements needed as a result of dilapidations before the trade began would in any event be regarded as capital (*Bidwell* v. *Gardiner* (1960) 39 T.C. 31).

The fact that the expenditure is essential, *e.g.* on coastal protection or a new access road, does not make it allowable. Such expenditure would normally be capital (*Avon Beach & Cafe Ltd.* v. *Stewart* (1950) 31 T.C. 487, *Pitt* v. *Castle Hill Warehousing Co. Ltd.* [1974] S.T.C. 420).

3–018   The argument about an improvement element being treated as capital often arises in relation to the drainage of fields. The replacement of mole drainage by tile or polythene pipe drainage would count as an improvement on which agricultural buildings allowances could be claimed (see Chapter 5). The replacement of tile drainage by polythene pipe drainage however should be claimed as a mere repair using equivalent modern materials and therefore wholly deductible on the authority of *Conn* v. *Robins Brothers Ltd.* (1966) 43 T.C. 266 as mentioned above (para. 3–016). A tenant farmer will often write the cost off over three or four years by including a reducing proportion of the cost in the annual valuation of tenant's rights.

Plant leasing

3–019   The Revenue is likely to inquire into the true nature of up-front leasing payments. For example, a farmer who wishes to buy a combine at a retail price of £90,000 may be offered a discount of £22,500 which can be added to the amount given in part-exchange for his old combine, or can be paid to him in cash or can be deducted from the cost of a new combine. He may add it to the part-exchange value and use the total to pay the first instalment on a leasing agreement. The Revenue does not appear to challenge the large amount of the initial leasing payment. It is understood that provided that it is treated as income in the lessor's hands it feels that there is no loss of revenue.

Specific disallowables

3–020   Other items specifically disallowed include non-trade expenses and taxes: I.C.T.A. 1988, s.74(e); capital expenditure generally: I.C.T.A. 1988, s.74(f); interest forgone on expenditure incurred: I.C.T.A. 1988, s.74(h); a general provision for bad debts, although bad debts actually written off or a specific provision for identified debts considered likely ultimately to become bad is deductible: I.C.T.A. 1988, s.74(j). Expenditure recoverable under an insurance claim cannot be claimed as an expense: I.C.T.A. 1988, s.74(k) and (l), nor can annuities or other annual payments such as patent royalties, mining rents or wayleaves paid under deduction of tax: I.C.T.A. 1988, s.74(p) and (q), apart from interest: I.C.T.A. 1988, s.74(m).

Interest

Interest paid to a non-resident which exceeds a normal commercial rate **3–021** of interest is specifically disallowed as a trading expense: I.C.T.A. 1988, s.74(n), but otherwise interest incurred for trading purposes will normally be deductible as a trading expense. Where a proprietor's capital and current accounts are overdrawn, however, the Revenue might argue that part of the interest paid relates to personal drawings and is therefore disallowed as a private expense under I.C.T.A. 1988, s.74(b). Interest paid to non-residents normally has to be paid under deduction of tax under I.C.T.A. 1988, ss.82 and 349 as does interest paid by a company or partnership with a corporate partner: I.C.T.A. 1988, s.349(2)(a) and (b) except to a bank carrying on business in the U.K.: I.C.T.A. 1988, s.349(3).

Interest in respect of loans to acquire let property are considered in Chapter 9.

Interest on a loan to purchase plant and machinery by a partner entitled to capital allowances qualifies for relief, but only in respect of interest paid within the three years following the year of assessment in which the debt was incurred. A loan to acquire plant or machinery only partly used for a trade or profession is eligible for interest relief on the appropriate part. Where an employee borrows money for the purchase of plant or machinery required for the employment and qualifying for capital allowances, the loan is eligible for interest relief for a period of up to three years from the end of the year of assessment in which the debt was incurred. A loan to purchase plant or machinery used partly for the employment qualifies for interest relief on the appropriate part. Unpaid purchase consideration counts as a loan. Apportionment is permissible for loans which only partly qualify for interest relief: I.C.T.A. 1988, s.359(1)–(4), 367(3), (4).

Property owned by a partner and occupied rent free by the partnership may be bought with the assistance of a loan, the interest on which is paid by the partnership. This is treated as rent paid by the firm, and rent received covered by interest paid by the partner owning the property, SP4/85.

Further aspects of loan interest are dealt with in Chapter 20. **3–022**

Donations

Charitable donations are normally disallowed by I.C.T.A. 1988, s.577(8) **3–023** but small local donations for charitable purposes or to local trade associations are deductible under Extra Statutory Concession B7 (1988).

Entertaining

Entertaining expenditure, including expenditure on capital assets used **3–024** for entertainment, is not an allowable deduction as a trading expense or for capital allowances purposes: I.C.T.A. 1988, s.577.

Entertaining by employees is usually disallowed in the employer's computation although a round sum allowance to cover entertaining and other expenses would be an allowable expense of the trader and would be charged as income on the employee: I.C.T.A. 1988, s.577(1)(b) and Revenue Booklet I.R. 480, para. 17(8). The employee could then claim any allowable expenses. Bona fide staff entertainment is not disallowed as business entertainment. Gifts are disallowed as trading expenses apart from advertising items, such as calendars, which cost no more than £10 per recipient and do not consist of food, drink or tobacco: I.C.T.A. 1988, s.577(8).

Redundancy payments

3–025   Redundancy payments are normally allowable as a trading expense either under I.C.T.A. 1988, ss.579, 580 or, on discontinuance of farming, as extended by the Revenue Statements of Practice SP1/81 (March 10, 1981) and SP11/81 (November 6, 1981).

Specific deductions

3–026   Other important statutory deductions include payments for technical education related to the trade paid to a recognised university or college.

Fees paid to an agricultural college for a farmer's son's education was allowed in *Wickwar* v. *Berry* (1963) 41 T.C. 33: I.C.T.A. 1988, s.84.

Where a premium is paid on the original grant of a lease of up to 50 years, as opposed to on the assignment of an existing lease, part of the premium received is taxable under Schedule A on the recipient in accordance with I.C.T.A. 1988, ss.34 or 35 (see Chapter 9). In such cases a deduction may be claimed by the payer of the premium for that proportion taxed on the recipient as income spread equally over the period of the lease I.C.T.A. 1988, s.87. In other cases a premium or lump sum to vary the terms of a lease would not be deductible, *Rolli Estates Ltd.* v. *East Africa Income Tax Commissioners* [1961] 1 W.L.R. 329, *Whitehead* v. *Tubbs (Elastics) Ltd.* [1984] S.T.C. 1, *Tucker* v. *Granada Motorway Services Ltd.* [1979] S.T.C. 393, *MacTaggart* v. *B & E Strump* (1925) 10 T.C. 17.

Bad debts recovered are brought into charge to tax when received: I.C.T.A. 1988, s.94.

3–027   Where there is an artificial cessation on a change of partners under I.C.T.A. 1988, s.113 or where a company ceased to trade, a deduction may be claimed against future profits for bad debts in respect of transactions arising before the change: I.C.T.A. 1988, s.89.

3–028   Incidental costs of obtaining loan finance are allowable under I.C.T.A. 1988, s.77 and discounts on bills of exchange under I.C.T.A. 1988, s.78. Contributions to local enterprise agencies are allowable under I.C.T.A. 1988, s.79 and approved profit sharing schemes under I.C.T.A. 1988, s.85. A deduction may be claimed for remuneration of employees seconded to charities and educational establishments under I.C.T.A. 1988, s.86.

Case law

3–029   There has over the years been extensive case law on the allowability of certain expenses and the following cases could be relevant to a farming trade.

Loans and guarantees by farmers would not normally be within their 3–030
trade and any losses arising would not be deductible as a trading expense
(*I.R.C.* v. *Hagart & Burn-Murdoch* (1929) 14 T.C. 433), although relief
may be available as a capital loss under C.G.T.A. 1979, s.136.

Misappropriations by employees will normally be deductible but misap- 3–031
propriations by directors would not (*Curtiss* v. *J & G Oldfield Ltd.* (1925) 9
T.C. 319 and *Bamford* v. *A.T.A. Advertising Ltd.* (1972) 48 T.C. 359).

Interest incurred on a loan used for the purpose of a trade is deductible 3–032
so long as the trade continues. If there are two trades, one of which ceases,
the interest on that portion of the loan ceases to be deductible (*Olin
Energy Systems Ltd.* v. *Scorer* [1982] S.T.C. 800). It may be helpful that all
farming in the U.K. is a single trade: I.C.T.A. 1988, s.53.

Legal expenses on capital matters such as raising capital (*Texas Land and* 3–033
Mortgage Co. v. *Holtham* (1894) 3 T.C. 255) would be disallowed except in
connection with the issue of loan stock under I.C.T.A. 1988, s.77. Company formation expenses are disallowed (*Kealy* v. *O'Mara (Limerick) Ltd.*
[1942] I.R. 616) as are those of reorganising the share capital (*Archibald
Thomson, Black & Co. Ltd.* v. *Batty* (1919) 7 T.C. 158 and *Montreal Coke
& Manufacturing Co.* v. *Minister of National Revenue* [1944] 1 All E.R.
743).
　Legal costs in respect of licences are capital (*Southwell* v. *Savill Brothers
Ltd.* (1901) 4 T.C. 430) as would be the costs of obtaining planning permission (*ECC Quarries Ltd.* v. *Watkis* [1975] S.T.C. 58, see SP 4/78
(November 6, 1978) and legal expenses in connection with the acquisition
or sale of freehold property, or on the granting of a lease. Legal expenses
on the renewal of an existing lease would however be deductible.
　The costs of **accountancy advice** in the preparation and submission of tax 3–034
computations is deductible but the cost of an appeal before the Commissioners is not: *Allen* v. *Farquharson Brothers & Co.* (1932) 17 T.C. 59,
Smith's Potato Estates Ltd. v. *Bolland* (1948) 30 T.C. 267 and *Rushden
Heal Co. Ltd.* v. *Keene* (1948) 30 T.C. 298.

Remuneration paid must be shown to be reasonable having regard to the 3–035
services provided (*Copeman* v. *William Flood & Sons Ltd.* (1940) 24 T.C.
53, *Dollar & Dollar* v. *Lyon* [1981] S.T.C. 333, *Stott & Ingham* v. *Tre-
hearne* (1924) 9 T.C. 69 and *Johnson Brothers & Co.* v. *I.R.C.* (1919) 12
T.C. 147).
　If wages are to be paid to a wife not only must they be reasonable for the
services rendered but they must actually be paid and the P.A.Y.E and
national insurance dealt with properly (*Thompson* v. *Bruce* (1927) 11 T.C.
607, *Moschi* v. *Kelly* (1952) 33 T.C. 442).

Gratuities made in accordance with normal practice to retiring 3–036
employees are deductible (*Smith* v. *Incorporated Council of Law Reporting
for England and Wales* (1914) 6 T.C. 477) as are pension payments under
approved superannuation schemes under I.C.T.A. 1988, ss.590–596. A
payment to set up a staff pension fund in excess of the amount approved by

the Superannuation Funds Office would be capital (*Atherton* v. *British Insulated and Helsby Cables Ltd.* (1925) 10 T.C. 155).

3–037 **Rent payable** is, not surprisingly, a proper deduction for tax purposes but so is rent for premises no longer required (*I.R.C.* v. *Falkirk Iron Co. Ltd.* (1933) 17 T.C. 625) and the rent differential if premises no longer required are sub-let at less than the rent payable (*Hyett* v. *Lennard* (1940) 23 T.C. 346). However, a lump sum payment for the cancellation of a lease is a capital payment (*Mallett* v. *Staveley Coal & Iron Co. Ltd.* (1928) 13 T.C. 772, *Cowcher* v. *Richard Mills & Co. Ltd.* (1927) 13 T.C. 216, *Union Cold Storage Co. Ltd.* v. *Ellerker* (1938) 22 T.C. 547 and *Dain* v. *Auto Speedways Ltd.* (1959) 38 T.C. 525). A premium on the grant or assignment of a lease is capital subject to any relief under I.C.T.A. 1988, s.87 (*MacTaggart* v. *B & E Strump* (1925) 10 T.C. 17) as is a payment to modify the terms of a lease (*Tucker* v. *Granada Motorways Services Ltd.* [1979] S.T.C. 393).

Rent paid at a proper commercial rate to a partner by a farming partnership would be deductible (*Heastie* v. *Veitch & Co.* (1933) 18 T.C. 305).

It is not unusual for a farmer's wife to buy land in her name financed by a loan. If she is not a partner she bears the loan interest which she can claim only against the rent she receives. Where the interest paid exceeds the rent received the excess interest is not allowable but may be carried forward: I.C.T.A. 1988, s.355(1)(b) and (4).

If, however, rent is paid for what is in reality the acquisition of the freehold by instalments the additional rent is a capital payment (*I.R.C.* v. *Land Securities Investment Trust Ltd.* (1969) 45 T.C. 495, *Littlewoods Mail Order Stores Ltd.* v. *McGregor* (1969) 45 T.C. 519).

3–038 **Payments to trade associations** and purchasing and marketing co-operatives would normally be deductible (*Guest Keen & Nettlefolds Ltd.* v. *Fowler* (1910) 5 T.C. 511, *Thomas* v. *Richard Evans & Co. Ltd.* (1927) 11 T.C. 790).

3–039 **Travelling expenses** purely for the purpose of the trade are deductible (*Edwards* v. *Warmsley Henshall & Co.* (1967) 44 T.C. 431). Travel from the home to place of business is private (*Newsom* v. *Robertson* (1952) 33 T.C. 452) even if there are business calls on route (*Sargent* v. *Barnes* [1978] S.T.C. 322).

In the case of a farmer, however, it is normal to live on or very close to the premises and if the home is the base the whole of the business travelling costs would be allowable (*Horton* v. *Young* (1971) 47 T.C. 60).

3–040 It is, however, necessary for the expense to be wholly and exclusively for the purpose of the business and if there is a dual purpose, strictly speaking the whole of the cost has to be disallowed (*Bowden* v. *Russell & Russell* (1965) 42 T.C. 301). If the prime purpose of the trip is a business purpose and any non-business purpose is purely incidental the non-business element may be ignored (*Bentleys Stokes & Lowless* v. *Beeson* (1952) 33 T.C. 491). The expenses incurred by a farmer to visit Australia with a view to farming there was disallowed in *Sargent* v. *Eayrs* [1973] S.T.C. 50. It is arguable that where a tour is organised by, *e.g.* the National Farmers Union or the Country Landowners Association specifically to visit farms and study farming methods, in the U.K. or abroad, the cost should be fully deductible, although the costs of an accompanying spouse would probably be disallowed.

The excess cost of lunching away from home was disallowed in *Caillebotte* v. *Quinn* [1975] S.T.C. 265. If, however, a business trip involves an overnight stay the allowable costs would include the travelling, hotel accommodation and food (*Nolder* v. *Walters* (1930) 15 T.C. 380).

Protective clothing used specifically for the purpose of the trade is deductible but the additional costs of normal clothing as a result of the exceptional wear and tear as a result of farming would not be allowable (*Woodcock* v. *I.R.C.* (1977) 51 T.C. 698, *Hillyer* v. *Leeke* [1976] S.T.C. 490, *Ward* v. *Dunn* [1979] S.T.C. 178, *Mallalieu* v. *Drummond* [1983] S.T.C. 665).   **3–041**

Medical expenses would not normally be deductible as they are partly for the private purpose of getting well and therefore fail on the grounds of duality of purpose (*Murgatroyd* v. *Evans-Jackson* (1966) 43 T.C. 581, *Norman* v. *Golder* (1944) 26 T.C. 293, *Prince* v. *Mapp* (1969) 46 T.C. 169).   **3–042**

Fees for services, even if paid to an associated service company or partnership, should be deductible provided that they are reasonable for the services provided and represent only a modest profit on the cost of the service organisation (*Stephenson* v. *Payne Stone Fraser & Co.* (1967) 44 T.C. 507).   **3–043**

Hire-purchase payments are split between the capital amount equal to the cash price on which capital allowances would be claimed and an interest element which is written off as a trading expense over the life of the agreement, either on a time basis or an actuarial method related to the remaining amount outstanding under the agreement: *Darngavil Coal Co. Ltd.* v. *Francis* (1913) 7 T.C. 1.   **3–044**

Removal expenses, *e.g.* on moving to a new farm, are normally allowed in practice but could be disallowed following *Granite Supply Association Ltd.* v. *Kitton* (1905) 5 T.C. 168.   **3–045**

Payments for technical assistance were allowed in *Paterson Engineering Co. Ltd.* v. *Duff* (1943) 25 T.C. 43.   **3–046**

Where an asset of an enduring nature is acquired as a result of a payment, it is likely to be regarded as a capital payment, *I.R.C.* v. *Pattison & Others* (1959) 38 T.C. 617, *Atherton* v. *British Insulated & Helsby Cables Ltd.* (1925) 10 T.C. 155 (a leading case on the distinction between a capital and a revenue expense).

Examples of typical farm trading and profit and loss accounts showing the main expense headings are shown below:

3–047 EXAMPLE

Arable farming partnership
Trading and Profit and Loss Account

	1988		1987	
	£	£	£	£
SALES				
Sheep and wool		2,325		2,501
Peas		24,290		19,776
Wheat		204,074		200,884
Barley and oats		40,787		40,857
Rape		52,178		62,013
Contract work		2,230		1,197
		325,884		327,228
COST OF SALES				
Opening stock	173,539		181,730	
Purchases				
Seed	14,118		15,761	
Spray	20,481		24,012	
Livestock	1,468		1,427	
Fertiliser	25,034		29,659	
	234,640		252,589	
Closing stock	187,596		173,539	
		47,044		79,050
GROSS PROFIT		278,840		248,178
FARM OPERATING COSTS				
Wages	32,767		32,537	
Light, heat, petrol and oil	8,800		12,017	
Hire of plant and haulage	255		1,447	
Repairs and maintenance:				
Plant and machinery	15,950		15,529	
Premises	16,543		25,954	
Licences	612		2,183	
Sundry fare costs	963		—	
Rents	6,486		5,993	
Rates and water	3,055		2,690	
carried forward	85,431	278,840	98,350	248,178

brought forward	85,431	278,840	98,350	248,178
Insurance	5,743		2,681	
Telephone	691		629	
Training costs	—		520	
General expenses	2,833		2,762	
		94,698		104,942
FARMING PROFIT		184,142		143,236

OTHER OPERATING COSTS

Professional fees	2,345		8,089	
Depreciation:				
Tractors and combines	46,554		29,633	
Plant and equipment	15,592		12,907	
Motor vehicles	2,761		905	
Profit on disposal of assets	(11,476)		(4,945)	
	55,776		46,589	
Bank interest	57		51	
Sundry	—		(21)	
		55,833		46,619
		128,309		96,617

OTHER INCOME

Rent and wayleaves	11,991		13,000	
Interest received	4,199		10,762	
Grants receivable	138		—	
		16,328		23,762
NET PROFIT		£144,637		£120,379

Shared among partners

A	—	6,000
B	63,640	60,621
C	57,855	48,039
D	23,142	5,719
	£144,637	£120,379

3–048 EXAMPLE

Mixed farming company
Trading and Profit and Loss Account

	1988		1987	
	£	£	£	£
Sales during year				
Milk		168,420		160,616
Cattle		27,699		45,323
Corn		168,454		245,830
Rape		48,484		30,694
Potatoes		127,610		62,325
Other		—		247
		540,667		545,035
Cost of Sales				
Opening valuation	119,185		122,295	
Cattle purchases, semen and A.I. charges	2,285		16,567	
Vets' fees and medicines	5,254		1,863	
Purchased foodstuffs	61,926		43,243	
Seeds	44,667		26,558	
Fertilisers and sprays	82,433		72,542	
Contract hire and cartage	650		2,506	
Marketing costs	6,701		9,420	
Crop sundries	21,164		6,446	
Stock sundries	7,208		3,123	
Wages and social security costs	107,856		89,154	
Vehicles and equipment running costs	48,557		56,027	
Electricity and water	11,851		8,295	
Property repairs and maintenance	26,892		8,178	
Contract charges	2,388		15,898	
Insurance	6,847		4,385	
Rent and rates	7,720		7,028	
Sundry expenses	1,746		1,835	
Depreciation of fixed assets	67,828		57,573	
	633,158		552,936	
Closing valuation	142,902	490,256	119,185	433,751
carried forward		50,411		111,284

brought forward		50,411		111,284
Administrative costs				
Directors' salaries	25,500		31,700	
Directors' social				
security costs	2,288		2,519	
Directors' pension				
premiums	4,316		4,316	
Auditors'				
remuneration	2,200		2,200	
Accountancy,				
secretarial and				
professional services	12,991		7,076	
Telephone	1,291		1,255	
Office administration				
and expenses	4,426	53,012	2,854	51,920
		(2,601)		59,364
Other operating income				
Rents received	3,744		2,109	
Wayleaves	106	3,850	101	2,210
		1,249		61,574
Interest received				
Bank deposit interest		4		3
		1,253		61,577
Interest payable				
Bank interest and				
charges	24,251		18,198	
HEGG interest	8,246	32,497	695	18,893
TRADING PROFIT (LOSS)				
FOR THE YEAR		£(31,244)		£42,684

Farm profit averaging

One of the hazards of farming is the weather. This affects both crop   **3–049**
yields and the productivity of livestock on the one hand and costs on the
other. This means that a good season both increases the income and
reduces the costs, whereas a bad season has the opposite effect. Thus the
profits of a farmer can fluctuate substantially from year to year and with
progressive rates of tax he could well be paying at the basic rate one year
and at the higher rate the next. Relief is therefore introduced by I.C.T.A.
1988, s.96 which enables the profits from farming or market gardening for
two years to be aggregated and divided by two for income tax purposes,
provided that certain conditions are met. The relief is not available for corporation tax purposes: I.C.T.A. 1988, s.96(11).

It is necessary to make a claim for the section to apply: I.C.T.A. 1988, s.96(1) and such claim must be made by notice in writing to the inspector within two years of the end of the second year of assessment to which the claim relates, or by the end of the year of assessment following that in which an adjustment is made to the profits: I.C.T.A. 1988, s.96(8).

3–050 Where the profits of two consecutive years of assessment are such that one does not exceed 70 per cent. of the other, or are nil, the profits for each year are adjusted to one-half of the aggregate profits of the two years: I.C.T.A. 1988, s.96(2).

If the profits for one year exceed 70 per cent. of the profits for the other but are less than 75 per cent. of those profits, marginal relief is available: I.C.T.A. 1988, s.96(3). It is necessary to subtract the lower-year profit from the higher and multiply the result by three. From this figure is deducted three-quarters of the higher-year profit; the resultant figure is added to the lower-year profit and subtracted from the higher-year profit.

The second year amended assessment may be further amended by an averaging claim for the next year.

3–051 If the profits of one year are more than 75 per cent. of the profits of the other no averaging relief is available. Nor is relief available in respect of a year of assessment in which the trade is set up and commenced or permanently discontinued, or is treated as such as a result of a partnership change under I.C.T.A. 1988, s.113, s.96(4)(b). It is not possible to make a claim for an earlier year of assessment where a claim has already been made for a later year: I.C.T.A. 1988, s.96(4)(a). For example, if a claim has been made for 1987–88 and 1988–89 it would not then be possible to make a claim for 1986–87 and 1987–88, but a claim for 1988–89 and 1989–90, using the amended figure for 1988–89 as a starting point, may be made.

For the purpose of this section a loss merely counts as a nil profit and the normal loss relief is not affected; I.C.T.A. 1988, s.96(5)(a). Any adjustments to the penultimate and antepenultimate years on cessation are calculated ignoring the farm averaging provisions: I.C.T.A. 1988, s.96(5)(b). If the profit of either year is adjusted for any other reason the claim is in the first instance deemed not to have been made and there is then a further opportunity for making a claim by the end of the tax year following that in which the adjustment takes place: I.C.T.A. 1988, s.96(5)(c) and (8). Averaging provisions are available for partnership profits on a claim made jointly by all the partners who are individuals, including all those who were partners before and after any change in the partnership where a I.C.T.A. 1988, s.113 continuation election has been made: I.C.T.A. 1988, s.96(6). Personal representatives sign for a deceased partner.

3–052 Averaging relief is given before deducting losses, capital allowances or charges on income except those given by way of deduction in computing the profits: I.C.T.A. 1988, s.96 (7). An averaging claim made in time enables any other claim for any year of assessment affected by the averaging claim to be made, revoked or amended within the time limit for submitting an averaging claim: I.C.T.A. 1988, s.96(9). This does not however enable an averaging claim which has been made and agreed to be revised. All consequential amendments are to be made: I.C.T.A. 1988, s.96(10).

Claim for this purpose does not include an election, *e.g.* for taxation of wife's earnings under I.C.T.A. 1988, ss.287, 288 up to 1989–90 but the

Revenue can extend the time–limit: C.C.A.B. Press Release T.R. 386, March 31, 1980, para. 31.

For the purposes of farm profit averaging, farming includes the intensive rearing of livestock or fish on a commercial basis for the production of food for human consumption: Extra Statutory Concession A29 (1988). **3–053**

EXAMPLE

Profits as adjusted for tax purposes
assessable (a) 1987–88		£7,300
(b) 1988–89		£10,000
(c) 1989–90		£6,500
(b) − (a) = £10,000 − £7,300 =		£2,700
£2,700 × 3 =		£8,100
(b) × $\frac{3}{4}$ = £10,000 × $\frac{3}{4}$ =		£7,500
£8,100 − £7,500 =		£600

amended assessment 1987–88
£7,300 + £600 =   £7,900

provisionally amended assessment
1988–89
£10,000 − £600 =   £9,400
original assessment 1989–90   £6,500 (less than 70%
                                                  ——— of £9,400)
                                                  £15,900

amended assessment 1988–89
$\frac{1}{2}$ × £15,900 =   £7,950
provisionally amended assessment
1989–90
$\frac{1}{2}$ × £15,900 =   £7,950

Market gardening

Market gardening, like farming, is deemed to be a trade within I.C.T.A. **3–054** 1988, s.53(1). Market garden buildings and works qualify for agricultural buildings allowances in view of C.A.A. 1968, s.69.

The distinction between market gardening and farming is for most purposes academic, except that: (a) there is no provision that all market gardening in the U.K. be treated as a single trade, unlike farming: I.C.T.A. 1988, s.53(2); and (b) for the purposes of the business expansion scheme the trade of farming but not market gardening is excluded from relief in respect of an issue of shares after March 13, 1984, under I.C.T.A. 1988, s.297(2)(j), although the provisions about interests in land in I.C.T.A. 1988, ss.294–296 limit the application of the scheme.

There have been a number of cases on the distinction between farming **3–055** and market gardening which was of importance when farming was assessed under Schedule B and market gardening was not. The leading case is *Monro & Cobley* v. *Bailey* (1932) 17 T.C. 607 in which it was stated at p. 666:

    (a) the statutory "garden" must be within reasonable limits a defined unit of occupation having a degree of permanence and continuity;

    (b) the "splitting" of a farm into two separate units of the kind indicated in subparagraph (a) is only legitimate if such a division is present in fact;

    (c) a finding that a farm is "worked as a single mixed unit" prima facie means that it is one in management, cultivation, labour, business accounting and so on and therefore that it is a single farming unit, and if so, that fact must necessarily exclude the possibility of "splitting";

    (d) if any given unit of occupation is worked in part as a farm and in part as a garden, but is not susceptible of splitting under subparagraph (b), it must be held to be either wholly a farm, or wholly a garden; it cannot be held to be a garden unless gardening is the dominant purpose of the whole; and such a conclusion is not legitimate, unless the farming part is found as a fact to be truly a necessity of and for the garden—as it was in Monro's case;

    (e) but odd fields or fields scattered over a farm amongst what are plainly farm fields cannot be added up arithmetically into a "garden," whatever crops they carry, either for the purpose of splitting or for the purpose of subparagraph (d): in other words, such arable fields in a farm, being usually subject to rotation, sometimes bearing vegetables or fruit crops of kinds which may also regularly be seen growing in market gardens, do not constitute a unit "occupied" as a garden.

3–056   The remaining cases on this point are somewhat difficult to reconcile. In *Dennis* v. *Hick* (1935) 19 T.C. 219 vegetable crops consisted of 263 out of 350 acres and the activity was held to be market gardening. In *Kerr* v. *Davis* (1939) 22 T.C. 515 vegetables were grown on 115 acres out of 120 but the activity was held to be farming, and a similar decision was reached in *Williams* v. *Rowe* (1939) 22 T.C. 508, even though the crops grown were vegetables, largely cultivated by hand. In *Bomford* v. *Osborne* (1939) 23 T.C. 642 the arable land in a farm of 550 acres amounted to 229 acres of which 213 were devoted to the growing of vegetables and fruit and the remaining 16 acres to the growing of wheat. It was held that the whole of the arable land constituted a market garden with the wheat acreage being merely ancillary to the market gardening. The remaining area of the farm was devoted to normal farming activities and dealt with at that time under the normal Schedule B assessment.

    In the case of *David Lowe & Sons Ltd.* v. *I.R.C.* (1941) 24 T.C. 105 the distinction between market gardening and farming was held to be the intensity of cultivation and, as in that case normal agricultural methods were used, it was held that the business constituted farming. This is supported by *Roberts* v. *Barter's Executors* and *Roberts* v. *Williamson* (1944) 26 T.C. 201 in which it was held that intensively cultivated watercress beds amounted to market gardening. In *Cross* v. *Emery* (1949) 31 T.C. 194 the taxpayer's activities were apportioned between market gardening and fruit farming as a separate activity.

National insurance planning

National insurance benefits are funded by means of contributions pay- **3–057** able to the Secretary of State, supplemented by the Treasury. The four classes of contributions under the Social Security Act 1975, s.1 are:

(1) Class 1 contributions, earnings-related and in two forms—primary contributions from employees, and secondary contributions from employers and anyone paying earnings.
(2) Class 2 contributions, payable by the self-employed at a flat rate (see s.7).
(3) Class 3 contributions, purely voluntary and paid to provide or make up entitlement to benefits (see s.8).
(4) Class 4 contributions, payable by the self-employed on profits.

The continued increase in contribution rates has led those bearing the burden to seek ways in which to minimise the cost. As planning becomes more sophisticated it is to be expected that the Department of Social Security (DSS) will clamp down on the more obvious loopholes such as the payment of salaries in kind which are not currently chargeable to national insurance. Since October 6, 1985 the Upper Earnings Limit (UEL) has not applied to Class 1 secondary contributions payable by the employer, currently at a rate of 10.45 per cent. of gross earnings. There are, however, a number of areas where planning may be possible.

Payments in kind

"Earnings" for the purposes of NIC are defined as "remuneration or **3–058** profit derived from employment." There are, however, a number of exceptions specified in the Social Security (Contributions) Regulations 1979, reg. 19, *e.g.* "any payment in kind (other than a security) or by way of the provision of board or lodging or of services or other facilities." NIC are therefore payable whenever an employee is remunerated with cash or payments which are capable of being converted into cash.

A few examples of exempt "payments in kind" are considered below: **3–059**

(1) Where an employer supplies a company asset for the use of the employee, it is exempt: the most obvious example is a company car but this applies to any asset, for example a television or video equipment or indeed a flat.
(2) Luncheon vouchers which are not convertible into cash are exempt.
(3) Where an employer meets a liability of the employee the current DSS view is that if an employee enters into a contract with a third party, for the supply of goods or services, and the employer actually settles the liability on the employee's behalf this is not a payment in kind and gives rise to a NIC liability. An example of this is the employee's telephone bill where the contract is between British Telecom and the employee but is paid by the employer. One way to

avoid this is for the employer to enter into the contract directly with the supplier. This is perhaps impractical for telephone payments but is quite simple in other instances where the employer, say, gives the employee a free holiday as an incentive prize or bonus. If the employee selects the holiday but the employer contracts with the travel company to supply the chosen holiday to the employee this is probably exempt from NIC as a payment in kind.

(4) An employee's contributions to a company pension scheme do not reduce earnings for NIC purposes but the employer's contributions to an approved scheme are considered a payment in kind and therefore are not chargeable to NIC (nor do they give rise to an income tax liability). A pension scheme is therefore a very effective way of providing an employee with a tax and NIC free benefit in kind. If a "net pay" scheme is being operated for P.A.Y.E. purposes the employees' national insurance contributions should still be calculated on the gross pay.

3–060   A word of caution on the subject of benefits concerns the principle established in the case of *Heaton* v. *Bell* (1969) 46 T.C. 211 which may also apply to NIC although this has not been conclusively established. The case concerned an employee who sacrificed part of his salary as compensation to his employer for the private use of a company car. As part of the arrangements the employee had the option to surrender the car and revert to his normal gross salary. It was held that as the value of the car could be converted into cash the gross salary was taxable; this was not therefore a way of reducing the tax liability of the employee. In so far as the decision is applicable for NIC purposes, liability thereunder can be avoided by ensuring there are no convertible rights available.

Cash payments in lieu of benefits are not exempt, irrespective of whether the benefit itself might have been.

Company directors

3–061   Where a director receives annual fees, in addition to his regular salary, these are treated as paid when the accounts are approved in general meeting (*i.e.* at the AGM), so that if it is possible to hold two AGMs in one tax year NIC will be saved where the director's fees exceed the upper earnings limit. The saving is limited to the employee's contribution as the employer's contribution is based on the employee's total earnings. Worthwhile NIC savings can still be made but this can affect a director's entitlement to contributory state benefits.

As dividends are not regarded as earnings for the purposes of NIC, major NIC savings can be made for a shareholding director and the company by paying dividends instead of remuneration.

3–062   However, there are a number of implications which should be considered:

(a) the directors may not be eligible for state benefits or pensions which are contributory, although this may be remedied to a certain degree by making Class 3 voluntary contributions or ameliorated by a mix of dividends and remuneration.

(b) other shareholders will be entitled to receive dividends, unless special classes of share are held, but this is not always possible where I.C.T.A. 1988, ss.138, 139 (share incentive schemes) are in point;

(c) the payment of dividends can increase the value of the shares;

(d) no contributions will be made by the directors towards the company pension scheme;

(e) dividends can only be paid if the company has available profits which can be distributed. There may therefore be some advantage in rewarding directors with a mixture of dividends and remuneration;

(f) advance corporation tax is payable on the amount of the dividend, credit for which will not be obtained immediately;

(g) the individual will receive the dividend, with a credit for the advance corporation tax, equivalent to basic rate tax, and will only have further tax to pay if he is a higher rate taxpayer. This liability will be payable on December 1 in the year following the tax year of receipt. Accordingly, careful planning can result in a significant cash flow advantage accruing to the individual, at the expense of accelerating the corporation tax payments.

(h) remuneration can sometimes be more effective where, *e.g.* a wife's earnings election would be beneficial (up to 1989–90).

(i) if the taxable profits of the company are between £150,000 and £750,000 its marginal rate of corporation tax is (1989–90) 37.5 per cent. so remuneration, which is a deductible expense, may be preferable to a dividend which is not.

The final decision as to whether a dividend, remuneration or a mixture of both is paid will depend on a number of factors including the impact on cash flow.

The self-employed option

Another method of achieving substantial NIC savings, both for the employer and the worker, is to have services supplied by self-employed workers. This arrangement can be far from easy to set up, bearing in mind other tax considerations, and it is important that the DSS and the Revenue accept that the worker is self-employed. **3–063**

Failure to satisfy the authorities that the self-employment is genuine means that if the issue is raised at some later date there could be a substantial liability for unpaid NIC and income tax under P.A.Y.E., with the possibility of penalties. **3–064**

Job sharing

It will be appreciated that weekly earnings of less than £43 (1989–90) do not attract contributions by either the employee or employer. Earnings from two or more employments with the same or an associated employer have to be aggregated. Employers may be able to reduce their employment costs by engaging part-time staff. **3–065**

Sundry payment points

3–066
1. If an individual has more than one job he may obtain deferment of the Class 1 employee contributions by applying to the DSS on form CF379, explained in leaflet NP28.
2. An individual may also apply for deferment if he is both employed and self-employed during the tax year, application being made on DSS form CF359 as explained in leaflet NP18.

Anti-avoidance

3–067   An important final aspect is the anti-avoidance regulations found in the Social Security (Contributions) Regulations 1979, S.I. 1979 No. 591:

Regulation 3   — relates to earnings periods;
Regulation 21   — relates to abnormal pay practices: these have not been defined by the DSS and the regulation is not used very often;
Regulation 22   — relates to irregular or unequal payments.

There is also the possibility that the *Furniss* v. *Dawson* principle (see Chapter 1) may apply to NIC, but this seems unlikely.

Class 4 national insurance contributions

3–068   A self-employed individual, whether a sole trader or partner, pays in addition to Class 2 flat rate contributions a variable contribution which is based on his taxable profits between lower and upper limits, and is collected through the income tax system under the Social Security Act 1975, s.9. One half of the Class 4 contribution is deductible for income tax purposes on making a claim: I.C.T.A. 1988, s.617(5). Class 4 contributions are levied on profits assessable under Schedule D, Cases I and II. The sum due is payable at the same time as the related income tax and the amount payable is shown separately on the notice of assessment.

3–069   Profits for this purpose are after deducting:

(1) Capital allowances deductible in arriving at the trading profit (Social Security Act 1975, Sched. 2, para. 2(a)(i)).
(2) Capital allowances given by way of discharge or repayment of tax (Social Security Act 1975, Sched. 2 para. 2(a)(ii)).
(3) Where a Schedule D, Case I loss has been set against employment or investment income the benefit of that relief is preserved for Class 4 purposes and consequently can be deducted fully from future farming profits (Social Security Act 1975, Sched. 2, para. 3(1)(a)).
(4) Should there be an excess of capital allowances over trading income relief is given for Class 4 purposes for that excess (Social Security Act 1975, Sched. 2, para. 3(1)(b)) by carry forward to reduce future income.
(5) Similarly Class 4 relief is granted for the carry forward of Schedule D, Case I losses (Social Security Act 1975, Sched. 2, para. 3(1)(c)).

(6) Class 4 relief is also given for terminal loss relief (Social Security Act 1975, Sched. 2, para. 3(1)(d)).

(7) Capital allowance balancing charges increase the figure on which contributions are levied (Social Security Act 1975, Sched. 2, para. 2(b)).

CHAPTER 4

Losses

4–001   As with profits, the provisions in I.C.T.A. 1988, s.53(2) that all farming carried on by any particular person or partnership or body of persons shall be treated as one trade means that a loss on any particular farm or farming activity is brought into the aggregate results of the farmer for the year. Only if the overall result, as adjusted for tax purposes, shows a loss do the provisions in this chapter come into force.

Unincorporated businesses

4–002   Where a loss is incurred in a trade under Schedule D, Case I the accounts need to be adjusted for tax purposes in exactly the same way as if the accounts showed a profit: I.C.T.A. 1988, s.382(4). Depreciation, entertaining and other disallowable items are deducted from the loss. The accounts and tax computations would be subject to normal Revenue enquiries in order to ensure that the correct adjusted loss figure is arrived at for tax purposes. If the accounts did not correctly reflect the results they would be rejected (*Hood Barrs* v. *I.R.C. (No. 4)* (1967) 46 A.T.C. 448).

Set-off against general income

4–003   The loss arising in the year of assessment may be set against the "total" taxable income for that year: I.C.T.A. 1988, s.380(1). For this purpose, "total income" means the aggregate taxable income before personal allowances but after deducting charges, *e.g.* allowable payments under covenants, court orders, etc.: I.C.T.A. 1988, s.835. For this purpose charges are not limited to those allowable at both the basic and the higher rates. The loss cannot be set against capital gains.

Actual basis

4–004   Technically the loss is calculated on the actual basis, namely that of the year ended April 5. Where necessary the profits and losses of accounting periods are apportioned on a time basis: I.C.T.A. 1988, s.72. In a corporation tax case it was held that if it were possible to arrive at a reasonable allocation of profit or loss into the appropriate years then the time apportionment basis should not apply: I.C.T.A. 1988, ss.70, 72 (*Marshall Hus & Partners Ltd.* v. *Bolton* [1981] S.T.C. 18). As the wording of the relevant legislation for income tax purposes is similar to that considered in the corporation tax case it is conceivable that, where there is sufficient information profits and losses should be allocated according to when they arose. In practice both the Revenue and the taxpayer are normally happy to time apportion the results. The apportionment of the results could mean that no loss relief is available to set against total income. If the loss is computed by splitting accounts, there is no forfeiture of any part of the loss even where the amount to be deducted from total income is restricted owing to a profit in the previous or succeeding accounting year.

Accounts basis

It is normal Revenue practice, however, to treat the accounting year as **4–005** coterminous with the tax year ending on the subsequent April 5. Should the accounts basis apply then a loss claim related to (say) the year ended December 31, 1988, would be treated as for the year ended April 5, 1989. If the penultimate and antepenultimate years are adjusted to the actual basis as a result of a cessation it is not normally Revenue practice to disturb loss relief already granted.

EXAMPLE

A farmer who had carried on business for some years had the following results, as adjusted for tax purposes:

		£
Profit – year ended October 5, 1988		5,000
Loss – year ended October 5, 1989		(18,000)
Year 1988–89 – actual basis:		
Profit 6/12ths of £5,000	2,500	
Loss: 6/12ths of (£18,000)	(9,000)	
Net loss available	£(6,500)	

The unrelieved loss of £11,500 (£18,000 less £6,500) which falls outside the tax year cannot be set against general income for 1987–88 but will be available for use in 1989–90.

If the accounts basis applies the whole of the loss of £18,000 for the accounting year ending in 1989–90 would be available to set against income for the year 1989–90 or 1990–91.

Compulsory actual basis

The actual basis must be used in the following cases: **4–006**

(a) The first three tax years of the business. This includes the three years commencing with a deemed cessation and recommencement.
(b) The fourth year where the first three years mentioned in paragraph (a) are assessed by reference to the actual profits.
(c) The year immediately following a year where a claim for loss relief was calculated on the actual basis to April 5.
(d) The year of cessation.
(e) Where accounts are prepared to irregular dates.
(f) Where the taxpayer requests that the strict actual basis applies.

Where the taxpayer has the choice of whether or not to compute the loss relief by reference to either basis, an accounting date falling just after April 5 would be useful. This is because there is greater scope for placing the relief into the tax year which is most beneficial.

Following year set-off

4–007 As an alternative, or in addition, to setting the loss against income of the tax year in which it arises it may be carried forward for one year and set against the total income of the following year of assessment, provided that the taxpayer is carrying on the same business in that later year: I.C.T.A. 1988, s.380(2). Consequently, it could possibly be tax beneficial to delay the date of cessation of a trade so that the activity is being undertaken in the following tax year thereby enabling the loss to be set against the income of two tax years. For example, if trading ceased on April 10, 1989, rather than March 31, 1989, the loss for 1988–89 could be set against total income of 1988–89 and 1989–90 subject to the income and loss levels. The taxpayer cannot allocate the loss relief between the two tax years as he wishes— however, until April 5, 1990, there are certain options to restrict the loss to his income and exclude the income of his spouse. Should the taxpayer wish to have the loss set against the income of the two tax years concerned then the maximum relief for the second year is the extent to which the loss exceeds the total income of the first year.

4–008 Before a decision is made to delay the cessation to a later tax year as a means of maximising the loss relief, the tax effect on the penultimate and antepenultimate years must be considered.

In the case of a continuing trade the relief is normally taken as the loss for the accounting period in the tax year. This means that the loss for the year ended April 30, 1989, would be set against the total income for 1990–91 where advantage is taken of this carry forward provision. This is the year for which the accounting period would be the normal basis period; thus a profit for the year ended April 30, 1989, would be assessed in 1990–91 on the usual preceding year basis. It may be appropriate to apportion the loss to the tax years concerned.

Where the loss has been claimed against the profits for the following year of assessment it is given in priority to any loss arising in that second year of assessment and this would normally be beneficial.

EXAMPLE

A farmer had the following results:

	£
Year ended April 30, 1988 loss	(15,000)
Year ended April 30, 1989 loss	(12,000)
Total income 1989–90	20,000

A claim is made for the 1988 loss to be set against the income, not of 1988–89, but of the following year 1989–90. In addition, a claim is made for the 1989 loss to be set against the income of 1989–90. These losses are used in the following order:

	£	£
Year 1989–90		
Total income		20,000
Less: Loss for year ended 30/4/88		(15,000)
		5,000
Loss for year ended 30/4/89	(12,000)	
Deduct: Utilised	5,000	(5,000)
Balance carried forward	£(7,000)	
Taxable		Nil

The balance of the 1989 loss can be set against the total income of 1990–91 if the trade is carried on in that later year.

Because of the preceding year basis of assessment it will be seen that a loss is effectively carried back against the profits of the preceding accounting period. For example, a profit for the year ended April 30, 1988, would be assessed in 1989–90 and a loss for the year ended April 30, 1989, would normally be relieved in the year 1989–90, that is against the profits for the year ended April 30, 1988, and any other taxable income for 1989–90.

Time-limits for set-off claim

A claim for relief must be made within two years of the end of the year of **4–009** assessment in which the loss relief is to be given: I.C.T.A. 1988, s.380(1). Therefore, if the loss relief is to be set against the total income of the year following the year of loss there are effectively three years in which to lodge a claim.

EXAMPLE

A farmer, who is continuing to trade, has incurred a loss for the year ended April 5, 1989. If the loss is to be set against total income then claims must be made within the following time limits:

Year of loss relief 1988–89 — claim by April 5, 1991.
Year of loss relief 1989–90 — claim by April 5, 1992.

Order of set-off

4–010   Where the loss is to be set off against total income it is used in the following order: I.C.T.A. 1988, s.382(2).

(1) As a deduction from the taxpayer's income of the corresponding class, then
(2) As a deduction from the taxpayer's other income, then (until April 5, 1990):
(3) As a deduction from the income of the taxpayer's spouse of the corresponding class, then
(4) As a deduction from the other income of the taxpayer's spouse.

From April 6, 1990, married couples will be taxed separately, so there will not be the opportunity to set a loss off against the spouse's income.

In order to determine the "income of the corresponding class," it is necessary to ascertain, had a profit rather than a loss arisen, whether that profit would have been earned or unearned income. Consequently, in most cases the loss will be set off against earned income before unearned income.

EXAMPLE

Farmer Jones has incurred a trading loss of £15,000. In 1989–90 he has earned income of £6,000 and investment income of £2,000, while his wife has earned income of £5,000 and investment income of £3,000. The loss will be set off against total income in the following order:

Order of loss set off	*Loss* £	*Income* £
Trading loss of Farmer Jones	15,000	
1st: Farmer Jones' earned income		6,000
Less: loss claim	(6,000)	(6,000)
Balance of loss	9,000	
2nd: Farmer Jones' investment income		2,000
Less: loss claim	(2,000)	(2,000)
Balance of loss	7,000	
3rd: Mrs. Jones' earned income		5,000
Less: loss claim	(5,000)	(5,000)
Balance of loss carried forward	2,000	

Balance of loss brought forward	2,000	
4th: Mrs. Jones' investment income		3,000
Less: loss claim (remainder)	(2,000)	(2,000)
	=====	———
Net income		£1,000
		=====

In the unusual situation where the profits of the trade would be regarded as unearned income a different order of set-off would apply because the "income of the corresponding class" would be different. For example, in the case of a sleeping partner or a farmer who was also a non-working Lloyd's underwriter who made a loss in that trade, that loss would be set first against investment income, then against earned income, then (until April 5, 1990) against the spouse's investment income, and finally against the spouse's earned income: I.C.T.A. 1988, s.382(2).

Restricted set-off

4–011 It is possible to restrict a claim (up to 1989–90) to the income of the person sustaining the loss without extending the claim to the spouse's income. For the loss to be restricted in this way it is necessary for the claim to make it clear that the loss is not to be set against the income of the claimant's spouse. Unless this is the case then the loss will be set against the income of both husband and wife. Therefore, care should be taken to ensure that the possibility of restricting the loss is considered before the claim is submitted. It is not possible to set the loss against part only of the income of a spouse or to use only part of a loss. It will often make sense to restrict the claim in this way to avoid a loss of personal allowances which cannot be carried forward and would otherwise be lost: I.C.T.A. 1988, s.382(1). In this connection it needs to be remembered that certain personal allowances given to the husband can be transferred to his wife if he has no assessable income.

Consequently, where a husband has incurred a trading loss, which is to be set against only his own total income to prevent the forfeiture of personal allowances, it might be useful to create income for the wife. This could possibly be achieved by paying a salary to her, so long as it can be justified commercially. Alternatively, capital could be transferred to her thereby giving her investment income.

4–012 Under certain circumstances repayments of income tax resulting from loss relief should be accounted for to the wife either as a result of case law or statute. Where a repayment of tax arose owing to a wife's farming losses set against her dividend income it was held that the refund obtained must be accounted for to her (*re Cameron deceased* (1965) 42 T.C. 539). This was the position even though the husband and wife were not separately assessed and the fact that the income of his wife was correctly to be aggregated with that of the husband: I.C.T.A. 1988, s.279. It is now provided that under certain circumstances a repayment of tax attributable to a wife's

Schedule E income resulting from losses and certain other reliefs must be made by the Revenue direct to the wife: I.C.T.A. 1988, s.281.

Should there be in force for the year of claim (up to 1989–90) an election for the separate taxation of wife's earnings and the corresponding income of the wife in relation to a loss incurred by her is earned income, then the loss can only be set against her earned income. The loss cannot be set against her unearned income or her husband's income. However, if the corresponding income related to a loss incurred by the wife is unearned, or the loss was incurred by the husband, the loss cannot be set against the earned income of the wife: I.C.T.A. 1988, s.287(7).

No double relief

4–013   Where relief has been given for the loss it cannot be given a second time, even where the accounting period in which the loss arises falls into more than one year of assessment, *e.g.* on commencement: I.C.T.A. 1988, s.382(3) *I.R.C.* v. *Scott Adamson* (1952) 17 T.C. 679.

As already stated, it is only possible for the loss to be set off against income for the following year under I.C.T.A. 1988, s.382(2) where the trade, profession or vocation continues. For this purpose a partnership change which is treated as a discontinuance under I.C.T.A. 1988, s.113(1) is not regarded as giving rise to a different trade for a continuing partner: I.C.T.A. 1988, s.380(3).

Where the appropriate election has been made under I.C.T.A. 1988, s.54, for the results of a woodland to be assessed under Schedule D, any loss arising is treated as if it were a loss sustained in a trade: I.C.T.A. 1988, s.380(4) (Chapter 12). (Such an election cannot be made after March 14, 1988.)

Carry forward

4–014   A loss which has not been relieved by set-off against total income under I.C.T.A. 1988, ss.380, 382, or otherwise may be carried forward against future profits: I.C.T.A. 1988, s.385(1) and (2), *Westward Television* v. *Hart* (1968) 45 T.C. 1. However, if a loss of an opening accounting period relating to the commencement of a new business is used when calculating the amount assessable then the loss to carry forward is restricted, as shown by the following two illustrations:

EXAMPLE

Mr. Smith commenced to farm on June 1, 1986. His first accounts ran to March 31, 1987, and his next set of accounts ran to March 31, 1988. For the first 10 months there was a loss of £12,200 and the profit for the next period of one year was £56,400.

The assessments are as follows:

Year of Assessment	Basis period		Sched. D. Case I assessment
			£ £
1986–87	Actual basis — 1/6/86 to 31/3/87		Nil
1987–88	First 12 months — 1/6/86 to 31/5/87		
	(i) 1/6/86 to 31/3/87	loss	(12,200)
	(ii) 1/4/87 to 31/5/87:		
	£56,400 × 2/12ths	profit	9,400 Nil
	Loss to carry forward		£(2,800)
1988–89	Previous year's profit		56,400
	Less: loss brought forward		(2,800) 53,600

As profits of £9,400 have been set against the loss in 1987–88 the amount of loss to carry forward has been correspondingly reduced.

EXAMPLE

Mr. Brown commenced farming on October 6, 1986 and prepared his first accounts to July 5, 1987. The results were as follows:

Period of nine months ended July 5, 1987	profit	£9,000
Year ended July 5, 1988	loss	(£12,000)

The assessments are:

Year of assessment	Basis period	Sched. D. Case I assessment
		£ £
1986–87	Actual basis — 6/10/86 to 5/4/87	
	£9,000 × 6/9ths	6,000
1987–88	First 12 months 6/10/86 to 5/10/87	
	(i) 6/10/86 to 5/7/87	9,000
	(ii) 6/7/87 to 5/10/87	
	(£12,000) × 3/12ths	(3,000) 6,000
1988–89	First 12 months — 6/10/86 to 5/10/87	
	As for 1985–86	6,000

The loss used is £6,000 (£3,000 in each of the years 1987–88 and 1988–89) and therefore he has loss relief of £6,000 to carry forward.

It will be observed from the two previous examples that relief for a loss is **4–015** generally only given once. However, in the opening years an item of allowable expenditure can be taken into account more than once where it merely reduces a profit, but only to the extent that the expenditure reduces a profit to nil. The reason why there is an inconsistency of treatment between the use of loss relief and expenditure set against profits is simply that there are two separate provisions.

EXAMPLE

Mr. Green commenced farming on June 1, 1987. His results are:

	£
7 months to December 31, 1987 — loss	(6,000)
Year ended December 31, 1988 — profit	21,000

No election is made to have the assessments for the first three years of assessment computed by reference to the actual results.

Year 1987–88
(Actual results for the period 1/6/87 to 5/4/88)

Loss to 31/12/87	(6,000)
Profit 1/1/88 to 5/4/88:	
£21,000 × 3/12ths	5,250
Assessable	£Nil
Losses relieved	£5,250

Year 1988–89
(Results for the first 12 months to 31/5/88)

Loss to 31/12/87	(6,000)
Profit 1/1/88 to 31/5/88:	
£21,000 × 5/12ths	8,750
Assessable	£2,750
Losses relieved	£6,000

Year 1989–90
(Results for the year ended 31/12/88)

Profit	21,000
Less: loss relief	Nil
Assessable	£21,000

In the example, relief for the loss sustained has been secured against profits of £5,250 for 1987–88 and £6,000 for 1988–89. This means that the loss of £6,000 has resulted in total relief of £11,250. However, owing to the statutory wording of the provisions for determining the assessments for the first two years, the Revenue cannot make a restriction in the relief being set off.

4–016   If a claim is made for the loss of £750 (£6,000 less £5,250) in 1987–88 to be set against total income that claim will affect the position for 1988–89 as follows:

	£
Year 1988–89	
(Results for the first 12 months to 31/5/88)	
Loss to 31/12/87	(6,000)
Less: set against total income	750
Net loss available	(5,250)
Profit to 31/5/88:	
£21,000 × 5/12ths	8,750
Assessable	£3,500

EXAMPLE

Mr. Gray commenced farming on January 6, 1987, and incurred a loss of £5,000 in his first accounting period ended April 5, 1987. For the subsequent year ended April 5, 1988, he has a profit of £16,000.

Assessments:

	£	
Year 1986–87		
(loss to 5/4/87)	Nil	
Year 1987–88		
(net profit for y/e 5/1/88)	£	
Loss to 5/4/87	(5,000)	
Profit for 9 months to 5/1/88:		
£16,000 × 9/12ths	12,000	£7,000

If a claim is made for the 1986–87 loss of £5,000 to be set against total **4-017** income, the loss cannot be used in the computation for 1987–88: I.C.T.A. 1988, s.382(3), *I.R.C.* v. *Scott Adamson* (1952) 17 T.C. 679 with the result that the profit assessable for 1987–88 becomes £12,000.

Relief is given against the first available profits and, to the extent unrelieved, is carried forward indefinitely so long as the activity does not cease: I.C.T.A. 1988, s.385(3).

EXAMPLE

Mr. Blue has profits for the years ended April 30, 1986 and 1987, of £10,000 and £6,000 respectively but has incurred a loss of £12,000 for the year ended April 30, 1988. A profit of £15,000 is made for the year ended April 30, 1989. If loss relief is to be given against 1987–88 then the carry forward loss is calculated as follows, assuming, for the sake of simplicity, that there is no other income:

Year 1987–88   £
Assessable income
(based on y/e 30/4/86)   10,000
Less: loss relief for y/e 5/4/88 on statutory basis
£12,000 × 11/12ths = £11,000 but restricted to   10,000

	£
y/e 30/4/87 £6,000 × 1/12th	500
y/e 30/4/88 £12,000 × 11/12ths	(11,000)
Maximum statutory loss	£(10,500)

Assessable   £Nil

There is a balance of loss of £2,000 (£12,000 less used in 1987–88 £10,000). This loss cannot be set against total income for 1988–89 because on the statutory basis there is no net loss:

	£
y/e 30/4/88 (£12,000) × 1/12th	(1,000)
y/e 30/4/89 £15,000 × 11/12ths	13,750
Profit	£12,750

4–018 Consequently, the loss of £2,000 can only be carried forward and set against future profits of the same trade. The Revenue will usually maintain that the first trading profits against which the loss can be set are those for 1990–91. The reason for this is simply that the loss for the year ended April 30, 1988 forms the basis period for 1989–90 and therefore 1990–91 is the first subsequent year for loss relief purposes. This means that there remain the following assessable profits:

	£
Year 1988–89 (y/e 30/4/87)	6,000
Year 1989–90 (y/e 30/4/88)	Nil

It must be appreciated that a loss incurred in the first year of a new business will not be available for relief for some years if the loss is carried forward.

EXAMPLE

Mr. Yellow has a loss for his first year ended April 30, 1989. There will be nil assessments for 1988–89, 1989–90 and 1990–91. Consequently, if the loss is carried forward it is not until 1991–92 that relief is obtained against trading profits.

Suspension of trade

The cessation of trading which prevents the carry forward of losses **4–019** should not be confused with a suspension of trading which does not debar the carry forward of losses. This point is brought out in the following tax cases:

(1) A contracting company shut its premises and for some years it had no contracts and no plant. However, the directors tried to secure work for the company. Eventually, contracts were obtained. It was held that trading had not ceased (*Kirk & Randall Ltd.* v. *Dunn* (1924) 8 T.C. 663).

(2) By way of contrast, where a company ceased trading activities, sold all its assets, informed the Revenue that it would be wound up, remained dormant for some years and then commenced a similar trade it was held that the old trade had ceased (*Goff* v. *Osborne & Co. (Sheffield) Ltd.* (1953) 34 T.C. 441).

(3) A brick manufacturer shut its factory, sold its assets and paid off its employees. Some months later it recommenced manufacturing bricks at a new location. It was held that the trade continued throughout notwithstanding the interval (*Robroyston Brickworks Ltd.* v. *I.R.* (1976) 51 T.C. 230).

(4) Where two complementary businesses merged it was held that for tax purposes the result was a new business (*George Humphries & Co.* v. *Cooke* (1934) 19 T.C. 121), but an election for continuation may be permitted under I.C.T.A. 1988, s.113(2) in view of SP9/86 issued by the Revenue on December 10, 1986.

(5) Where, conversely, a business is divided into two or more complementary businesses, such as on the break up of a family farming partnership, two or more new businesses will result and no election for continuation would be possible unless one of the businesses carried on after the division is so large in relation to the rest as to be recognisably "the business" as previously carried on; SP9/86, para. 5.

Where, under the rules before the abolition of Schedule B, an election was made before March 15, 1988, to have the results of a woodland taxed under Schedule D under I.C.T.A. 1988, s.54, the losses could be carried forward as if they were losses of a trade: I.C.T.A. 1988, s.385(6) (Chapter 12).

Should a farmer dispose of an industrial building (such as a fertiliser pro- **4–020** cessing unit) after the cessation of the trade, any balancing charge is assessable under Schedule A or Schedule D, Case VI (C.A.A. 1968, s.12). Technically, unused trading losses cannot be carried forward and deducted from that balancing charge. However, it is provided by extra-statutory concession that the losses and any unused capital allowances can be carried forward and deducted in these circumstances (Extra-Statutory Concession B19 (1988). The same treatment applies to balancing charges related to commercial buildings in enterprise zones, and will presumably be applied for agricultural buildings allowances under F.A. 1986, s.56 and Sched. 15 (see Chapter 5). Double relief for the same loss is prohibited: I.C.T.A. 1988, s.385(7).

Time-limits for carry forward claim

4–021   A claim to carry forward a loss must be made within six years of the end of the year of assessment in which the loss arises, and a claim to set off against the profits of the same trade for a particular year must be made within six years of the end of that year: I.C.T.A. 1988, s.385(8). Where tax computations are submitted showing a loss to carry forward and relieve under these provisions, that is treated by the Revenue as a claim.

Capital allowances

4–022   Where a claim has been made to set off losses against total income, the losses may be increased by the appropriate capital allowances related to the activity or such allowances may be used to convert a profit into a loss relievable against total income. The capital allowances can only be used in this way if the loss is available to be offset against other income. As with losses set against total income, so the capital allowances can be relieved for the tax year of loss and the following year if the activity is being carried on in that second year: I.C.T.A. 1988, s.383(1).

EXAMPLE

For the year ended December 31, 1988, Mr. Red has a trading profit of £15,000. The capital allowances related to that basis period are £21,000. A loss claim is made for 1988–89 and will be £6,000 (£15,000 less £21,000). Should Mr Red's total income for 1988–89 be £5,000 the balance of £1,000 loss can be set against his total income for 1989–90: I.C.T.A. 1988, s.383(3).

This claim must be made within two years of the end of the tax year for which relief is to be given. The capital allowances that may be included in the loss claim are those for the year of assessment for which the claim is made, excluding allowances brought forward from an earlier year. For the purpose of this claim capital allowances given by discharge or repayment of tax (C.A.A. 1968, s.71), and not by deduction in computing trading profits, are ignored. To the extent that capital allowances have been set against total income they cannot be carried forward. For example, if in the case of an established business, the loss for the year ended April 30, 1989, is claimed against total income of 1989–90, the capital allowances to be added to the loss would be those for 1989–90, based on the expenditure for the year ended April 30, 1988, which would normally be the basis period of assessment for 1989–90. Relief may also be claimed for capital allowances for the year of assessment for which the year of the loss is the basis period. In this example the accounting period to April 30, 1989, would normally be assessed for 1990–91 and the capital allowances which would otherwise be given in 1990–91 by reference to expenditure in the year ended April 30, 1989, could be added to the loss for the year ended April 30, 1989, and relieved in 1989–90: I.C.T.A. 1988, s.383(5)(a).

Cessation relief

In the case of a cessation of a trade where capital allowances might **4–023** otherwise go unrelieved, any brought forward allowances are to be deducted from the profits of the tax year before the allowances of that tax year thereby maximising the allowances for that final year potentially available for relief against total income: Extra-Statutory Concession A8 (1988).

EXAMPLE

A farmer ceases business in 1989–90. His profits for that last year are £6,000 and the related 1989–90 capital allowances are £4,000. The unused capital allowances brought forward from earlier years are £5,000. Without the extra statutory concession there would be no loss in the last year available to set against total income; but some relief might be claimable under the terminal loss provisions. The concession permits the loss relief for the final year to be calculated as follows:

	£
Year 1989–90:	
Profits	6,000
Less: Capital allowances brought forward	(5,000)
	1,000
Less: Capital allowances of the final year	(4,000)
Loss that can be set against total income of the final year	£(3,000)

The capital allowances to be added to losses are those of the normal **4–024** basis period related to the tax year even where the loss itself is apportioned by splitting the accounting period in which the loss arises over the relevant tax years: I.C.T.A. 1988, s.383(3).

EXAMPLE

Mr. Violet starts farming on October 6, 1985. He claims loss relief against total income and his results as adjusted for tax are:

	£
Year ended October 5, 1986 profit	12,000
Year ended October 5, 1987 profit	6,000
Year ended October 5, 1988 profit	75,000
Capital allowances:	
Year 1985–86 £7,000	
Year 1986–87 £26,000	

Assessments:	
Year 1985–86	
£12,000 × 6/12ths (actual basis)	6,000
Less: capital allowances	7,000
Loss available	£(1,000)
Amount assessable	Nil
Year 1986–87	
First 12ths	12,000
Less: capital allowances of the year to cover	(12,000)
Amount assessable	Nil
Loss relief for 1986–87 on the statutory basis	
£12,000 × 6/12ths =	6,000
£6,000 × 6/12ths =	3,000
	9,000
Less: capital allowances for 1986–87	26,000
Loss — see below for comments	£(17,000)

The loss relief against total income is limited to £14,000 (£26,000 capital allowances less that set against the income of £12,000 for the year).

It will be noticed that to establish the extent of the loss attributable to capital allowances that can be set against total income it is necessary to prepare a separate calculation. However, the loss calculation does not alter the amount of the assessment.

The capital allowances used in the loss claim against total income are those related to the basis period: I.C.T.A. 1988, s.383(1).

4–025   Where the basis period applies for more than one year of assessment, the capital allowances used in the loss claim are given in the earlier year: I.C.T.A. 1988, s.383(5)(*a*). This is clearly shown by considering the position of a new trade.

EXAMPLE

Mr. Purple starts farming on September 1, 1985, and takes his accounts to August 31. No election is made for the first three years of assessment to be computed by reference to the actual basis under I.C.T.A. 1988, s.62.

Basis period for capital allowances	Year of assessment	Capital allowances for loss relief purposes
1/9/85 to 5/4/86	1985–86	1985–86
6/4/86 to 31/8/86	1986–87	1986–87
No basis period	1987–88	Nil
y/e 31/8/87	1988–89	1987–88
y/e 31/8/88	1989–90	1988–89

For the year 1987–88 in the example there should be no balancing charges or allowances related to relief given in taxing the trade but there may well be writing down allowances.

If those writing down allowances for 1987–88 exceed the profits, or if there are no profits, for that year, then those capital allowances cannot be used to create or augment the loss claim against total income. The non-effective allowances can only be carried forward and set against future balancing charges or trading profits.

Continuing the previous example, if a claim had been made for the first **4–026** three years of assessment of Mr. Purple, to be computed by reference to the actual profits, then the year for which capital allowances cannot be used in the loss claim is simply deferred to 1988–89.

EXAMPLE

Basis period for capital allowances	Year of assessment	Capital allowances for loss relief purposes
1/9/85 to 5/4/86	1985–86	1985–86
6/4/86 to 5/4/87	1986–87	1986–87
6/4/87 to 5/4/88	1987–88	1987–88
No basis period	1988–89	Nil
6/4/88 to 31//8/88	1989–90	1988–89
y/e 31/8/89	1990–91	1989–90

When considering whether or not to elect for the first three years of **4–027** assessment to be on the actual basis, the effect on the capital allowances computation should not be overlooked. If an election makes little difference to the Schedule D, Case I profit or loss such an election could have a significant effect on the relief for capital allowances.

Capital allowances may only be added to a loss to the extent that they are not needed to absorb balancing charges on disposals in the year of the loss. However, any unused capital allowances brought forward from earlier years should be set against balancing charges before the current year's allowances, in order to maximise the amount that could be set against total income: I.C.T.A. 1988, s.383(2).

It is possible to restrict the claim for certain capital allowances and this could be beneficial in preventing the loss of personal allowances. In addition, careful planning of plant and machinery purchases and sales could be beneficial.

In calculating capital allowances before that year's Finance Act is passed, the old tax rates may be applied; any subsequent amendment is made by repayment of tax or an additional assessment under Schedule D Case VI as necessary: I.C.T.A. 1988, s.383(11).

4–028 As has been explained above, the appropriate basis period for capital allowances for any loss claim is extended to include the basis period, the end of which falls in the year of assessment or coincides with the end of the year of assessment, *i.e.* the accounting year ended April 30, 1989 would be an additional capital allowances basis period for 1989–90 and not, as would normally be the case, for 1990–91: I.C.T.A. 1988, s.383(5)(*a*).

Any reference to capital allowances and balancing charges for a year of assessment are those relating to the normal basis period; thus an accounting year ended April 30, 1989, would be the basis of the capital allowances and balancing charges for 1990–91: I.C.T.A. 1988, s.383(5)(*b*). For this purpose capital allowances brought forward are ignored.

Non-effective capital allowances are those for which relief cannot be given because the profits are not large enough: I.C.T.A. 1988, s.388(5)(*c*).

Surplus capital allowances carried forward are dealt with on a first-in-first-out basis: I.C.T.A. 1988, s.383(5)(*d*). Where the total income is insufficient to absorb all the losses and capital allowances, losses are used up in preference to capital allowances which would be carried forward: I.C.T.A. 1988, s.383(6). This could be of importance in determining, on a cessation of trade, the amount of capital allowances brought forward and so eligible under Extra-Statutory Concession A8, referred to above.

Where capital allowances have been given against profits and subsequently allowed against total income, an additional assessment would be made to withdraw any double relief: I.C.T.A. 1988, s.383(7).

Capital allowances may be included in a loss claim for a profession or vocation as well as a trade, and also for a commercial woodland where, under the rules before the abolition of Schedule B, the election to deal with the results under Schedule D was made before March 15, 1988: I.C.T.A. 1988, s.388(12).

Restrictions on right of set-off

4–029 A loss can be set off against total income of the year or the next year following, as opposed to being carried forward against future profits of the same trade, only if it can be shown that the loss arose in a trade which was being carried on on a commercial basis and with a view to the realisation of profit: I.C.T.A. 1988, s.384(1), (2). However, if there has been a change in the way in which the trade is carried on, it is the situation which prevailed at the end of the year of assessment which is important: I.C.T.A. 1988, s.384(3). Similar provisions apply in the year of commencement and cessation: I.C.T.A. 1988, s.38(4). If it can be shown that there was a reasonable expectation of profit, this is conclusive evidence that the trade was carried on with a view to the realisation of profits: I.C.T.A. 1988, s.384(9).

This commercial basis provision does not apply where the loss is to be carried forward against profits of the same trade.

Farming losses

If a loss arises from a trade of farming or market gardening and there **4–030** was also a loss in each of the preceding five years of assessment, the loss for the current year cannot be set against total income but can only be carried forward against future profits in the same trade: I.C.T.A. 1988, s.397(1) which means any farming trade in the U.K.: I.C.T.A. 1988, s.53(1). When considering the profit or loss position it is strictly necessary to apportion results on the statutory basis to ascertain whether or not the five-year rule in transgressed.

The Revenue considers that capital expenditure on accommodation for cattle, dairy equipment, grain stores and land drainage are not relevant in considering whether or not activities are profitable.

There is no restriction if the farmer shows that:

(a) the *whole* of his farming activities (in the year of claim following five or more years of loss) are of such a *nature* and carried on in such a *way* as would, if they had been undertaken by a competent farmer, have justified a reasonable expectation of profits in the future, *but*

(b) if a competent farmer had undertaken *those activities* at the *beginning* of the five or more years of loss he could not reasonably have expected profits until *after* the end of the year of claim: I.C.T.A. 1988, s.397(3).

This entails detailed long-term plans prepared at the beginning of a period of expected losses for a long-term activity, such as breeding a pedigree herd from scratch, showing budgeted income and expenditure ending in a profit some time after the end of the five-year period. It is understood however that by concession the Revenue is prepared to extend the period from five years to 11 years in the case of a pedigree stock breeder.

It should be noted that if (b) does not apply and a *profit* in the year of **4–031** claim could reasonably have been expected, but it shows an unforeseen loss because of a hazard or catastrophe, use of that loss is restricted even though any competent farmer would have suffered it. On the other hand, if a loss was expected, the fact that it is much larger does not matter.

If the farming or market gardening is part of and is ancillary to a larger trade, the restriction does not apply: I.C.T.A. 1988, s.397(4).

Interestingly, the Revenue seems to accept that if after, say, four years, the farm is sold and a new farm purchased, the five-year cycle starts again, notwithstanding it is treated as the same trade under I.C.T.A. 1988, s.53.

The loss is arrived at before capital allowances or balancing charges. If **4–032** the loss for a year of assessment is not available for set off, neither are the related capital allowances, namely those for the basis year which is the subject of the loss: I.C.T.A. 1988, s.397(5) and (6). Only capital allowances which fall to be given in taxing the trade are so restricted. These would normally be capital allowances on plant for the farming business, and agricultural buildings allowances because for expenditure on or after April 1, 1986 these are given in taxing a trade under F.A. 1986, s.56(1) and Sched. 15, para. 11.

EXAMPLE

Mr. Orange carries on the trade of farming and has the following results:

Year ended March 31	(Loss)/ Profit	Capital allowances	Interest received	Assessed/ (repaid)	Loss carried forward
	£	£	£	£	£
1982	(500)	500	1,700	700	Nil
1983	(1,500)	1,700	2,500	(700)	Nil
1984	(1,800)	1,200	2,500	Nil	500
1985	(2,500)	3,000	1,800	Nil	4,200
1986	(4,000)	800	2,000	Nil	7,000
1987	(6,000)	2,000	1,500	1,500	15,000

The losses and capital allowances in the year ended March 31, 1987, cannot be set against other income for that year (as there have been losses in that and the five preceding years), but may be carried forward against future income from the trade.

The results continue:

	£	£	£	£	£
1988	1,000	3,000	2,500	500	15,000
1989	(4,500)	2,000	3,000	(500)	18,000

The restriction is lifted by the profit (before capital allowances) for the year ended March 31, 1988.

4–033   Where an owner-occupied farm incurs a loss, the cost of maintenance, repairs and insurance of the agricultural land and buildings can be claimed against total income, even though the losses are restricted: Extra-Statutory Concession B5 (1988) and I.C.T.A. 1988, s.33. So far as such expenditure relates to the farmhouse, only one-third can be claimed, leaving two-thirds regarded as domestic.

20   The loss is computed under the Schedule D, Case I rules: I.C.T.A. 1988, s.397(7). As stated above there is no restriction on losses for the first five years of the trade provided that it is carried on on a commercial basis and this includes the position so far as partners are concerned on a deemed cessation on a change in partners: I.C.T.A. 1988, s.397(8), (9) but only for the new partners, not for continuing partners: I.C.T.A. 1988, s.397(10). It is not possible to transfer the business between husband and wife or to a company under their control in order to obtain the extension of the first five-year relief: I.C.T.A. 1988, s.397(10).

Clearly, it is important to ensure that in at least one year in any six consecutive years there is a trading profit thereby taking the farmer outside the restrictions. It might be possible to achieve this end by taking one or more of the following steps:

(1) Defer repairs for a year.
(2) Do not pay a salary to a wife and other working relatives for that year.
(3) Avoid paying a bonus to staff.
(4) Reduce expenses generally.
(5) Minimise bank interest payable.
(6) Sell surplus livestock.

Overseas losses

A Schedule D, Case V loss arising from deficiencies of income from **4–034** farming overseas or lettings of overseas farms may be carried forward in full and set off against future income from the same farm (Extra-Statutory Concession B25 (1988)).

Terminal losses

Where farming is permanently discontinued, a loss incurred during the **4–035** last 12 months can be deducted from the profits of the activity charged to tax in the three tax years before the final year. The loss is referred to as a terminal loss: I.C.T.A. 1988, s.388(1), (2). The relief comprises the whole of the loss arising in the year of cessation and the proportion of the loss of the previous tax year beginning 12 months before the date of discontinuance: I.C.T.A. 1988, s.388(6)(a) and (c). To this may be added the relevant capital allowances for the final year of assessment: I.C.T.A. 1988, s.388(6)(b) and the appropriate fraction of the preceding year's capital allowances for the period beginning 12 months before the cessation: I.C.T.A. 1988, s.388(6)(d).

Should the calculation produce a trading profit for either the final year or the proportion of the previous tax year falling within the 12-month period then that profit is deemed to be nil.

The terminal loss may be carried back against the profits from the same trade for the three tax years preceding the year of cessation: I.C.T.A. 1988, s.388(1). The relief is given against the latest year first: I.C.T.A. 1988, s.388(3).

Terminal loss relief cannot reduce the taxable income below that necessary to cover annual charges, unless they were incurred for trading purposes, so that, had tax been collected in view of an insufficiency of income held in charge to tax additional loss relief would have been available: I.C.T.A. 1988, ss.388(5), 349, 350, 387.

The capital allowances to be included in a terminal loss claim are those **4–036** given by way of deduction in taxing the trade and not those given by way of discharge or repayment of tax (such as agricultural buildings allowances for expenditure incurred before April 1, 1986): I.C.T.A. 1988, s.388(7).

The terminal loss claim may include allowances given by way of discharge or repayment of tax for the tax year in which the cessation takes place and for the appropriate fraction of these allowances for the preceding tax year, namely the allowances applicable to the last 12 months, by making a claim under I.C.T.A. 1988, s.389(6), (7).

Where an election was made before March 15, 1988, under the rules before the abolition a Schedule B, to have the results of a commercial woodland assessed under Schedule D any terminal loss was deemed to be a trading loss for purposes of relief: I.C.T.A. 1988, s.389(8) (Chapter 12).

4–037   The terminal loss relief is not limited to the appropriate spouse's income where there is an election for the separate taxation of wife's earnings in the year of cessation but not in the year to which the terminal loss may be carried back: I.C.T.A. 1988, s.287(3). From 1990–91 the income of husband and wife is treated separately, so any loss incurred by the husband cannot be set against the income of his wife, and vice versa.

EXAMPLE

Mr. Indigo ceased to farm after many years. His results for the last four years and nine months are as follows:

			£
Year ended December 31, 1985	Profit		6,000
Year ended December 31, 1986	Profit		7,000
Year ended December 31, 1987	Profit		600
Year ended December 31, 1988	Loss		(2,400)
Nine months ended September 30, 1989	Loss		(3,600)
Year of assessment			

	£	
1986–87 Capital allowances	2,000	Non-trade charges 150
1987–88 Capital allowances	1,900	Non-trade charges 200
1988–89 Capital allowances	1,800	
1989–90 Capital allowances	450	

Calculation of the terminal loss relief:

	£	£
1989–90 6 months (£3,600) × 6/9ths		(2,400)
Add: Capital allowances		(450)
1988–89 6 months (£3,600) × 3/9ths	(1,200)	
(£2,400) × 3/12ths	(600)	(1,800)
Add: Capital allowances		
(£1,800) × 6/12ths (see below)		(900)
Total		£(5,550)

There is no need to restrict the £900 capital allowances relief because only £600 of the £1,800 capital allowances for 1988–89 have been used to reduce the assessment for that year to nil. Had the profits been such that the surplus capital allowances were less than £900 then that lower figure would be used.

The assessments are revised as follows to take account of the terminal loss **4–038** relief:

Year	Summary £	Terminal loss relief £	Revised amount assessable £
1989–90	Nil	N/A	Nil
1988–89 Profits	600		
Less: Capital allowances	600		
	Nil	N/A	Nil
1987–88 Profits	7,000		
Less: Capital allowances	1,900		
	5,100	4,900	200
1986–87 Profits	6,000		
Less: Capital allowances	2,000		
	4,000	650	3,350
		£5,550	

The £200 is kept in charge to tax for 1987–88 owing to the non-trade charges.

This illustration only shows how terminal loss relief is calculated and utilised so it makes no attempt to consider other possible loss claims.

New business losses

Where a loss arises in the first four years of any trade, a claim may be **4–039** made within two years of the end of the year of assessment in which the loss arises: I.C.T.A. 1988, s.381(1) and (7) to carry the loss back against total income for the three years of assessment preceding that in which the loss arises. The relief is given against the earliest year first: I.C.T.A. 1988, s.381(2). Relief cannot also be given under any other provision: I.C.T.A. 1988, s.381(3).

The trade must have been carried on on a commercial basis with a view to profits within a reasonable period of time: I.C.T.A. 1988, s.381(4). The Revenue takes the view that the test is an objective one and not subjective. This is more difficult to establish than where the loss is to be set against general income of the year of loss.

It is understood that at an appeal hearing, in a case where the Revenue had refused loss relief under the carry back provisions but was prepared to offer a deduction against total income of the current year, the inspector made the following additional points:

(1) In the Revenue's view, to satisfy the loss carry back provisions, not only must the commercial nature of the trade be clear but (in normal circumstances at least) profits should actually be realised within 12 months of commencement. In other words, a delay of more than 12 months would usually mean that profits were not being realised within a reasonable time thereafter.

(2) "Profit" means profit arrived at on normal accountancy principles (after deductions for interest, depreciation, etc.), and not earnings as adjusted for taxation purposes.

(3) In order to sustain a claim to relief in a case where losses arise in the early periods of trading, the proprietor should be able to show that, at the time he commenced business, he had available to him the results of some form of research or enquiry demonstrating the likely viability of the project. This would satisfy the objective test, which the mere subjective appraisal of a hopeful entrepreneur would not.

These points show the attitude of certain inspectors to loss carry back claims. Therefore, it is desirable to ensure that the taxpayer meets the requirements. For example, in order to achieve the "profit" mentioned in (2), it might be possible to adopt a suitable accounting policy which nonetheless accords with good accounting principles.

4–040 Under the rules existing before April 6, 1988, woodlands normally assessable under Schedule B and brought within Schedule D, Case I by an election made before March 15, 1988, could not be included in such a loss claim. The reason for this is that forestry is not of itself a trade (*Coates* v. *Holker Estates Co.* (1961) 40 T.C. 75) (Chapter 12), merely a deemed trade for certain loss relief purposes under I.C.T.A. 1988, s.380(4).

4–041 A husband and wife who are living with each other cannot obtain an extension of the four-year period by transferring a business and claiming the activity as a new business of the transferee spouse: I.C.T.A. 1988, s.381(5). However, where an individual succeeds to a trade on the death of the other spouse this restriction does not apply and the first four-year period recommences, enabling loss relief to be carried back to previous years and relieved against the income of the deceased. The position is the same should the transfer take place when both spouses are not living together but in this situation care must be exercised on the entitlement to the benefit of any tax reduction. From 1990–91 the income of husband and wife is treated separately, so any loss incurred by the husband cannot be set against the income of his wife, and vice versa.

The restrictions to the four-year period do not apply in relation to a loss sustained in a trade which has been commenced separately from the activity undertaken by the individual's spouse. For example, a husband runs a sandwich bar and the wife subsequently opens a sandwich bar as a separate trade: the wife should be entitled to relief under this loss provision for her first four years. This will also apply in the case of farming. Although it is provided that all the farming carried on by any particular person shall be treated as one trade: I.C.T.A. 1988, s.53(2), neither this provision, nor the general aggregation provision under which, until April 5, 1990, the income of a married woman is deemed for income tax purposes to be her husband's income and not her income: I.C.T.A. 1988, s.279(1), is sufficient to deem her activities to be those of her husband.

Where there is a cessation and commencement under the partnership provisions: I.C.T.A. 1988, s.113(1) this event does not enable the continuing partners to claim the relief in respect of losses for the four years following the change which triggered the technical cessation. However, any new partner can claim the relief: I.C.T.A. 1988, s.381(6).

The legislation does not specify how relief under this provision fits in **4–042** with loss relief due against total income of the year of loss and the year immediately following. The Revenue accepts that the taxpayer can decide the order in which the loss relief is to be given under the separate provisions. It the relief is to be given under these carry back provisions, only the balance remaining is available to set against other income or carry forward. Alternatively, the order of priority could be reversed. If the taxpayer does not set out the order of the claims then the matter will be decided on the basis of which claim was made first. The loss cannot be split so that part is used under one provision and part under the other provision (*Butt* v. *Haxby* [1983] S.T.C. 239). Only when the loss claim under one of the provisions has been exhausted can the individual proceed to deduct the balance of the loss from total income under the other provisions.

The normal loss set-off provisions apply, allowing, until 1989–90, a claim to be restricted to the income of the person sustaining the loss and excluding the income of the spouse. If the loss for a particular year is restricted to the income of the spouse incurring the loss then that restriction must operate for all the years of assessment against which that year's loss is carried back under these provisions. It is not possible to restrict the relief in this manner for one of the earlier years to which the loss is carried but not another. However, as each year of loss requires a separate claim the restriction could apply to a loss for one year but not for another year.

An earned income loss is set first against earned income, then against investment income and then (until 1989–90) against the spouse's earned income followed by the spouse's investment income. This order of set-off against each spouse's income is reversed in the case of an unearned income loss. Similarly under the opening years' provision a loss cannot be allowed more than once. A loss is computed in the same way as a profit under Schedule D, Case I.

Capital allowances may be added to the losses. The allowances so **4–043** included are those for the year of loss so far as they have not been relieved: I.C.T.A. 1988, s.383(4). Where allowances are carried backwards under these provisions any necessary adjustments are made to the allowances carried forward, if necessary by additional repayment or assessment: I.C.T.A. 1988, s.383(7)(*b*). Careful planning of plant and machinery purchases and sales could be most beneficial. For example, to bring forward a purchase of plant from the fifth year into the fourth year would enable the capital allowances concerned to be available to be carried back against general income if there were a loss in the fourth year.

EXAMPLE

Mr. Carmine commenced farming on October 6, 1985. The results for the first four years were as follows:

		£
Year ended October 5, 1986	loss	(2,000)
Year ended October 5, 1987	loss	(3,000)
Year ended October 5, 1988	loss	(4,000)
Year ended October 5, 1989	profit	6,000

Capital allowances for the first four years were:

	First year £	*Writing down* £	*Total* £
Year 1985–86	2,200	1,000	3,200
Year 1986–87	Nil	2,000	2,000
Year 1987–88	Nil	1,200	1,200
Year 1988–89	Nil	900	900

Total income for previous years

	£
Year 1982–83	3,500
Year 1983–84	5,000
Year 1984–85	5,750

Losses available under I.C.T.A. 1988, s. 381 are:

Year 1985–86 £

Case I loss £2,000 × 6/12ths	1,000	(actual basis)
Add: Capital allowances	3,200	
	£4,200	

Year 1986–87

Case I loss £2,000 × 6/12ths plus £3,000 × 6/12ths	2,500	(actual basis)
Add: Capital allowances	2,000	
	£4,500	

Year 1987–88

Case I loss (£3,000 × 6/12ths plus £4,000 × 6/12ths)	3,500	(actual basis)
Add: Capital allowances	1,200	
	£4,700	

Year 1988–89

Case I loss y/e 5/10/88 (£4,000 × 6/12ths)	(2,000)	
Case I profit y/e 5/10/89 (£6,000 × 6/12ths)	3,000	
	1,000	
Less: Capital allowances	900	
	£ 100	

Therefore no loss relief is due for 1988–89 in view of the net profit.

The loss relief is allocated as follows:

	Year 1985–86 £	Year 1986–87 £	Year 1987–88 £
Maximum relief available	4,200	4,500	4,700

Set against total income:

Year 1982–83	3,500		
Year 1983–84	700	4,300	
Year 1984–85		200	4,700
	£4,200	£4,500	£4,700

Pre-trading expenditure

Where revenue expenditure is incurred by a person within five years **4–044** before the trade commences, the expenditure is deemed to be a loss sustained in the tax year in which the activity is first carried on by him: I.C.T.A. 1988, s.401. This means that the taxpayer does not secure the benefit of effectively deducting the expenditure a number of times over under the special rules for computing the amount assessable for the opening years of the business.

It is necessary to show that the expenditure would have been allowable as an expense of the trade had it been incurred after the commencement of the business. Consequently, for the expenses to be allowed under this provision they must meet all the usual tests set out in the tax legislation and the decisions enshrined in case law. For example, the legal and professional costs related to the acquisition of a farm will remain unrelievable. However, relief would normally be due for remuneration of employees and the payment of rent, heating and lighting bills incurred before the commencement of farming.

Relief for this deemed loss can be secured in the normal way for an early year loss by deduction from total income or carry forward. However, a claim for relief under this provision is to be made separately from any other claim for loss relief and without regard to any other loss: I.C.T.A. 1988, s.401(2).

Where a significant amount of pre-trading expenditure has been incurred, the year of assessment in which the trade starts could greatly affect the tax reduction arising from a claim for the resulting loss to be set against total income. For example, it might be within the taxpayer's power to commence trading towards the end of one tax year or at the beginning of the next. Before reaching a decision it is necessary to consider not only the extent of the tax reduction created by the loss claim but also the amount of any repayment supplement receivable and the cash flow situation arising from the repayment of tax.

It was possible until March 14, 1988, for an election to be made which resulted in commercial woodlands being assessed under Schedule D rather than under Schedule B (which was abolished from 1988–89): I.C.T.A. 1988, s.54. However, such an election would not have caused the pre-trading expenditure to qualify for relief under these provisions. The reason for this is that the section applies to a trade and was not specifically extended to cover woodlands (*Coates* v. *Holker Estates Co.* (1961) 40 T.C. 75) (Chapter 12).

Excess charges

4–045 Where an assessment to tax has been made in respect of charges on income which have been paid out of profits or gains not brought into charge to tax, the amount so assessed is treated as if it were a loss in the trade for the purpose of carry forward against future profits from the same trade: I.C.T.A. 1988, s.387(1). This applies only where the excess charges have been made wholly and exclusively for the purposes of a trade, have not been charged to capital, and where the charges on income are ultimately borne by the person so assessed: I.C.T.A. 1988, s.387(2).

Under I.C.T.A. 1988, s.387(3) these provisions do not apply to:

(a) excess annual interest paid to a person whose usual place of abode is outside the U.K.;
(b) rents paid to a non-resident under deduction of tax;
(c) easements in connection with radio relay services;
(d) capital sums paid in respect of patent rights;

(e) copyright royalties paid to a person whose normal place of abode is outside the U.K.; or

(f) manufactured dividends.

EXAMPLE

A farmer pays patent royalties on a Supa Dupa Silo System of £1,000 gross per year. His adjusted profits are:

	£
Year ended April 5, 1988	600
Year ended April 5, 1989	5,000

Assessments:

Year 1988–89

Schedule D, Case I	600
Assessment on excess charges	400
Total amount assessed	£1,000

The £400 assessed can be carried forward as a trading loss.

1989–90

Schedule D, Case I	5,000
Less: Excess charges assessed in previous year	(400)
Net amount assessable	£4,600

Where interest is paid for the purposes of a trade, profession or vocation, but is allowed as a charge on income under I.C.T.A. 1988, s.390 rather than as an expense, for example, as a loan applied in acquiring an interest in a partnership under I.C.T.A. 1988, s.362(1) and (2), the unrelieved interest may be treated as a loss arising in the trade for the purposes of carry forward against future profits from the same trade and for terminal loss relief.

Transfer of business to a company

Where a farming business carried on by a sole trader, or by individuals in **4–046** partnership, has been transferred to a company in consideration solely or mainly of the allotment of shares in the company to that individual or individuals, then the former proprietor is entitled to relief for unused losses sustained before the transfer, subject to certain other conditions being met: I.C.T.A. 1988, s.386. The taxpayer must be the beneficial owner of the shares throughout the year of assessment in respect of which the relief

83

is claimed—for the first year from the date of transfer to the following April 5. Similarly, the company must carry on the business throughout the year of assessment for which the relief is claimed—for the first year from the date of transfer to the following April 5.

Losses carried forward and retainable business charges carried forward under I.C.T.A. 1988, s.386(1) and (2) are available for relief. The income against which the relief can be given is that derived by the individual concerned from the company, whether as dividends on the shares or otherwise. The loss is first deducted from assessable income such as remuneration and then from other income such as dividends. A claim needs to be made within six years after the end of the year of assessment for which the claim is made.

Capital allowances unrelieved before the transfer cannot be carried forward and that relief is wasted. It may therefore be better to waive all or a proportion of the allowances.

4–047 The Revenue in practice normally allows the former proprietor to dispose of up to 20 per cent. of his shares without disallowing the relief. However, before making the disposal it is recommended that the proposed transactions are cleared with the inspector.

EXAMPLE

Mr. Ultramarine has farmed for several years and his results were:

		£
Year ended April 5, 1984	Profit	10,000
Year ended April 5, 1985	Profit	12,000
Year ended April 5, 1986	Profit	5,000
Year ended April 5, 1987	Loss	4,500
Period to October 5, 1988	Loss	300

The trade is transferred on October 5, 1988, to Nuts Ltd. for shares. Between October 5, 1988, and April 5, 1989, he was entitled to no director's fees but for the following year ended April 5, 1990, he was entitled to fees of £4,000 and he received a dividend of £750, *i.e.* £1,000 gross. The assessments on him are as follows:

	£
Year 1984–85 preceding year basis	10,000
Year 1985–86 preceding year basis	12,000
Year 1986–87 preceding year basis	5,000
Year 1987–88 actual basis	Nil

The losses to carry forward under this provision are £4,800. Mr. Ultramarine can deduct the loss from the director's remuneration of £4,000 leaving £800 to be deducted from the dividends received. For the purposes of this illustration any entitlement to terminal or other loss relief has been ignored.

Retirement annuity and personal pension premiums

Where an individual is assessable under Schedule D, Case I he can nor- **4–048** mally secure tax relief for the payment of retirement annuity or personal pension premiums subject to certain limits and restrictions.

In order to compute the relief it is necessary to calculate "net relevant earnings" and as would be expected such earnings are after deducting such items as capital allowances and losses: I.C.T.A. 1988, ss.623(6)(c), 646(2)(d). If losses and capital allowances are to any extent set against income other than relevant earnings for a year, then for the purpose of ascertaining net relevant earnings only, the profits of later years must be reduced to the same extent, taking the subsequent years in chronological order: I.C.T.A. 1988, s.623(7), 646(5), (6).

It is important to note that this restriction can arise only if the individual concerned has claimed retirement annuity relief for the year in which losses and capital allowances have been deducted from income other than relevant earnings.

Repayment supplements

If as a result of a claim for losses or other reliefs a taxpayer becomes **4–049** entitled to a tax repayment of at least £25 (£100 if a company) he may also be entitled to a repayment supplement which is calculated as simple interest from the relevant starting date to the end of the tax month in which the repayment order is issued (I.C.T.A. 1988, ss.824(1), 825(2)).

The rate of interest is fixed by Treasury Order and changes frequently.

The supplement includes repayments arising from tax credits on dividends: I.C.T.A. 1988, s.824(2).

The starting date from which the repayment supplement runs depends on the circumstances.

If the tax was paid more than one year after the end of the year of assessment for which it was payable, the repayment supplement runs from the end of the year of assessment in which the tax was actually paid: I.C.T.A. 1988, s.824(3)(*a*). If, however, the tax was paid earlier than that, the repayment supplement runs from the end of the year following the year of assessment to which the repayment relates: I.C.T.A. 1988, s.824(3)(*b*).

Where the tax has been paid in two or more years of assessment the repayment is treated as a repayment of the most recently paid tax as far as possible, thus minimising the repayment supplement: I.C.T.A. 1988, s.824(3).

A repayment supplement is tax free: I.C.T.A. 1988, s.824(8). There are powers to make regulations with regard to repayment supplements and such regulations have been made for Schedule E repayments: I.C.T.A. 1988, s.824(5)(a); S.I. 1975 No. 1283.

A tax month begins on the sixth of one month and ends on the fifth of the following month: I.C.T.A. 1988, s.824(10).

Strategy

4–050   In terms of planning which claims should be made for loss relief it is often helpful to calculate the value of the loss relief in terms of tax and any repayment supplement due, discounted at a reasonable rate of interest to the present value. This enables the two most important factors to be taken into account. First, money now is worth more than money at some time in the future and therefore the earlier that the relief can be claimed the more advantageous the claim. Second, losses that can be relieved at a high rate of taxation are more valuable than those relieved at a lower rate so the loss relief should be computed not merely in terms of the relief but in terms of the consequent reduction of tax payable. If the taxpayer is liable to penalties and interest owing to fraud, wilful default or neglect for a particular year, then to claim the loss as relief for that year could reduce such penalties and interest and therefore maximise the value of the losses.

Partnerships

Loss relief—general

4–051   Although the partnership assessment is a joint assessment on the firm under I.C.T.A. 1988, s.111, each partner is entitled to claim loss relief for his share of the losses as he wishes. He may therefore claim relief for set-off against income in exactly the same way as if the loss arose from farming carried on by him personally.

Relief against total income

4–052   Each partner may claim for his share of the loss to be set against his total income for the year in which the loss arises: I.C.T.A. 1988, s.380(1). If the activity is being undertaken in the following year then his share of the loss can be set against his total income of that subsequent year: I.C.T.A. 1988, s.380(2). If a partner is contemplating retirement just before April 5, and for his last year there is an allowable loss he should consider the implications of remaining as a partner until after April 5, thereby enabling the loss to be set against his total income of either tax year. For the purpose of ascertaining the relief to be given against total income, where the concessional accounts year basis applies, the loss sustained in an accounting period is allocated in accordance with the basis on which the partners allocate losses in the year of assessment in which the accounting period ends and not as the loss is allocated in the accounting period. This can create inequity.

EXAMPLE

Mr. Burnt, Mr. Umber and Mr. Gamboge are in partnership sharing profits and losses equally for the year ended October 5, 1988, when a loss of £9,000 is incurred. For the subsequent year ended October 5, 1989, the ratio is 3:2:1. Assuming that the partners wish the loss to be set against total income and that the accounts year basis applies, then the loss of £9,000 is regarded as the loss for the year ended April 5, 1989. The relief is allocated as follows:

	£
(i) Period 6/4/88 to 5/10/88 (£9,000 × 6/12ths)	4,500
(ii) Period 6/10/88 to 5/4/89 (£9,000 × 6/12ths)	4,500
Total	£9,000

	Firm £	Mr. Burnt £	Mr. Amber £	Mr. Gamboge £
(i) Equally	4,500	1,500	1,500	1,500
(ii) 3:2:1	4,500	2,250	1,500	750
Loss relief	£9,000	£3,750	£3,000	£2,250
Actual share of loss	£9,000	£3,000	£3,000	£3,000

New business

Where a loss is incurred in any of the first four years of assessment dur- **4–053** ing which the farming is carried on, a claim can be made for the loss to be carried back against the total income for the three years of assessment preceding that in which the loss is sustained. This relief is given to partners in a partnership, each of whom can claim separately in relation to his share of the partnership loss: I.C.T.A. 1988, s.381(6). A partnership change which is treated as a cessation does not cause there to be a new business for the purpose of this loss relief. Therefore, all the partners can claim relief in relation to the losses sustained for the first four years of assessment of a new business. However, when those four years have passed the relief is only available to a new partner joining the firm.

Terminal loss relief

Where a partner leaves the firm and that is the occasion of a cessation, **4–054** whether actual or deemed, then he is entitled to claim terminal loss relief: I.C.T.A. 1988, s.388. The relief in question is given against his share of the available profits for past years. If the event is not treated as a discontinuance then the departing partner cannot claim terminal loss relief. Nor can a continuing partner claim this particular relief. The loss relief can be given in a period falling before a previous discontinuance. Not only is the previous change ignored when ascertaining the available profits against which the relief can be given, but also the earlier change is ignored when calculating the loss of the last 12 months: I.C.T.A. 1988, s.389(4)(b).

Carry forward

4–055   A partner can carry forward losses against his share of profits from the same trade or profession in future periods: I.C.T.A. 1988, s.385. It is considered that the loss should be allocated among the partners for this purpose on the same basis that the loss is allocated in the accounting period in which the loss arose. Consequently, if each partner in the previous example were to wish to have the loss for the year ended October 5, 1988, carried forward, they would each have future relief for a loss of £3,000 (£9,000 shared equally). If a partner dies or retires without having fully used up his share of the loss relief then his relief is wasted.

Order of set-off

4–056   The order of set-off for a partnership loss is the same as for a loss by a sole trader: I.C.T.A. 1988, s.382(1).

A partner's share of capital allowances may be added to his share of loss for the purpose of set-off against total income but only with the written consent of all persons who were partners between the end of the year of loss and the making of the claim: I.C.T.A. 1988, s.383(9), (10). Where the claim relates to a loss for a period falling before an event treated as a permanent discontinuance, no written consent is required from those persons who only became partners after that event.

Allocation among partners

4–057   Partnership losses are allocated among the partners in accordance with the partnership agreement which may specify a different ratio for sharing losses from that in which profits are divided. In the absence of an agreement to the contrary the losses would be shared equally among the partners.

If there is an overall adjusted profit for the partnership, but because of prior charges such as salaries to equity partners and interest on capital, certain partners may suffer accounting losses; those partners would not be entitled to loss relief. The profit-sharing partners would have the adjusted partnership profit apportioned amongst them in accordance with their share of profit in the year of assessment.

Example

Mr. Black, Mr. White and Mr. Vermilion are in partnership sharing profits and losses one-half, one-third and one-sixth respectively. The accounts for the year ended April 5, 1989, show:

			£
Net trading loss (before prior charges) (assume no adjustment needed for tax)			6,000

Add

	Interest	*Salary*	
	£	£	
Mr. Black	400	2,000	
Mr. White	200	1,000	
Mr. Vermilion	300	3,000	
	£900	£6,000	6,900
Overall loss per accounts			£12,900

Division of loss:

	Interest	*Salary*		*Balance*	*Total*
	£	£		£	£
Mr. Black	400	2,000	(1/2)	(6,450)	(4,050)
Mr. White	200	1,000	(1/3)	(4,300)	(3,100)
Mr. Vermilion	300	3,000	(1/6)	(2,150)	1,150
	£900	£6,000		£12,900	(£6,000)

4–058 In the normal course of events each partner would be eligible for loss relief for the figure shown in the total column as being attributable to him. However, in the example Mr. Vermilion has a "profit" and the combined losses of Mr. Black and Mr. White amounted to £7,150 (£4,050 plus £3,100). In view of the real loss incurred the "profit" of £1,150 for Mr. White is treated as nil, and the loss relief of £6,000 is allocated between Mr. Black and Mr. White as follows:

				£
Mr. Black	£6,000 ×	$\dfrac{£4,050}{£7,150}$	=	3,399
Mr. White	£6,000 ×	$\dfrac{£3,100}{£7,150}$	=	2,601
Loss relief for tax purposes				£6,000

4–059 Similarly, there could be an adjusted profit but a loss is allocated to a partner. In such a situation the loss is not allowable and the profit is allocated to the partners in the ratio in which they participated in the overall profit.

EXAMPLE

Mr. Gold, Mr. Silver and Mr. Buff are in partnership. There is a profit for the year ended April 5, 1989, of £18,000 that is assessable in 1989–90 on the previous year basis. This is allocated to partners in accordance with how the partners divide their profit for the year ended April 5, 1990.

Year 1989–90

	Salary £	Interest £		Balance £	Total £
Mr. Gold	9,300	2,000	(1/3)	(1,000)	10,300
Mr. Silver	8,000	1,000	(1/3)	(1,000)	8,000
Mr. Buff	200	500	(1/3)	(1,000)	(300)
	£17,500	£3,500		£(3,000)	£18,000

Mr. Buff has apparently sustained a loss of £300. However, as there is a profit overall it follows than no loss relief is due to him. Also, Mr. Gold and Mr. Silver have profits totalling £18,300 (£10,300 plus £8,000). Consequently, the assessable profit of £18,000 is reallocated as follows:

$$\text{Mr. Gold} \quad £18,000 \times \frac{£10,300}{£18,300} \quad = \quad 10,131$$

$$\text{Mr. Silver} \quad £18,000 \times \frac{£8,000}{£18,300} \quad = \quad 7,869$$

Amount assessable £18,000

The problems highlighted in the last two examples could call for an equitable adjustment among the partners through a tax equalisation account.

Miscellaneous losses

4–060   Where a loss arises on a transaction, such as furnished lettings on surplus farm cottages, on which, had it given rise to a profit the assessment would have been under Schedule D, Case VI, the loss may be deducted from any profits or gains arising from any transaction assessable under Schedule D, Case VI: I.C.T.A. 1988, s.392(1).

To the extent that the loss is not used against Case VI profits in the current year it may be carried forward against any future Case VI profits from whatever source: I.C.T.A. 1988, s.392(1). Case VI losses cannot be set against Case VI income of previous years; nor can Case VI losses be set against general income. A partner may claim his own proportion of the total partnership Case VI loss and this can be set against any Case VI profits of the partnership for the same or future years: I.C.T.A. 1988, s.392(2).

Losses carried forward are used against the first available profit: I.C.T.A. 1988, s.392(3).

The claim for loss relief must be made within six years from the end of the year of assessment in which the loss arises: I.C.T.A. 1988, s.392(5). If the tax computation sent to the Revenue clearly shows the loss to carry forward then that is treated as a formal claim.

Certain Case VI losses are not relievable against general Case VI profits: I.C.T.A. 1988, s.392(4). Such losses are those arising to a third party in connection with the grant of a lease at a premium where a profit would have been taxable under Case VI: I.C.T.A. 1988, s.34(6), (7) or on the assignment of a lease granted at an under value: I.C.T.A. 1988, s.35 or on the sale of land with a right to reconveyance: I.C.T.A. 1988, s.36. These provisions are of an anti-avoidance nature as a loss could only arise in a purely artificial transaction.

In the case of holiday lets, although the income is regarded as earned **4–061** income and roll-over and retirement relief is available, the assessment is nevertheless under Case VI. However, the losses are relieved against any other income in the same way as trading losses: I.C.T.A. 1988, ss.503, 504, see Chapter 9.

Where a casual loss arises that may be relieved under Case VI the first matter to consider is whether or not there is an argument that there is a trading activity so that the loss could be relieved under Schedule D, Case I against total income as explained in Chapter 1. If a trading intention can be proved, the Revenue does not have the option of assessing under Case VI (*Pearn* v. *Miller* (1927) 11 T.C. 610).

If the loss can only be relieved under Case VI it is obviously necessary to have some Case VI income against which to set the loss. Any Case VI income will suffice.

This means that it might be possible, for example, to invest in property which is let furnished, or to invest in certificates of deposit or roll-up funds, to generate Case VI income. It is possible to elect for the property rental portion of letting rent to be assessed under Schedule A instead of Case VI, but obviously it is not desirable to make such an election where the object is to maximise the Case VI income in order to absorb a loss: I.C.T.A. 1988, s.15(1)(4).

As mentioned, it might be possible to manufacture Case VI income by placing money on a certificate of deposit, instead of a normal bank deposit account, and assigning the certificate immediately before the payment of interest. Under the anti-avoidance provisions the profit on the sale of the deposit is taxed not as interest but as a casual profit under Case VI:

I.C.T.A. 1988, s.56. However, care needs to be taken to ensure that the profit does not fall to be assessed under Schedule D, Case I.

Capital losses

4–062 These are dealt with in Chapter 15.

Companies

4–063 The result shown in the accounts have to be adjusted for tax purposes in the same way as for an unincorporated business. It is ineffective to prepare a single set of accounts covering several years in order to average profits and losses, because the Revenue may allocate profits to the appropriate period on an actual rather than on a time basis (*Marshall Hus & Partners Ltd.* v. *Bolton* [1981] S.T.C. 18).

The company accounts are adjusted for tax purposes where a loss is shown, in exactly the same manner as where a profit results: I.C.T.A. 1988, s.393(7). Disallowable items such as depreciation or entertaining: I.C.T.A. 1988, s.577, are deducted from the loss and income not taxed under Schedule D, Case I, such as interest received, rents, franked investment income and capital profits, are added to the loss shown by the profit and loss account. For corporation tax purposes "profits" refer to income and chargeable gains: I.C.T.A. 1988, s.6(4)(*a*).

Carry forward

4–064 The first option with a trading loss for an accounting period is to carry the loss forward and set it against the first available taxable profits from the same trade in future years: I.C.T.A. 1988, s.393(1).

Such losses can be carried forward without time limit so long as the trade does not cease. While there may be periods of inactivity it may still be possible to show that the trade had not ceased, *e.g.* there is merely a suspension of trading owing to a temporary hiatus in operations (*Robroyston Brickworks Ltd.* v. *I.R.C.* [1976] S.T.C. 329). However, it is necessary to make a claim within six years of the end of the accounting period: I.C.T.A. 1988 s.393(11).

EXAMPLE

Classic Farms Ltd.

The adjusted trading results of Classic Farms Ltd. were as follows:

		£
Year ended March 31, 1987	Loss	(20,000)
Year ended March 31, 1988	Profit	16,000
Year ended March 31, 1989	Profit	12,000

Computation

Year ended 31/3/87	Loss available to carry forward	£20,000

Year ended 31/3/88 Less	Profit Loss brought forward	16,000 16,000
	Assessment	£Nil
	Loss carried forward	£4,000

Year ended 31/3/89 Less	Profit Loss brought forward	12,000 4,000
	Assessment	£8,000

The Board of Inland Revenue has, however, agreed to accept informal **4-065** claims, under I.C.T.A. 1988, s.393(1), provided that it is clear from the taxation computation that loss relief is being claimed against future profits and not otherwise. This could be done by describing it as "loss for carry forward": SP/C11 which reads as follows:

> "Following discussions with the allied accountancy bodies the Board of Inland Revenue has agreed to accept claims of an informal character [under T.A. 1988, s.393(1)] provided that it is clear from the way in which the computation of the amount of loss is formulated that relief is being claimed against future profits of the company and not otherwise; for this purpose it will be sufficient if such words as "loss for carry forward" appear against the amount of the loss as computed. A similar working practice will be adopted for claims to loss relief carried forward in relation to income tax under Cases I, II and VI of Schedule D."

Set off

Alternatively a claim may be made within two years of the end of the **4-066** accounting period to set off the loss against other income and assessable capital gains for the accounting period: I.C.T.A. 1988, s.393(2).

EXAMPLE

Rustic Farms Ltd.

The taxation computation of Rustic Farms Ltd. for the year ended June 30, 1989, showed the following:

	£
Adjusted trading loss	(10,000)
Interest received	4,000

Net rental income	8,000
Chargeable gains realised	6,000

The corporation tax profit for the year is calculated as follows:

	£
Chargeable gain assessable	6,000
Interest	4,000
Rental income	8,000
	18,000
Less trading losses	10,000
Assessment	£8,000

Carry back

4–067   A further possibility is to elect within two years of the end of the accounting period to carry any part of the loss which cannot be set against current total profits against profits of the preceding accounting period: I.C.T.A. 1988, s.393(2). The same trade must have been carried on in the preceding accounting period. However, the loss set-off is not restricted to profits of the trade, but may be deducted from total income and chargeable gains: I.C.T.A. 1988, s.6(4)(*a*). If, however, the accounting period during which the loss arises is less than 12 months the loss can only be carried back against a similar proportion on a time basis of the profits of the previous accounting period: I.C.T.A. 1988, s.393(3). The loss cannot be set against profits for an accounting period during which the company was not carrying on the trade. For losses arising from first year capital allowances see Chapter 5.

EXAMPLE

Mystic Farms Ltd.

Adjusted trading results

		£
Year ended September 30, 1988	Profit	20,000
9 months ended June 30, 1989	Loss	(16,000)

The loss may be relieved as follows:

	£
Corresponding proportion of previous accounting period £20,000 × 9/12ths	15,000
Loss for 9 months to 30/6/89	16,000
Loss to carry forward	£1,000

The position for the year ended 30/9/88 then becomes:

	£
Profits	20,000
Less loss brought back (restricted)	15,000
Revised assessment	£5,000

Order of set-off

The order in which losses may be claimed is usually known as the order **4–068**
of set-off.

In the first instance it is necessary to compute the total profits from all
sources, including the assessable proportion of chargeable gains: I.C.T.A.
1988, s.393(2).

In determining the figure of total profits, capital allowances on plant and
machinery or agricultural buildings used in a trade are treated as an
expense of the trade and deducted from trading income (C.A.A. 1968,
ss.73(2), 6(1), 46(1), 66, 67(4)). The trading income is also reduced by any
trading losses and excess trading charges brought forward from previous
accounting periods: I.C.T.A. 1988, s.393(1) and (9)(*b*).

When the total profits have been computed (I.C.T.A. 1988, s.6(4)(*a*)),
trading losses for the accounting period or carried back from future
accounting periods are deducted first: I.C.T.A. 1988, s.393(2). Next are
deducted surplus capital allowances given by way of discharge or repayment (C.A.A. 1968, s.74(3)). Such allowances would be to the lessor of an
industrial building or plant as they are given primarily against the appropriate income from property or leasing and only the surplus is deducted from
the total profits (C.A.A. 1968, ss.74(1), 46(2), 68(2)).

Charges on income for the accounting period are next deducted:
I.C.T.A. 1988, s.338(1) and finally group relief: I.C.T.A. 1988, s.407(1).
Group relief is not itself affected by losses brought back from subsequent
accounting periods, I.C.T.A. 1988, s.407(1) which would therefore have to
be restricted if the available profits had already been reduced by group
relief.

In the case of an investment company, for example owning but not oper- **4–069**
ating a farm, management expenses for the year and any excess management expenses and charges brought forward from earlier years are allowed
as a deduction from total profits: I.C.T.A. 1988, s.75(1) and (2).

Farming losses

Farming losses may be subject to special restrictions as in the case of an **4–070**
unincorporated business: I.C.T.A. 1988, s.397(2). If a trading loss, before
capital allowances, has been incurred in an accounting period it cannot
usually be set against other income if there have been such losses throughout the preceding five years before the beginning of the accounting period:
I.C.T.A. 1988, s.397(5). In these circumstances the loss can only be carried
forward and set against future profit from farming. The restriction does not
apply where the trade forms part of and is ancillary to a larger trading

undertaking: I.C.T.A. 1988, s.397(5). It must be remembered that all farming activities are treated as a single trade: I.C.T.A. 1988, s.53(1).

The restrictions for farming losses is intended to limit the relief available to "hobby farms." However, if it can be shown from inception that the farming is being carried on with a reasonable expectation of profits in the long term and that a competent farmer would have incurred losses in the accounting period and in the previous five years, but would expect to show profits in the future, the losses may be available for relief against other income in the normal way: I.C.T.A. 1988, s.397(3), in the same manner as for an unincorporated business, see para. 4–030.

It is specifically provided that loss relief will be unrestricted if the trade commenced within the five years prior to the accounting period, I.C.T.A. 1988, s.397(8), (9), always assuming it can be shown that the trade is undertaken with a view to profit: I.C.T.A. 1988, s.397(4) and (5). There are provisions preventing a reconstruction or transfer to a connected company from restarting the five-year period: I.C.T.A. 1988, s.397(8)–(10).

4–071   If, in view of these provisions, it is desirable to make a profit as adjusted for tax but before capital allowances in one year to overcome the five-year rule, it might be possible to defer repairs for a year, not pay a salary to a wife, avoid a bonus to staff or otherwise reduce the expenses or increase the income for that year. After one year of profit the restriction would not apply until there had again been a loss in an accounting period and in each of the prior five years: I.C.T.A. 1988, s.397(2).

EXAMPLE

Gristick Farms Ltd.

The company had the following results from commencement of business on January 1, 1987.

Years ended December 31	1987 £	1988 £	1989 £
Trading profits before capital allowances	20,000	—	50,000
Trading loss before capital allowances	—	(56,000)	—
Capital allowances on plant and machinery	(8,000)	(9,000)	(10,000)
Deposit interest received	2,000	2,400	3,200
Patent royalties received from silage plant design	12,000	14,000	16,000
Charges on income (trading)	(6,000)	(6,800)	(7,200)
Rents receivable (net)	4,000	5,000	6,400
Chargeable gains	—	6,000	3,000

Assessable profits were:

Trading profits		20,000	—	50,000
Less capital allowances		8,000	—	10,000
		12,000		40,000

Less

Trading loss b/fwd	(7,600)		
Charges b/fwd (6,000 + 6,800)	(12,800)*		20,400
			19,600
Deposit interest		2,000	2,400	3,200
Patent royalties		12,000	14,000	16,000
Rents receivable		4,000	5,000	6,400
		30,000	21,400	45,200
Add chargeable gains		—	6,000	3,000
		30,000	27,400	48,200
Trading loss for 1988	(56,000)		
Plus capital allowances	(9,000)		
	(65,000)		
Relieved against 1987	(30,000)	(30,000)	
Balance of trading loss (1988)	£(35,000)		(35,000)
Trading losses carried forward			£(7,600)
Charges carried forward as trading loss		£6,000*	£6,800
Current year's charges			(7,200)
Assessable profit			£41,000

Note (*) Beware of loss relief for non-trading charges such as charitable deeds of covenant which can be relieved against current total profits. These could be displaced by losses of current or future periods which may be set against total profits. Non-trading charges cannot be carried forward and set against future profits. It might be beneficial to claim losses by carry forward against future profits from the same trade rather than lose relief in respect of such charges if the amounts were substantial.

Overseas trading losses

4-072   If a company carries on a separate trade overseas through, for example, a farm outside the U.K. managed and controlled from overseas, the profits of that branch would be assessed to tax under Schedule D, Case V as income from a foreign possession. Corresponding losses from such a trade are allowable only against future profits from the same trade: I.C.T.A. 1988, s.393(5). In practice, however, foreign branches are usually managed and controlled from the U.K. and so the profits are assessed and the losses relieved under Schedule D, Case I.

Non-commercial losses

4-073   Trading losses cannot be set against other profits unless the trade is part of a statutory function or is undertaken on a commercial basis with a view to realising an overall profit. It is not in fact necessary to make a profit in a trade if it can be shown that at the end of the accounting period the business was being carried on with a reasonable expectation of gain: I.C.T.A. 1988, s.393(5) and (6).

EXAMPLE

Fantasy Farms Ltd.

A company carring on the trade of farming, had the following results from commencement for the years ended December 31:

	Farming results before capital allowances		Capital allowances	Deposit interest	Assessment	Losses carried forward
	Loss	Profit				
	£	£	£	£	£	£
1982	2,000	—	1,000	4,000	1,000	—
1983	6,000	—	1,600	5,000	(1,000)	1,600
1984	4,000	—	2,000	4,400	—	3,200
1985	5,000	—	10,000	2,000	—	16,200
1986	6,000	—	8,000	20,000	6,000	16,200
1987	5,800	—	4,000	14,000	12,000	24,000
1988	—	2,000	10,000	10,000	2,000	24,000
1989	10,000	—	5,000	12,000	(2,000)	25,000

Notes

1983 Loss claim against 1982 profits: I.C.T.A. 1988, s.393(2).

1985 It might have been desirable to disclaim any first year allowances in this year, which would have increased the capital allowances for later years.

1987 Loss disallowed as there was a loss in that year and in the five prior years: I.C.T.A. 1988, s.397(2). However, capital allowances include agricultural building allowances of £2,000 which may be set against other income (C.A.A. 1968, ss.74(1), 46(2), 68(2)).

1988 Restriction no longer applies as there is a profit before capital allowances.

1989 Loss claim against 1988 profits.

Terminal losses

If a company ceases to trade it may claim terminal loss relief within six **4-074** years of the cessation: I.C.T.A. 1988, s.394(6). The loss for the last 12 months of trading which is ascertained by apportioning accounting periods if necessary: I.C.T.A. 1988, s.394(3) can, where not otherwise relievable, be carried backwards against the trading profits only for the preceding five-year period or such lesser time as the company may have carried on the trade: I.C.T.A. 1988, s.394(1), (2). For the purpose of terminal loss relief any capital allowances given by way of discharge or repayment, such as agricultural buildings allowances in respect of expenditure before April 1, 1986, are treated as an addition to the trading loss as if they were a trading expense, I.C.T.A. 1988, s.394(4) and C.A.A. 1968, s.68(2), provided that they cannot be set against the agricultural income.

EXAMPLE

S.A. Farms Ltd.

There were the following trading results for the periods ending with cessation of trading on June 30, 1989. There was no other income.

	£
Year ended 31/12/85 profit less capital allowances	2,000
Year ended 31/12/86 profit less capital allowances	16,000
Year ended 31/12/87 profit less capital allowances	4,000
Year ended 31/12/88 loss plus capital allowances	(6,000)
Six months ended 30/6/89 loss plus capital allowances	(24,000)
Year ended 31/12/88 loss	6,000
Carried back against profit year ended 31/12/87	4,000
Carried forward	£2,000

The loss for the last 12 months is:

$$£24,000 + (\frac{6}{12} \times £2,000) = £25,000$$

The terminal loss of £25,000 is relieved as follows:

Profits for year ended 30/6/88		£
$\frac{6}{12} \times £4,000$ (1987) less loss relieved £4,000 (1988)	=	Nil
$\frac{6}{12} \times$ loss (1988)	=	Nil
		Nil

Profits for year ended 30/6/87

$\dfrac{6}{12} \times$ £16,000 (1986)   =   8,000

$\dfrac{6}{12} \times$ £4,000 (1987) less loss relieved £4,000 (1988)   Nil

---

                                                                                            8,000
Less terminal loss   25,000

Terminal loss carried back   £17,000

Profits for year ended 30/6/86

$\dfrac{6}{12} \times$ £2,000 (1985)   1,000

$\dfrac{6}{12} \times$ £16,000 (1986)   8,000

Less terminal loss as above   9,000
                                                                                           17,000

Unrelieved loss   £8,000

The terminal loss relief is therefore limited to the available trading profits of £17,000.

Restrictions on carry forward of unused relief

4–075   Any loss incurred by a company is taken to consist of:

(1) the current year's trading loss, ignoring capital allowances;
(2) capital allowances as reduced by any profits for that period; and
(3) stock relief brought forward less any residual profit after deducting the capital allowances previously referred to.

Stock relief under F.A. 1981, Sched. 9 cannot be carried forward for more than six years (F.A. 1981, Sched. 9, para. 17).

However, where a loss was attributable in part to first year allowances and a three-year carry-back claim was made, I.C.T.A. 1988, s.393(4), such losses were treated as capital allowances; the remainder following the above order of set-off.

Losses carried forward and used against future profits from the same trade, I.C.T.A. 1988, s.393(1) are set off as follows:

(1) capital allowances brought forward from accounting periods ended after November 13, 1980;
(2) new scheme stock relief on a last in first basis;
(3) losses brought forward from accounting periods ended after November 13, 1980, including those reinstated having previously been used against franked investment income, I.C.T.A. 1988, s.242(5) but ignoring capital allowances and stock relief related to accounting periods ended after November 13, 1980;
(4) any other losses, capital allowances and reliefs.

There is therefore a requirement for records of losses and reliefs to be kept in order that the constituent parts of any composite loss may be identified.

Miscellaneous losses

Trading losses may also be set against franked investment income under **4–076** I.C.T.A. 1988, s.242.

Strategy

A farming company faced with the task of making the most beneficial **4–077** use of a trading loss would normally start on the assumption that the best course of action is to seek relief for a loss as soon as possible. This has the cash flow advantage either of obtaining an immediate tax repayment or of reducing or eliminating the next tax payment. On any discounted cash flow basis the net present value of loss relief is greatest the sooner that relief is available.

Unfortunately, it is not always most advantageous to take the relief for losses at the earliest possible opportunity. If, for example there were a substantial increase in the rate of corporation tax, the tax saved by carrying the loss forward against future income might be materially greater than carrying it back against income taxed at a lower rate.

Excess charges

Annual payments may be set against the total of the profits including **4–078** chargeable gains for the accounting period: I.C.T.A. 1988, s.338(1). Excess charges, *i.e.* the amount in excess of the taxable profits for the accounting period, may be added to trading losses and carried forward for loss relief purposes against future profits from the same trade provided that the charges were made wholly and exclusively for the purposes of the trade or business carried on by the company: I.C.T.A. 1988, s.393(9). Such excess charges may not however be carried back and set against the profits of the preceding accounting period.

Casual losses

As in the case of an unincorporated business, casual losses where the **4–079** corresponding profit would be assessed under Schedule D, Case VI may only be set against Case VI profits from whatsoever source in the same or any subsequent accounting period on an appropriate claim being made: I.C.T.A. 1988, s.396. Similarly a claim may not be made for a loss arising from a lease granted at a premium, an assignment of a lease granted at an under-value or on sale of land with a right to reconveyance: I.C.T.A. 1988, s.396(2). The claim must be made within six years after the end of the accounting period in which the loss was incurred even though the relief cannot be given until after the end of that six-year period owing to the absence of Case VI profits: I.C.T.A. 1988, s.396(3).

Group relief

4–080   If the farming company is part of a group of companies, losses may be surrendered to group companies with profits in the same accounting period under the provisions of I.C.T.A. 1988, s.402. Loss relief can be lost as a result of changes in the ownership of the company coupled with major changes in the nature of the trade during a three-year period under the provisions of I.C.T.A. 1988, s.768. Relief can also be lost as a result of company reorganisations. These are complicated areas of taxation outside the scope of this book. Readers are referred to "Utilising Personal Tax Losses and Reliefs," "Utilising Company Tax Losses and Reliefs" and "Tax Aspects of Company Reorganisations" by Eastaway and Magrin, all published by the Institute of Chartered Accountants in England and Wales, for further details on the more complicated areas of loss relief.

Maintenance expenses of owner-occupied farms not carried on on a commercial basis

4–081   Where loss relief is precluded because a farm is not carried on on a commercial basis or the loss is restricted under I.C.T.A. 1988, s.397 an owner-occupier farmer may nonetheless claim one-third of the cost of maintenance, repairs and insurance of his agricultural land and buildings under Extra Statutory Concession B5 (1988).

CHAPTER 5

Capital Allowances

Introduction

As explained in Chapter 3 the depreciation of capital assets is not allow- **5–001** able for tax purposes. However, certain capital expenditure, for example on plant and agricultural buildings and works, qualifies for statutory depreciation allowances known as capital allowances.

Capital allowances are usually allowed as a deduction in arriving at the taxable profits (C.A.A. 1968, s.70(2)). Claims for capital allowances should in strictness be made in the tax returns, but are usually included in the tax computations submitted to the Revenue together with the accounts. The normal period of six years for making a claim is allowed.

Should the capital allowances due exceed the profits then the excess may be carried forward and utilised against profits of the same trade for future years until the trade ceases: I.C.T.A. 1988, s.383. Balancing charges may be the subject of a separate assessment.

Capital allowances are available to trades such as farming and to the occupiers of woodlands where an election has been made before March 15, 1988, for assessments to be raised under Schedule D instead of Schedule B (C.A.A. 1968, s.70(7)) (Chapter 12).

Certain capital allowances are given by way of discharge or repayment of tax as opposed to a deduction in arriving at the profits of the trade. For example, allowances for expenditure before April 1, 1986, on agricultural buildings and allowances to lessors of industrial buildings, have to be deducted from income of the appropriate class, *e.g.* agricultural rents (C.A.A. 1968, s.71(1)). Any excess allowances may be carried forward against income of the same class in future years, or by election, set off against other income for the year of assessment. Such an election has to be made within two years from the end of the year of assessment (C.A.A. 1968, s.71(2)). The appropriate claim is made to the inspector under the normal procedure.

103

Basis periods

5–002 The basis period for capital allowances is the same as the basis period for income tax except that in the opening and closing years of assessment special rules apply (C.A.A. 1968, s.72).

Where two basis periods overlap, the period common to both falls in the first basis period only. If there is a gap between basis periods for succeeding years of assessment then the interval is taken into the second basis period unless that second period is the year of cessation, in which case it is added to the first basis period (C.A.A. 1968, s.72(2)).

A full year's writing-down allowance is given for each basis period, except for the opening period and the period of cessation, in which case a proportionate part of the writing-down allowance on a time basis is given. This applies whether or not the capital allowances basis period is, under these rules, a greater or lesser period than 12 months.

EXAMPLE

A. E. Warn commenced farming on December 29, 1979 and ceased August 17, 1987. Accounts are made up to April 30 each year.
No election is made for the "actual" basis of assessment to apply for 1980–81 and 1981–82.

Situation "A"

	Capital allowances basis period	Income tax basis period
1979–80	29/12/79–5/4/80	29/12/79–5/4/80
1980–81	6/4/80–28/12/80	29/12/79–28/12/80
1981–82	no basis period	29/12/79–28/12/80[i]
1982–83	29/12/80–30/4/81	1/5/80–30/4/81
1983–84	1/5/81–30/4/82	1/5/81–30/4/82
1984–85	1/5/82–30/4/83	1/5/82–30/4/83
1985–86	1/5/83–30/4/84	1/5/83–30/4/84
1986–87	1/5/84–5/4/87[ii]	1/5/84–30/4/85
1987–88	6/4/87–17/8/87	6/4/87–17/8/87

Notes

(i) Probable Revenue selection, although it had power to choose any 12-month period ending in 1980–81 (I.C.T.A. 1970, s.115(2)(b)).
(ii) The gap from May 1, 1985, to April 5, 1987, is added to the basis period for 1986–87 because the following year is that of cessation.

5–003 Had the taxpayer elected for the actual basis of assessment for the two years following commencement, I.C.T.A. 1988, s.62, and the Revenue had applied the "actual" basis as a result of the discontinuance, I.C.T.A. 1988, s.63, the capital allowances basis periods would be as follows:

Situation "B"

	Capital allowances basis period	Income tax basis period
1979–80	29/12/79–5/4/80	29/12/79–5/4/80
1980–81	6/4/80–5/4/81	6/4/80–5/4/81
1981–82	6/4/81–5/4/82	6/4/81–5/4/82
1982–83	no basis period	1/5/80–30/4/81
1983–84	6/4/82–30/4/82	1/5/81–30/4/82
1984–85	1/5/82–30/4/83	1/5/82–30/4/83
1985–86	1/5/83–5/4/86[(i)]	6/4/85–5/4/86
1986–87	6/4/86–5/4/87	6/4/86–5/4/87
1987–88	6/4/87–17/8/87	6/4/87–17/8/87

(i) The gap from May 1, 1983, to April 5, 1985, is added to the basis period for 1985–86 because that is not the year of cessation.

In both situations "A" and "B" a full year's writing-down allowance will be made for all tax years other than the years of commencement and cessation, namely 1979–80 and 1987–88. Initial and any first-year allowances together with balancing allowances and charges will fall into the appropriate basis period in which the expenditure was incurred, proceeds were received or other event occurred.

Another situation where there can be periods of overlap and gaps arises if there have been changes in accounting dates.

The basis period for allowances other than those given against income assessed under Schedule D, Cases I or II is the year of assessment in which the expenditure was incurred (C.A.A. 1968, s.72(4)).

In the case of companies, capital allowances are given by reference to the chargeable accounting period as corporation tax is charged on a current year basis.

Plant and machinery—definition

What precisely counts as plant is often not immediately apparent. Plant has been held to include: **5–004**

(a) law books (*Munby* v. *Furlong* [1977] S.T.C. 232);
(b) movable office partitioning (*Jarrold* v. *John Good and Sons Ltd.* (1962) 40 T.C. 681);
(c) a dry dock (*I.R. Comrs.* v. *Barclay, Curle and Co. Ltd.* (1969) 45 T.C. 221);
(d) a swimming pool (*Cooke* v. *Beach Station Caravans Ltd.* [1974] S.T.C. 402);
(e) grain silos (*Schofield* v. *R. and H. Hall Ltd.* [1975] S.T.C. 353);
(f) loose tools, knives and lasts (*Hinton* v. *Maden and Ireland Ltd.* (1959) 38 T.C. 391);
(g) a horse (*Yarmouth* v. *France* (1887) Q.B.D. 647);

(h) certain electrical installations, but not all (*Cole Brothers Ltd.* v. *Phillips* [1982] S.T.C. 307;

(i) decorative light fittings and murals in a hotel (*I.R.C.* v. *Scottish and Newcastle Breweries Ltd.* [1982] S.T.C. 296);

(j) decorative panels (*Leeds Permanent Building Society* v. *Proctor* [1982] S.T.C. 821);

(k) option cancellation payments (*International Drilling Co. Ltd.* v. *Bolton* [1983] S.T.C. 70);

(l) the sterling cost of a foreign currency instalment purchase (*Van Arkadie* v. *Sterling Coated Materials Ltd.* [1983] S.T.C. 95).

5–005   Assets held not to be plant include:

(a) prefabricated buildings (*St. John's School* v. *Ward* [1975] S.T.C. 7); plant in the buildings did not qualify because there were no records apportioning the cost between plant and buildings);

(b) a garage canopy (*Dixon* v. *Fitch's Garage Ltd.* [1975] S.T.C. 480);

(c) a ship used as a floating restaurant (*Benson* v. *Yard Arm Club Ltd.* [1979] S.T.C. 266);

(d) an inflatable tennis court cover (possibly wrongly as insufficient evidence before Commissioners) *Thomas* v. *Reynolds* [1987] S.T.C. 135;

(e) finance charges for an oil rig (*Ben-Odeco Ltd.* v. *Powlson* [1978] S.T.C. 360);

(f) wallpaper pattern books of a retailer (*Rose and Co. (Wallpapers and Paints) Ltd.* v. *Campbell* (1967) 44 T.C. 500);

(g) a football spectators' stand (*Brown* v. *Burnley Football & Athletic Club Ltd.* [1980] S.T.C. 424);

(h) suspended ceilings (*Hampton* v. *Forte Autogrill Ltd.* [1980] S.T.C. 80);

(i) shop fronts, floor and wall tiles (*Wimpy International Ltd.* v. *Warland*, *Associated Restaurants Ltd.* v. *Warland* [1988] S.T.C. 149;

(j) plant does not include own labour;

(k) decoration was held not to be plant in *Mason* v. *Tyson* [1980] S.T.C. 284.

The definition in *Yarmouth* v. *France* (1887) 19 Q.B.D. 647 at 658 that plant and machinery is used by a businessman for carrying on his business, not his stock in trade which he buys or makes for sale, but all the goods and chattels, fixed or movable, live or dead, which he keeps for permanent employment in his business, was approved in *Wimpy International Ltd.* v. *Warland* [1988] S.T.C. 149. In the *Wimpy* case it was decided that in order to establish whether or not an item constituted plant, three essential questions had to be posed:

(1) Did the item in question form the premises or part of the premises or place in or upon which the business was conducted?

(2) If the answer to (1) was "no," was the item used for the purposes of the particular trade in question?

(3) If the answer to (2) was "yes," was the item part of the business's stock in trade?

If the answer to (3) was "no," the item was plant.

Special purpose buildings

It should be noted that where an agricultural building is a special build- **5–006** ing to house plant, such as a special milking shed with a sunken floor to enable electric milking machines to be used, it is normally possible to claim capital allowances as plant on the whole of the expenditure rather than agricultural buildings allowance on the building element. Where, however, a building is able to be used for a variety of purposes, such as a barn, only the agricultural buildings allowance would be available, for the expenditure on the building itself, although there could be a substantial element of plant in the overall cost on which capital allowances could be claimed.

Plant and machinery, general pool system

The pool system applies to expenditure on plant and machinery incurred **5–007** after October 26, 1970 (F.A. 1971, s.40).

Under this system the written down value of plant brought forward from the preceding basis period is added to expenditure on plant in the current period.

However, cars and plant not used wholly for the purposes of the business and certain leased assets acquired after May 31, 1980, are not added to the general pool (F.A. 1980, ss.65 and 69). These excluded items are kept in a separate pool with broadly similar provisions applying to those relating to the general pool. In addition, for certain expenditure incurred after March 31, 1986, on plant which has an expected life of less than five years it is possible to elect for it to be dealt with separately (see para. 5–014).

From this total for the general pool is deducted the proceeds of assets sold in the period and a writing-down allowance of 25 per cent. is given on the balance.

Proceeds from the sale of excluded assets cannot be deducted from the **5–008** general pool. Should the sale proceeds exceed the original cost of the plant sold, then those proceeds are deemed to be equal to the original cost for capital allowance purposes (F.A. 1971, s.44(1)). If the basis period is for less than 12 months then the 25 per cent. writing-down allowance is reduced accordingly on a time basis. However, the time during the period when an asset was purchased or sold does not affect the capital allowance position, so long as the assets have been in use during the basis period or at some earlier time for the purpose of the business. The requirement that the machinery or plant had to be brought into use before the 25 per cent. writing-down allowance could be claimed has been abolished in respect of expenditure incurred in chargeable periods ending after March 31, 1985 (F.A. 1985, s.55).

If the sale proceeds exceed the pool value there is a balancing charge on the excess (F.A. 1971, s.44(3)). Only in the period ending with the permanent discontinuance of trading is there a balancing allowance equal to the whole of the expenditure so far unallowed less any proceeds of disposal, in relation to the general pool (F.A. 1971, s.44(2)). To the proceeds for the final period is added the market value of plant and machinery unsold at the date of cessation. Items scrapped do not create balancing allowances.

A cessation gives rise to a balancing charge or allowance. If the plant is **5–009** sold to a buyer who is entitled to capital allowances, the net sale proceeds are compared with the tax written–down value: F.A. 1971, s.44(6)(a), otherwise it is the market value that is compared with the tax written–down value: F.A. 1971, s.44(6)(b). If the partners each take over part of the partnership trade on a dissolution, the continuation elections in I.C.T.A. 1988, s.113(2) and F.A. 1971, Sched. 8, para. 13 are not available so there is a cessation. In such circumstances a balancing charge can be avoided by selling the plant at tax written–down value.

EXAMPLE

Capital allowances on plant.
Computation based on the accounts for the period from April 6 to December 31, 1987 (date of cessation)

	A	B	C	D	E	F
			Total	Sale		Or
	W.D.V.		qualifying	proceeds	Balancing	Balancing
Asset	brought		expenditure	(disposal	allowance	charge
heading	forward	Expenditure	(A + B)	value)	(C − D)	(D − C)
	£	£	£	£	£	£
Sundry plant	20,000	6,000	26,000	20,000	6,000	Nil
Motor cars	10,000	2,000	12,000	15,000	Nil	3,000

It should be noted that so far as plant is concerned a balancing allowance or charge on discontinuance is backdated to the date of cessation (F.A. 1971, s.44(6)(e)) unlike the position for agricultural buildings.

First year allowances

5–010   Most purchases of plant and machinery whether new or second-hand used to qualify for a first year allowance, except in the year of permanent discontinuance (F.A. 1971, s.41(1)). The expenditure had to be incurred during the basis period, which meant a binding contract must have been made and the amount due become payable. In the Revenue's view, which was not universally accepted (Hansard March 28, 1984), this happened at the end of the normal credit period. Furthermore, it was necessary for the asset to belong to the purchaser during the period, (*Stokes* v. *Costain Property Investments Ltd.* [1984] S.T.C. 204) although it was not normally necessary for the plant actually to be in use during the accounting period. Retention of title clauses are usually ignored for tax purposes (*Aluminium Industrie Vaassen BV* v. *Romalpa Aluminium Ltd.* [1976] W.L.R. 676, (C.C.A.B. Press Release September 22, 1976)). However, if the equipment was disposed of before it was brought into use for the purposes of the trade a first year allowance was not given, or if given, was withdrawn (F.A. 1971, s.41(2)). In such a case the loss on disposal was added to the general pool and qualified for writing-down allowances at the rate of 25 per cent. a year (F.A. 1971, Sched. 8, para. 4).

Plant acquired under a hire-purchase contract qualified in full for first year allowances if in use in the period, even though the full capital cost of the plant had not been paid (F.A. 1971, s.45(1)).

The first year allowance was 100 per cent. of the cost of the plant where the purchase took place before March 14, 1984 (F.A. 1971, s.42 and F.A. 1984, Sched. 12). For expenditure after March 13, 1984 the first year allowances were:

Expenditure incurred between—

March 14, 1984 and March 31, 1985	75 per cent.
April 1, 1985 and March 31, 1986	50 per cent.
After March 31, 1986	Nil

A writing-down allowance was not given for a period in which a first year allowance could have claimed.

Where a person incurs expenditure on the purchase of plant for use in his **5–011** business before the commencement of trading the expenditure is deemed to be incurred on the date when trading starts (F.A. 1971, s.50(4)). However, in ascertaining the rate of first year allowance, if any, this deeming provision was ignored and the date the expenditure was actually incurred applied (F.A. 1984, Sched. 12).

Plant purchased between March 14, 1984 and March 31, 1987 under a contract entered into before March 14, 1984 qualified for first year allowances at the rate of 100 per cent. (F.A. 1984, Sched. 12).

Similarly, expenditure on plant in development areas or in Northern Ireland which met certain conditions was eligible for 100 per cent. first year allowances.

Three-year carry back

Trading losses in a company attributable in whole or in part to first year **5–012** allowances could be carried back for a period of up to three years ending before the year of loss under I.C.T.A. 1988, s.393(4). The allowance could not be carried back under this provision to an accounting period before that in which the particular trade commenced.

As indicated, capital allowances may be withdrawn in whole or in part by a balancing charge if the plant were disposed of for a figure in excess of the pool value. However, further plant additions which do not qualify for first-year allowances would mitigate the balancing charge position.

The absence of first year allowances does not necessarily mean that tax relief in the form of capital allowances can only be obtained over a number of years in accordance with the 25 per cent. writing-down allowance on the reducing balance basis. The allowances not claimed increase the pool to carry forward and can be used to cover proceeds of disposals of plant in a later period thereby eliminating or reducing a balancing charge.

Plant acquired before October 27, 1970

Plant and machinery acquired before October 27, 1970, were dealt with **5–013** on a different basis (C.A. 1968, Chapter 2, Part I). Under that system initial and annual allowances were given in respect of each individual asset at various rates and a balancing charge or allowance was calculated when an asset was sold by comparing the written-down value with the disposal proceeds, item by item.

Plant and machinery purchased before October 27, 1970, were normally brought within the capital allowances pool for basis periods ending after April 5, 1976 (F.A. 1976, s.39) and therefore qualified for the 25 per cent. writing-down allowance. In such a situation balancing charges and allowances are calculated in accordance with the provisions relating to post October 27, 1970, expenditure.

It was possible to elect for the new provisions not to apply upon giving the appropriate election (F.A. 1976, s.39(5)). It was normally advantageous to adopt the pooling procedure unless the assets were already eligible for writing-down allowances in excess of 25 per cent., for example, certain agricultural equipment, or if the asset was likely to be sold at a loss compared with the written-down value and a balancing allowance was preferred to a deduction from the pool value of the sale proceeds.

The extension of the pooling provisions to pre-October 27, 1970 expenditure did not apply to assets subject to subsidies for wear and tear, cars where the writing-down allowance is subject to the £2,000 restriction or assets where the allowance is due to a lessor or lessee (F.A. 1976, s.39(4)).

De-pooling

5–014 Following the abolition of first year allowances from April 1, 1986, machinery and plant generally attract an annual writing-down allowance of 25 per cent. (reducing balance basis). This allows 90 per cent. of the cost of the machinery or plant to be written off for tax purposes over eight years. Writing off at this rate is an appropriate average for the machinery and plant of many businesses, but the working life of some assets, because of heavy use or rapid obsolescence, can be much shorter.

Therefore, a new arrangement deals with short life assets (F.A. 1985, s.57) which enables tax allowances to be brought into line with actual depreciation of machinery or plant when it is sold or scrapped within around five years of acquisition. A taxpayer who expects to dispose of an item of machinery or plant, acquired on or after April 1, 1986, at less than its written-down value within a period of five years beginning with the year of acquisition, can elect to have the capital allowances on that machinery or plant calculated separately from the main machinery or plant pool—"de-pooling."

An election to de-pool must be made within two years of the end of the accounting period in which the expenditure was incurred.

The separate calculation allows the balancing adjustment to be made on a disposal but if the machinery or plant has not been sold by the end of the five–year period, its tax written-down value is transferred to the main machinery or plant pool and thereafter dealt with as if it had never been de-pooled.

De-pooling is primarily intended for assets from which first year allowances were withdrawn following F.A. 1984. It does not apply to machinery or plant which is generally dealt with outside the main pool of expenditure such as cars and assets leased to non-traders.

The following example illustrates how de-pooling works.   **5–015**

EXAMPLE

Plant costing £100 is bought in June 1987 by a farmer whose accounting year is December 31. A de-pooling election is made. The capital allowance calculation is as follows:

		£
Basis periods		
Year ended December 31, 1987	Cost	100
	—writing-down allowance	25
		75
Year ended December 31, 1988	—writing-down allowance	19
		56

(i) An election has to be made by December 31, 1989.

(ii) If the plant is sold in the year ended December 31, 1990, for £10 a balancing allowance of £32 (£42–£10) will arise.

(iii) If the plant is sold instead for £50 a balancing charge of £8 (£50–£42) will arise.

(iv) If the plant is not sold by December 31, 1991, the tax written-down value of £24 is to be transferred to the general machinery or plant pool.

Hire–purchase

Under a hire-purchase or similar contract where the hirer shall or may   **5–016**
become the owner of the machinery or plant on completion of the contract the capital allowances are due to the hirer and not the lessor (F.A. 1971, s.45).

Jointly–owned assets

It is not uncommon in farming for neighbouring farmers to club together   **5–017**
to buy a combine harvester or tractor, etc., for their joint use. Normally such assets would be owned as tenants in common and each farmer would claim his expenditure as capital expenditure on plant.

Exceptionally one farmer may retain ownership and a person sharing the cost will claim capital allowances on his contribution under C.A.A. 1968, s.85 so long as the parties are unconnected, C.A.A. 1968, s.85(1A). If they are connected the owner will claim capital allowances on the cost and the contributor will not be eligible for any relief.

Leasing

If machinery is let in the course of trade, such as that of an agricultural   **5–018**
plant contractor, the lessor will usually be entitled to capital allowances in the normal way against his trading profits from leasing (C.A.A. 1968, s.70). Where the lessor acquires plant and machinery as an investment which is hired out by him he is entitled to capital allowances as if he were carrying on a separate trade of hiring out plant (F.A. 971, s.46(1)). The

allowances must be set primarily against the income from letting the plant (F.A. 1971, s.48(3)).

Capital allowances in respect of expenditure incurred after March 26, 1980, by an individual or partnership are not available for set-off against other income if the equipment is leased, unless in the course of a trade. In the course of a trade, for this purpose, means a trade carried on for a continuous period of at least six months and one to which the claimant has devoted substantially the whole of his time throughout the period: I.C.T.A. 1988, s.384(6).

Replacements

5–019   Instead of claiming capital allowances on plant and equipment it is possible to forgo relief in respect of the initial expenditure and claim the replacement cost as a trading expense (*Caledonian Railway Co. v. Banks* (1880) 1 T.C. 487). When the rates of capital allowances were relatively low this may have been a sensible option but with current rates of capital allowances the renewals basis is normally to be avoided. By concession the Revenue permits capital allowances on unrelieved expenditure where there is a change, for a complete class of assets, (Extra Statutory Concession B1 (1985)) from the renewals basis to the normal basis.

Motor cars

5–020   All cars, whether or not leased, that are acquired after May 31, 1980, other than those costing £8,000 or more, are kept in a separate pool for capital allowances purposes (F.A. 1980, s.69).

Capital allowances on private motor cars costing more than £8,000 are restricted. Each car is treated separately and the writing-down allowance is limited to £2,000 (F.A. 1971, Sched. 8, para. 10).

Tax relief for car rental payments on long-term hire vehicles costing more than £8,000 is restricted in respect of cars purchased after June 12, 1979. The restriction applies a proportionate disallowance to the rental paid on the basis of one half of the excess of the cost of the car over £8,000, divided by the total cost of the car (F.A. 1971, Sched. 8, para. 12). If at the end of the rental agreement the taxpayer receives a return of part of the rent paid by him, then in accordance with an unpublished concession the Revenue will not seek to tax that proportion of the amount returned that relates to the amount disallowed.

Where a car costing more than £8,000 is sold, there is a balancing allowance or charge on the car at disposal because each such car is in its own pool for capital allowance purposes. Owing to the rental restriction it now becomes less attractive from a tax point of view to lease expensive cars compared with an outright purchase, because with a purchase the cost is allowed in full over the life of the car whereas with leasing the rentals disallowed are permanently unrelieved.

Plant in buildings

5–021   It is sometimes difficult to determine whether expenditure on a building, for example, on certain fittings, is part of the building.

Plant and machinery installed in a building, whether an industrial building or not, will normally qualify for capital allowances applicable to plant provided that the necessary records are kept to identify the expenditure (*St. John's School* v. *Ward* (1974) 49 T.C. 524).

The Chairman of the Board of Inland Revenue made the following helpful comments, in reply to the C.C.A.B., which were published in August 1977:

"You have asked about the treatment of expenditure on the installation of the 'main services' in new hotels. You said that hitherto the Revenue had regarded such expenditure as expenditure on plant and machinery which therefore qualified for capital allowances, but that recently, in the light of a court case about school buildings, inspectors of taxes had begun to challenge this view and to suggest that no relief was due as such expenditure could be regarded as part of the cost of the building.

"There has in fact been no recent change of Revenue practice in this area. What has happened is that the recent case to which you refer—*St. John's School* v. *Ward*—has focused fresh attention on the distinction which has always had to be drawn for the purposes of capital allowances between expenditure on plant and machinery and expenditure on buildings. As a result, inspectors of taxes have no doubt recently been looking more critically at borderline expenditure. But there has been no change in our view of what falls on either side of the dividing line, and there is, of course, no question of any "attack" on the hotel industry. It is our job to apply the law as we understand it, and the treatment of hotels and restaurants is in this respect no different from that of any other business.

"It may be helpful if I summarise our practice in this area, which is based on the views expressed by the courts over the years. Expenditure on the provision of main services to buildings such as electrical wiring, cold water piping and gas piping is regarded as part of the cost of the building, and therefore as not qualifying for capital allowances. We do, however, regard as eligible for capital allowances expenditure on *apparatus* to provide electric light or power, hot water, central heating, ventilation or air-conditioning, and expenditure on alarm and sprinkler systems. Relief is also given on the cost of all hot water pipes, and on the cost of baths, wash basins, etc., although the St. John's School case suggests that the courts might regard such expenditure as part of the cost of the building. We do not, however, propose any change of practice in this respect. Finally, to complete the picture, and since you mentioned modernisation, I should say that expenditure on alterations to *existing* buildings which is incidental to the installation of plant or machinery qualifies for relief under a separate provision (section 45 of the Capital Allowances Act 1968)."

The following items connected with the construction of a building will usually qualify as plant. However, it has been held that normal light fittings in a department store are not plant (*Cole Brothers Ltd.* v. *Phillips* [1982] S.T.C. 307) as they perform no function but to provide light, although light fittings in a hotel fulfilled a function in providing an atmosphere conducive to the comfort and well-being of its customers and therefore qualified as plant (*I.R.C.* v. *Scottish & Newcastle Breweries Ltd.* [1982] S.T.C. 296). **5–022**

— Advertising signs
— Aerials
— Air compressors and services
— Air conditioning
— Air lines

— Architects and other professional fees related to a number of items including plant (part may qualify)
— Bicycle holders
— Blinds, curtains, blind boxes and pelmets
— Boiler plants and auxiliaries
— Boilers
— Burglar alarms
— Canteen fittings and equipment
— Capital contribution to a sewerage authority in the U.K.
— Cat walks
— Clock installations
— Compressed air plant and piping
— Computers and floor tiling connected therewith
— Conduit for security alarm systems
— Contribution to plant purchased by others (certain conditions must be met)
— Conveyor installations and equipment
— Cooking, conveying and servicing equipment
— Cooling-water systems for (i) drinking and (ii) air-conditioning
— Counters and fittings
— Crane gantries and hoists
— Data transmission installations
— Dispensers
— Disposal units with all live feeds, wastes and flues
— Distribution systems
— Dust extraction equipment
— Dynamos
— Electrically operated doors
— Electrically operated roller shutters
— Electrical sub-stations and generators
— Electrical wiring closely related to an accepted piece of plant, for example to smoke detectors
— Emergency lighting
— Extinguishers
— Fans
— Fire alarms
— Fire blankets
— Fire protection systems
— Fires
— Fire safety to comply with the requirements of the fire authority
— Fitted desks, writing tables and screens
— Floodlighting
— Floor covering
— Flooring (demountable)
— Gas installations after incoming main
— General control and supervisory systems
— Generators
— Goods lifts and doors
— Grain silos
— Grill work (removable)

— Hand driers
— Heating installation, fittings, pipes and radiators
— Hoses and hose reels
— Hot water services
— Immersion and instant water heaters
— Laundry equipment and services
— Letter-boxes
— Lighting protection systems
— Loose floor coverings and doormats
— Loose furniture
— Mechanical ventilation systems
— Mechanical vehicle barriers
— Movable partitions (demountable)
— Name plates
— Personnel-location and call systems
— Pipelines
— Pneumatic tube conveying systems
— Power cables
— Power installations
— Protective structures closely related to accepted items of plant
— Pumps
— Quarantine premises
— Racking, cupboards and shelving (removable)
— Refrigeration installations and cold stores
— Refrigeration plant
— Refuse collecting and disposal systems (including shutes and incinerators)
— Safes, night safes and enclosures
— Safety equipment, screens, etc.
— Sanitary installations, lavatories, urinals, pans, etc.
— Security devices
— Shafts
— Shower and baths
— Shutters
— Smoke detectors and heat detectors
— Soft furnishings
— Software purchased at the same time as the hardware for a computer
— Software with a life of more than two years—if capitalised and capital allowances claimed
— Special buildings which cannot be used as ordinary buildings, such as boiler house, concrete shells housing plant, milking parlours with sunken floors
— Special foundations or reinforced flooring for plant
— Special housing around plant
— Special lighting
— Sprinklers systems
— Staff lockers
— Steam services and condensate return systems
— Storage racks

— Storage tanks
— Switchboards
— Switchgear
— Telephone equipment and conduits
— Teleprinters
— Telex systems
— Thermal insulation *re* industrial buildings
— Towel dispensers
— Towel rails
— Transformers
— Turntables
— Vacuum cleaning installations
— WC partitions (if demountable)
— Wash basins including drains, etc.
— Water treatment and filtration
— Weighbridge
— Welfare equipment
— Wet and dry risers
— Window panels, lighting and sockets for a shop front

Formerly, relief was not technically due for plant and machinery where the expenditure on landlord's fittings was incurred by the tenant as the plant did not belong to the tenant (*Stokes* v. *Costain Property Investments Ltd.* [1984] S.T.C. 204). However, unless there is a tax avoidance involved, relief is usually given to the tenant by the inspector. Capital allowances can now be claimed where machinery or plant is installed in a building or on land after July 11, 1984 and becomes a fixture (F.A. 1985, s.59).

Estate management

5–023   Plant and machinery used for the purposes of estate management qualifies for capital allowances which are relieved as if they were expenses on maintenance, repairs and management of premises for Schedule A purposes (I.C.T.A. 1970, s.78).

Agricultural buildings allowances

Expenditure incurred on or after April 1, 1986

5–024   Although the initial allowance for agricultural buildings was abolished by F.A. 1985, s.62 and the rate of writing-down allowances reduced to 4 per cent., the system of agricultural buildings allowances was not otherwise changed. However, in respect of expenditure incurred after March 31, 1986, F.A. 1986, s.56 and Sched. 15 introduced a new code of allowances broadly comparable with the industrial buildings allowances (see para. 5–039), including provisions for optional balancing allowances and charges.

The definition of husbandry is widened to include any method of intensive rearing of livestock or fish on a commercial basis for the production of food for human consumption which therefore enacts Extra-statutory Concession B2 (1985). Expenditure claimed for agricultural buildings allowances cannot also qualify for machinery and plant allowances or any other form of capital allowances (F.A. 1986, s.56(5)).

Writing-down allowances

Where a person having a major interest in agricultural land or forestry **5–025** land who made an election before March 15, 1988, under I.C.T.A. 1988, s.54 for assessment under Schedule D, incurs capital expenditure on the construction of farmhouses, farm or forestry buildings, cottages, fences or other works such as water and electricity installations, shelter belts of trees, glass houses and reclamation of former agricultural land he is entitled to a writing-down allowance of 4 per cent. a year, straight line. Allowances are withdrawn if the buildings or works are not used for the purpose of husbandry or forestry. A major interest in land is the freehold or leasehold interest, but not a licence to occupy (F.A. 1986, Sched. 15, para. 1).

Expenditure qualifying for allowances

Only a maximum of one-third of reasonable expenditure on a farmhouse **5–026** is allowable. Expenditure must be incurred for the purposes of husbandry or forestry on the land in question, although apportionment is available for buildings and works used partly for agricultural or forestry purposes (F.A. 1986, Sched. 15, para. 2). Grants from the Government or local authorities must be deducted when arriving at the qualifying expenditure, C.A.A. 1968, s.84.

It is often desirable to plan for a house to be built for a farmer's son while he is an employee and thereby obtain agricultural buildings allowances on 100 per cent. of the cost, as a farm cottage. If the son becomes a partner it is possible to continue the claim on 100 per cent. of the original cost on the ground that there is already a farmhouse on the farm and the cottage retains its status as such notwithstanding that the occupier is now a partner, *I.R.C.* v. *Lindsay* (1953) 34 T.C. 289, *I.R.C.* v. *Whiteford* (1962) 50 T.C. 379.

Contributions

Where a contribution is made towards another person's capital expendi- **5–027** ture for the purposes of a trade carried on by the contributor or by a tenant of his, capital allowances can usually be claimed under C.A. 1968, s.85.

The expenditure must have been eligible for industrial buildings allowances, plant and machinery allowances, mine, oil wells, etc., allowances other than mineral depletion or agricultural buildings allowances.

Writing-down allowances are given as if the contribution was capital expenditure on the appropriate asset.

No allowance is available after July 8, 1976, if the contributor and recipient are connected persons within I.C.T.A. 1988, s.839, so full allowances are due to the owner.

Contributions to a sewerage authority in the U.K. for an asset used for the treatment of trade effluents qualifies for capital allowances. This does not include the cost of connection to main drainage (*Bridge House (Reigate Hill) Ltd.* v. *Hinder* (1971) 47 T.C. 182).

Meaning of "the relevant interest"

5–028   A relevant interest in land is the major interest held by the person incurring the expenditure and if more than one major interest is held the relevant interest is the one reversionary on the others. Where the relevant interest is a lease which becomes extinguished the interest into which it merges becomes the relevant interest, as for industrial buildings allowances under C.A.A. 1968, s.11 (F.A. 1986, Sched. 15, para. 3).

Transfers of relevant interest

5–029   Where a relevant interest is transferred then, unless there is an election for a balancing adjustment, the transferor is entitled to allowances up to the date of transfer and the transferee from that date takes over the transferor's entitlement to the allowances. If the transfer takes place part way through a chargeable period or basis period the allowance is apportioned as necessary. Where only part of the interest is acquired the allowances are apportioned. Where a lease comes to an end and a new lease is granted to a new tenant, a capital sum paid by him to the old tenant for the value of buildings and works may enable him to claim the old lessee's balance of allowances which would otherwise pass to the landlord (F.A. 1986, Sched. 15, para. 4).

EXAMPLE

Mr. Bowyer, who makes up his accounts to June 30 each year, built a new pig unit for £100,000 on September 30, 1986, which was sold to Mr. Gunston, who makes up his accounts to March 31 each year, on October 1, 1987, for £120,000.

Mr. Bowyer's acquisition was in the year ended June 30, 1987, which forms the basis period for the assessment for 1988–89 and he is entitled to a full year's allowance.

The chargeable period related to the acquisition by Mr. Gunston is in Mr Bowyer's year ended June 30, 1988, which forms the basis period for 1989–90 and he is entitled to a proportionate allowance to the date of sale, *i.e.* three-twelfths.

The chargeable period related to the acquisition so far as Mr. Gunston is concerned is the year ended March 31, 1988, which forms the basis period for 1988–89 and he is entitled to relief for the proportionate period from the date of acquisition, *i.e.* six-twelfths.

The allowances are therefore allocated as follows:

1988–89

		£
Mr. Bowyer	4% × £100,000	4,000
Mr. Gunston	4% × £100,000 × 6/12ths	2,000

1989–90

		£
Mr. Bowyer	4% × £100,000 × 3/12ths	1,000
Mr. Gunston	4% × £100,000	4,000

1990–91

		£
Mr. Bowyer		Nil
Mr. Gunston	4% × £100,000	4,000

The allowances due to Mr. Gunston cannot exceed the residue of expenditure, *i.e.* £100,000 − £4,000 − £1,000 = £95,000, subject to an election under F.A. 1986, Sched. 15, para. 7, below.

In this example, because of the different accounting dates, allowances are accelerated; in other cases they could be deferred although in every case full relief for the whole cost is eventually given.

Buildings, etc., bought unused

Where the asset on which the capital expenditure was incurred is sold **5–030** before being brought into use, the original expenditure ceases to qualify and the purchaser may claim allowances on the lesser of the amount paid by him for the relevant interest in land, or the amount of the original expenditure (F.A. 1986, Sched. 15, para. 5).

Balancing allowances and charges

If an agricultural building is destroyed the former owner may elect for a **5–031** balancing allowance or charge or if one is transferred the former owner and new owner may jointly elect for a balancing allowance or charge on the transferor. Where there is such a balancing event and election it is necessary to compare the demolition or sale proceeds with the written-down value. If they are less than the written-down value there is a balancing allowance and if they are more there is a balancing charge up to the amount of allowances previously given. Apportionment is possible on partial sales or where the expenditure only partly qualifies for allowances, such as expenditure on a farmhouse (F.A. 1986, Sched. 15, para. 6).

Balancing events

A balancing event occurs on the transfer of the relevant interest to **5–032** another person or on the demolition, etc., of the asset by the original owner. Such an event only gives rise to a balancing allowance or charge if an election is made within two years of the end of the accounting period in which the event takes place, or two years of the end of the year of assessment during the basis period for which the event takes place. If there is a transfer the election must be signed jointly by transferor and transferee and the transferee's allowance is based on the transferor's residue of expenditure less any balancing allowance or plus any balancing charge equally over the remainder of the original writing-down period of 25 years.

An election cannot be made if the taxpayer is not within the charge to U.K. tax or if the sole or main benefit of the election would be an increased allowance (F.A. 1986, Sched. 15, para. 7).

If Mr. Bowyer and Mr. Gunston made a joint election under these provisions Mr. Bowyer would suffer a balancing charge in 1989–90 of the allowance given (£4,000 instead of receiving a proportionate allowance) and Mr. Gunston would obtain allowances for the full original cost of £100,000. As, in this case, the net effect would be to defer some of the allowances it is unlikely that an election would be made unless for example Mr. Bowyer had excess loss relief to absorb the balancing charge.

Exclusion of land values, etc.

5–033   Qualifying expenditure excludes the value applicable to the land on which the building or work is erected (F.A. 1986, Sched. 15, para. 8).

Special provisions as to certain sales

5–034   Sales between parties under common control are treated as taking place at open market value whether or not this is so and it is not possible to make an election under C.A.A. 1968, Sched. 7, para. 4 to have the sale treated as made at the lower of market value or written down value (F.A. 1986, Sched. 15, para. 9).

Restriction of balancing allowances on sale of buildings

5–035   There is an anti-avoidance provision which applies where an interest is sold subject to a subordinate interest such as a lease and any two of the three parties involved (the vendor, the purchaser and the grantee of the subordinate interest) are connected and it appears that the transaction was effected in order to increase the allowances. In such a case any balancing allowance is restricted by adding to the sale proceeds any premium on the lease, or the amount for which the lease could have been sold had it been for a full commercial rent. This can only restrict the balancing allowance, not create a balancing charge and the new owner's allowances are not affected. Only the capital proportion of the addition is treated as part of the capital consideration, that is excluding any amount treated as rent under I.C.T.A. 1988, s.36. A payment for the variation of the terms of a lease is treated as a premium (F.A. 1986, Sched. 15, para. 10).

Manner of making allowances and charges

5–036   Capital allowances for agricultural buildings and works in respect of expenditure incurred since April 1, 1986, are made in taxing the trade, not by way of discharge or repayment of tax as applied to expenditure before that date under C.A.A. 1968, s.68. Allowances are made by way of discharge or repayment of tax where a trade is not being carried on (F.A. 1986, Sched. 15, para. 11).

Expenditure originally incurred before April 1, 1986

The agricultural buildings allowance was 10 per cent. a year of the cost **5–037** on a straight line basis for expenditure originally incurred before April 1, 1986, until the original cost was written off (C.A.A. 1968, s.68). There were no provisions for balancing allowances or charges. A purchaser of an agricultural building was entitled to the balance of allowances which would otherwise have been claimed by the vendor even if the property ceased to be used for agricultural purposes (C.A.A. 1968, s.68(4)). If the building was sold any excess over the original cost (not written down value) would realise a capital gain in the usual way (C.G.T.A. 1979, s.2(1)) but a sale below cost would not give rise to a capital loss (C.G.T.A. 1979, ss.29(5) and 34) where relief had already been given as capital allowances.

EXAMPLE

Halifax Farms

Halifax Farms sold an agricultural building on June 1, 1988.

	£
Cost 1/4/83	30,000
Sale proceeds	45,000
Chargeable gain	£15,000
Agricultural buildings allowance:	
Initial allowance at 20%	6,000
Years ended 31/12/83 to 1987 inclusive	
£30,000 at 10% a year for 5 years	15,000
Year ended 31/12/88	
10% × £30,000 × 5/12ths	1,250
Total allowances to Halifax Farms	£22,250
Purchaser's allowances available	
Year ended 30/9/88 (say)	
10% × £30,000 × 4/12ths	1,000
Years ended 30/9/89 to 1990	
£30,000 at 10% a year for 2 years	6,000
Year ended 30/9/91—balance	750
Total allowances to purchaser	£7,750

Initial allowance: agricultural and forestry buildings and works

An initial allowance was introduced for expenditure on agricultural and **5–038** forestry buildings and works in relation to expenditure incurred after April

11, 1978. The initial allowance was reduced to nil in respect of expenditure incurred after March 31, 1986 (F.A. 1985, s.62). The initial allowance was 20 per cent. given in respect of the chargeable period in respect of which the expenditure was incurred and the writing-down allowance remained at 10 per cent. a year until the expenditure was written off, that is after eight years (F.A. 1978, s.39(1)). The allowance in the year of the expenditure was therefore 30 per cent. Where the initial allowance was not claimed or a reduced initial allowance was claimed the allowance was given on a straight line basis, still at 10 per cent. a year, until the expenditure had been written off.

It is provided that where expenditure on plant and machinery can qualify for agricultural buildings allowances under C.A.A. 1968, s.68 then no allowance is to be given for that expenditure under the plant and machinery provisions (C.A.A. 1968, s.50(2) and F.A. 1971, Sched. 8, para. 2(2)). Having regard to the foregoing, it is clear that if expenditure could qualify for relief under the agricultural buildings allowance legislation then, in strictness, no first year allowance nor 25 per cent. writing-down allowance could be claimed. For example, if expenditure was incurred on a grain silo, heating and ventilating plant forming part of an agricultural building then, according to the legislation the taxpayer should only be entitled to agricultural buildings allowances and not first-year allowances. The legislation did not permit the taxpayer to choose under which provisions to claim relief. However, notwithstanding the strict legal position, Inland Revenue, Technical Division confirmed that where an item clearly qualified as plant then it did not seek to apply the provisions strictly, thereby enabling the taxpayer to claim first-year and 25 per cent. writing-down allowances, where appropriate.

Industrial buildings

5–039   Industrial buildings allowances are granted in respect of industrial premises or structures such as factories and certain warehouses. However, commercial premises such as offices, retail farm shops and showrooms do not qualify. Farm buildings will usually attract agricultural buildings allowances but those used for certain activities, such as the packing or processing of eggs or the intensive rearing of livestock, may qualify for industrial buildings allowances instead. Expenditure on the building, but not on the land, may be eligible for a writing-down allowance of 4 per cent. a year straight line (C.A.A. 1968, s.2(2)). The writing-down allowance is 2 per cent. a year in respect of expenditure incurred before November 6, 1962.

The more recent rates of initial allowance are as follows:

Expenditure incurred	Rate of initial allowance
13/11/74 to 10/3/81	50%
11/3/81 to 13/3/84	75%
14/3/84 to 31/3/85	50%
1/4/85 to 31/3/86	25%
After 31/3/1986	Nil

In respect of expenditure incurred after March 10, 1981, the taxpayer could, at the time of making the capital allowances claim, have required the initial allowance to be a reduced amount (F.A. 1981, s.73).

The provisions regarding pre-March 14, 1984, binding contracts, expenditure qualifying for certain development grants and stage payments under a contract bridging March 31, 1985 to 1986, as explained previously, apply equally in determining the rate of initial allowance for expenditure incurred after March 13, 1984 (F.A. 1984, Sched. 12).

Balancing adjustments after December 17, 1980

Where a building is destroyed or demolished, or a relevant interest in it **5–040** is sold or disposed of, within 25 years of the building first being used (or 50 years where expenditure was incurred before November 6, 1962), there may be a balancing charge or allowance. If the sale proceeds are less than the written-down value there will be a balancing allowance of the difference. Should the proceeds be more than the written down value there will be a balancing charge equal to the excess (C.A.A. 1968, s.3). The balancing charge cannot exceed the allowances actually given (C.A.A. 1968, s.3(6)). It is worth noting that should the building be demolished then industrial buildings allowances can be claimed on the cost of demolition (C.A.A. 1968, s.4(11)). If there were a period of non-industrial use before a sale under a contract made after December 17, 1980, there would still be a balancing charge, as such a charge arises if there is a disposal or deemed disposal whether or not the building or structure is currently in use as an industrial building (C.A.A. 1968, s.3(1)). There is no balancing charge on the building permanently ceasing to be used as an industrial building, merely the loss of the annual writing-down allowance (C.A.A. 1968, s.2(1)(b)) unless it is sold or destroyed (C.A.A. 1968, s.3(3) to (5)). Any subsequent balancing allowance or charge would be scaled down by reference to any periods of non-industrial use.

Only the first disposal of a relevant interest during a period of non-use as an industrial building attracts a balancing allowance or charge. This is because no allowance would be given to the purchaser if the building were not used by him as an industrial building and consequently there would be none to withdraw.

On the disposal of a relevant interest in an industrial building under a contract entered into after December 17, 1980, the balancing allowance or charge is scaled down by reference to any period of time during which the building was not used as a qualifying building.

The rules which were introduced to cover disposals after December 17, **5–041** 1980, replace the previous provisions under which the scaling down was by reference to periods during which an annual allowance was not given. The previous provisions gave the desired result where there was no initial allowance but meant that where an initial allowance had been given there was a reduction in the balancing allowance or charge once the whole of the expenditure had been written off, even though the premises were still within their life as industrial buildings (*i.e.* 50 or 25 years) as there would be a period for which no annual allowances were given. More importantly however in the case of small workshops, where the legislation is by reference to the industrial buildings allowance provisions, there would be no balancing charge on the disposal of a small workshop under the old legislation. The reason for this is that with an initial allowance of 100 per cent. there would obviously be no annual allowance and if there were no annual allowance the balancing charges were automatically scaled down to nil (C.A.A. 1968, s.3(3) to (5)).

If the building is not constructed for the farmer but is purchased unused from a developer then the purchaser will be entitled to allowances based on the price paid by him and not on the developer's costs of construction (C.A.A. 1968, s.5(2)). However, the cost of the land is still excluded (C.A.A. 1968, s.17(1)). It should be noted however that a local authority may not be classified as a developer in cases where the authority has constructed an industrial building, because it may not be carrying on a trade.

Where for a temporary period the industrial building is not in use capital allowances can continue to be claimed as though the building were still in use (C.A.A. 1968, s.12).

Non-qualifying expenditure

5–042   Where an industrial building is planned it is worth noting that up to 25 per cent. of the construction costs may relate to non-industrial accommodation such as showrooms, offices (other than the offices of work managers, production managers, etc., which are industrial), or living accommodation without causing any part of the expenditure to cease to qualify for relief (C.A.A. 1968, s.7(4)). For expenditure incurred before March 16, 1983, the proportion was 10 per cent. and not 25 per cent. It is important however that such non-qualifying accommodation should be part of the basic structure and not a separate building. If the non-qualifying expenditure exceeds 25 per cent. of the total cost then the whole of the non-qualifying expenditure is ineligible for capital allowances and an apportionment is made.

Repairs to an industrial building which are disallowed as a revenue expense and treated as of a capital nature qualify for industrial buildings allowance (C.A.A. 1968, s.8) as does expenditure on preparing a site for plant and machinery (C.A.A. 1968, s.9).

Relevant interest

5–043   Relevant interest in relation to industrial buildings was so defined (C.A.A. 1968, s.11) that it was necessary to introduce anti-avoidance provisions to prevent the acceleration of industrial buildings allowances by disposing of the relevant interest in the building by means of a sale to a connected party where the seller retained a licence to occupy the premises for a long period at a low rent. The market value of the relevant interest retained was therefore negligible and a balancing allowance was claimed by subsequently selling that relevant interest. In respect of transactions after June 13, 1972, if a sale is made subject to retention of a subordinate interest, any balancing allowance will be withdrawn where the sale is to a connected person or if the grantee of the subordinate interest is connected with the vendor or purchaser. The restriction also applies where the parties are not connected, but the sole or main benefit which might be expected to accrue is the acceleration of industrial buildings allowances (F.A. 1972, s.69).

The effect of property prices over recent years has been that in many instances when a property is sold on the open market the whole of the industrial buildings allowance previously granted will be withdrawn by means of a balancing charge. If the cessation of business precedes the disposal of the building then in strictness any trading losses brought forward cannot be set against the balancing charge which is assessable under Schedule A or Schedule D, Case VI (C.A.A. 1968, s.12) as there is no provision deeming the disposal to take place at cessation, as applies for plant. However, by concession, the losses and unused capital allowances can be set against the balancing charge (Extra Statutory Concession B19 (1988), but not against any capital gain, (*Magnavox Electronics Co. Ltd. (in liquidation)* v. *Hall* [1986] S.T.C. 561).

Transfers other than sales and transactions between connected person

In respect of a contract for the disposal of a relevant interest in an indus- **5–044** trial building the market value is to be substituted if the relevant interest is transferred between connected persons as defined in I.C.T.A. 1988, s.839, where the obtaining of a greater allowance or the avoidance of a balancing charge was the object of the transfer. Under the previous legislation (C.A.A. 1968, Sched. 7, para. 1) the substitution of market value applied only in connection with a sale where there was common control or the purpose was to obtain an allowance. The amendment which covers transfers after March 10, 1981, is primarily aimed at the owner of an industrial building giving it to his wife immediately prior to resale in order to avoid a balancing charge (F.A. 1981, s.76).

Sale and lease back

A sale and lease back of premises could itself give rise to further prob- **5–045** lems in addition to the balancing charge if the rent was not at market value, I.C.T.A. 1988, s.779, as the deduction for rent paid would be limited to the market rent.

Lease and lease back

In order to prevent a balancing charge on industrial buildings it is com- **5–046** mon practice to avoid a sale and lease back and instead grant a long lease, say 999 years, and then lease back for a reasonable commercial period. The grant of a lease in such circumstances would not give rise to a disposal for industrial building balancing charge purposes (C.A.A. 1968, ss.11(3) and 3(1)). However, if the relevant interest is purchased and very shortly thereafter a long lease is granted the Revenue might contend that in reality the original expenditure was not of a capital nature with the result that no industrial buildings allowances are due. It is sometimes recommended that in such situations the value of the retained interest should be significant, say 25 per cent. of the value of the property acquired, thereby giving the taxpayer a greater opportunity to resist any Revenue attack.

Where after June 22, 1971, a short lease having no more than 50 years to run is sold and leased back for a term not exceeding 15 years, part of the sale consideration is treated as taxable income: I.C.T.A. 1988, s.780. That is $\dfrac{16 - n}{15}$

where n is the term of the new lease. If the sale and lease back is of trading premises the assessable amount is a trading profit otherwise it is taxable under Schedule D, Case VI.

5–047   In determining the period of the new lease it should be noted that it may be deemed to cease at the time of the first rent review under which the rent payable is reduced or at the date when either the lessor or lessee has power under the lease to determine it or to vary their obligations. This is to avoid a concentration of income into the early period. There are provisions to prevent a sale by one party and a lease back to an associated party.

Premiums on long leases

5–048   Where a long lease with a duration of more than 50 years is granted on or after February 15, 1978, the lessor and lessee of an industrial building may jointly elect for the grant of the lease to be regarded as a sale of the lessor's relevant interest to the lessee. This means that any capital sum paid by the lessee is regarded as the purchase price of the relevant interest enabling the lessee to claim industrial buildings allowances. The election has to be given by notice in writing within two years of the date on which the lease takes effect, which does not give a great deal of time (F.A. 1978, s.37).

The section does not apply where the lessor and lessee are connected or where the sole or main benefit which might be expected to accrue to the lessor is a balancing allowance on the disposal of the industrial building.

Residue of expenditure

5–049   Just as a balancing charge is assessable on the vendor of an industrial building only if the property is sold within the first 25 years (or 50 years where the original expenditure was incurred before November 6, 1962), the purchaser of a second-hand industrial building is only entitled to an allowance if his acquisition is within this period. The residue of expenditure, that is the cost not yet allowed plus the balancing charge on the vendor, is allowed to the purchaser over the remainder of the 25 or 50 year term (C.A.A. 1968, s.4).

Interest on sale

5–050   If a building subject to a mortgage is to be sold, the mortgage interest up to the date of sale should be paid by the vendor. If the mortgage is taken over cum interest, relief is not available to the vendor because the interest was not paid by him, nor is relief available to the purchaser to the extent that the interest relates to a period before his ownership of the property.

126

Enterprise zones

Industrial buildings allowances of 100 per cent. of the building cost are **5–051** given in respect of capital expenditure on industrial and commercial buildings where the expenditure is incurred or a contract entered into while the property is in an enterprise zone within the first 10 years of the site being so designated (F.A. 1980, s.74). Dwellings houses are specifically mentioned as not being eligible (F.A. 1980, s.74(4)). Expenditure on thermal insulation cannot also be claimed (F.A. 1975, s.14). The restriction on first year allowances to lessors does not apply to plant which is to be an integral part of a building in an enterprise zone (F.A. 1980, s.74 and Sched. 13).

There are about thirty enterprise zones. They have been set up to encourage business activity in the specified areas. Not only is the expenditure entitled to the 100 per cent. relief mentioned but there are in addition the following significant benefits:

(1) Business rates exemption on the industrial and commercial premises for the 10 years from the designation date. There remains liability for water rates.
(2) Industrial Development Certificates are not required.
(3) There is exemption from industrial training levies and from the obligations to supply information to Industrial Training Boards.
(4) There are certain Customs relaxations.
(5) Where development conforms with the published scheme for the enterprise zone concerned individual planning permission is not necessary.
(6) Government requests for statistical information are reduced and administration matters are dealt with more quickly.

The enterprise zones and their designation dates are:

Zone	*Designation date*
England	
Corby, Northants	June 22, 1981
Dudley	July 10, 1981
extended (Round Oak)	October 3, 1984
Glanford (Flixborough)	April 13, 1984
Hartlepool, Cleveland	October 23, 1981
Isle of Dogs (London Docklands)	April 26, 1982
Middlesbrough (Britannia) Cleveland	November 8, 1983
Nort-East Lancashire (Burnley, Hyndburn, Pendle and Rossendale)	December 7, 1983
North West Kent (Gillingham, Gravesend and Rochester-upon-Medway)	October 31, 1983
extended to Chatham	October 10, 1986
Rotherham, S. York	August 16, 1983
Salford/Trafford Park	August 12, 1981
Scunthorpe (excluding Glanford)	September 23, 1983
Speke (Liverpool)	August 25, 1981
Telford, Shropshire	January 13, 1984

Tyneside (Gateshead/Newcastle)   August 25, 1981
Wakefield, (Langthwaite Grange)
  W. Yorks   July 31, 1981
  extended (Dale Lane & Kinsley)   September 23, 1983
Wellingborough, Northants   July 26, 1983
Workington (Allerdale), West Cumbria   October 4, 1983

Wales
Delyn, Flint, Clwyd   July 21, 1983
Milford Haven Waterway   April 24, 1984
Lower Swansea Valley   June 11, 1981
  extended   March 6, 1985

Scotland
Clydebank/Glasgow   August 13 & 18, 1981
Inverclyde   March 3, 1989
Invergordon   October 7, 1983
Tayside (Arbroath and Dundee)   January 9, 1984

Northern Ireland
Belfast   October 21, 1981
Londonderry   September 13, 1983

Scientific research

5–052   Scientific research expenditure of a revenue nature is allowable as a trading expense (C.A.A. 1968, s.90) and capital expenditure on scientific research qualifies for a 100 per cent. deduction (C.A.A. 1968, s.91).

Scientific research is widely defined as any activity in the field of natural or applied science for the extension of knowledge (C.A.A. 1968, s.94(1)).

For the expenditure to qualify it needs to meet the test for scientific research *ab initio*. If it does not so meet the test when incurred then subsequent events cannot transmogrify the expenditure so that it meets the requirements for scientific research.

CHAPTER 6

Stock Valuations

Cost or market value

In order to prepare proper accounts, matching so far as possible income **6–001** with expenditure, to be able to arrive at the taxable profit, it is necessary to value the farming stock in trade and work-in-progress at each accounting date.

Stock is normally valued at the lower of cost and market value; cost is the historical cost of bringing the stock to its existing condition and location, and market value is the price at which the stock could be acquired in the normal market (*Brigg Neumann & Co. Ltd.* v. *I.R.C.* (1928) 12 T.C. 1191, *I.R.C.* v. *Cock Russell & Co. Ltd.* (1949) 29 T.C. 387). The lower of cost and market value may be applied to each individual item of stock.

Provided that the basis of valuation is consistent it is permissible to value stock at direct cost of production, that is labour and materials only, without any addition for overheads (*Ostime* v. *Duple Motor Bodies Ltd.* (1961) 39 T.C. 537).

A base stock method under which the minimum stock held is retained at **6–002** the original cost is not suitable for tax purposes (*Patrick* v. *Broadstone Mills Ltd.* (1953) 35 T.C. 44) nor is the LIFO (last in first out) method acceptable (*Minister of National Revenue* v. *Anaconda American Brass Co. Ltd.* [1956] 1 All E.R. 20). A market value calculated on the basis of the retail price less overall gross profit margin was held to be unsuitable for tax purposes in *BSC Footwear Ltd.* v. *Ridgway* (1971) 47 T.C. 495. If stock has previously been under valued and this is corrected for the closing stock the opening stock must similarly be adjusted (*Bombay Commissioner of Income Tax* v. *Ahmedabad New Cotton Mills Co. Ltd.* (1929) 8 A.T.C. 574) but not where the basis of valuation is changed (*Pearce* v. *Woodall-Duckham Ltd.* [1978] S.T.C. 372).

Livestock

In the case of growing animals which have to be valued for stock pur- **6–003** poses, that is where the herd basis (Chapter 7) does not apply, it is possible to arrive at the value for stock purposes on the basis of the farm costing records. However, where these are not sufficiently detailed to give accurate figures it is possible to use an estimated cost for home bred animals which has been agreed between the Revenue and the National Farmers' Union. This is based on the open market value of the animal and in the case of cattle it is 60 per cent. of such market value and in the case of sheep and pigs it is 75 per cent. of the open market value. There is no published text for this agreement; it was reached in correspondence and by discussion between the Revenue and the National Farmers' Union (Inland Revenue Press Office letter to the authors January 20, 1987) and is included in the Inspectors' Manual under "Particular Trades, etc.—farming."

The animal is valued at the appropriate percentage of the open market value at each accounting date until it reaches maturity. The figure of cost determined for any animal on the first valuation after reaching maturity fixes that animal's cost for the rest of its life and no subsequent valuation is necessary, unless below cost. An animal reaches maturity when it produces its first young or is available to service female animals. This estimated cost basis cannot be used where it would not result in a reasonable estimate of cost, for example in the case of pedigree stock.

This agreement was first made in 1942 and before October 1, 1972, cattle were also valued at 75 per cent. of the current market value. It is necessary to keep proper livestock records and the Revenue will, at the very least, request a reconciliation of the numbers of animals at the beginning of the year, purchases, births, sales and deaths and the number of animals in stock at the end of the year.

6–004 In the case of animals grown for eventual sale, it is permissible to take into account a proportion of profit earned to date in the stock valuation and this would be in accordance with the Statement of Standard Accounting Practice No. 9 which allows such treatment in the case of long-term work-in-progress (SP/B4). However, many farmers would regard it as more prudent to keep the value at cost of production.

Tillages, etc.

6–005 Growing crops, unexhausted manures and tillages, etc., should normally be valued as work-in-progress on the basis of the lower of cost and market value in the same way as stock. Costs would obviously be those of seeds, fertilisers, cost of ploughing, cultivation, drilling and unexhausted or residual manurial values. For cases where the normal value of tillages, unexhausted manures and growing crops does not exceed £7,000 (not £700 as sometimes quoted) at the accounting date and a detailed valuation is not available, there is, as with estimated costs of livestock, an agreement between the Revenue and the National Farmers' Union. Again, it has no official text. Under it a certificate that the value at the beginning of the accounting period did not differ materially from that at the end of the accounting period will be accepted in lieu of a detailed valuation. Even where the normal value exceeds £7,000 a valuation will not be pressed in every case and a similar certificate may be accepted after any enquiry necessary to establish its reasonable accuracy.

A valuation made on a full waygoing basis, which is the basis a valuer would use for the purpose of computing compensation due to a tenant from his landlord if he were giving up the tenancy at the accounting date and was entitled to commercial compensation, may be accepted if consistently applied.

Growing crops on a sale of a farm have to be accounted for separately **6–006** and are not regarded as part of the land (*Gunn* v. *I.R.C.* (1955) 36 T.C. 93). They are regarded as *fructes industriales,* that is the fruits of the farmer's labour. This is to be contrasted with the growing crop on the sale of an orchard which was held to be included in the land value in *I.R.C.* v. *Pilcher* (1949) 31 T.C. 314 as *fructes naturales,* that is the fruit of nature and therefore part of the land. Where an outgoing tenant is entitled to compensation for growing crops on a waygoing basis the so-called "harvest award" is treated as a sale of trading stock by the outgoing tenant and an allowable purchase of trading stock by the incoming tenant.

Harvested crops

In the case of harvested crops a valuation on the basis of 85 per cent. of **6–007** the current market value will be accepted by the Revenue as equivalent to cost.

Consumables

Most farmers will also have in store at the year end fuel, spares for plant **6–008** and equipment, seeds, fertiliser, feed stuffs and other items which require to be valued at the lower of cost or market value at the accounting date. If any of these have deteriorated or become obsolete an appropriate deduction should be made in the valuation.

It is important to ensure that the cut-off procedures are adequate so that where recently delivered items are included in stock the corresponding invoice has been included in purchases. Conversely, items bought just before the year end and included in purchases should be checked to ensure that they are included in the stock records, or have been included in values of tillages but that there has been no double counting.

Horses

The Revenue will accept that a working horse may be treated as either a **6–009** fixed asset or trading stock, but will resist any attempt to treat it as stock in one year and a fixed capital asset in another. A home bred horse transferred to fixed assets may be valued at a deemed cost of 85 per cent. of its current market value.

Working horses treated as fixed assets are normally dealt with on a replacement basis but a settled practice of making wear and tear allowances would not be disturbed.

Cessation

On cessation of trade the net sale proceeds of stock subsequently sold to **6–010** a person carrying on, or intending to carry on, a trade, are treated as a trading receipt at the date of cessation irrespective of when actually sold: I.C.T.A. 1988, s.100(1)(a). The costs of a stock valuation for the purposes of a sale are deductible from the proceeds. If the stock is retained and not sold the market value at the date of cessation is deemed to be a trading receipt under I.C.T.A. 1988, s.100(1)(b).

Change in basis

6–011   The basis of stock valuation must be consistent and this may entail recalculation of the opening stock on the same basis as the closing stock on the new basis (*Bombay Commissioner of Income Tax* v. *Ahmedabad New Cotton Mills Co. Ltd.* (1929) 8 A.T.C. 574. However, if stock values are written up on the new basis the Revenue is likely to resist any argument that any substantial increase in the opening stock between the new and old bases is tax free (*Pearce* v. *Woodall-Duckham Ltd.* [1978] S.T.C. 372).

The Revenue practice on a change in the basis of stock valuation was published in The Accountant on November 17, 1962, and amplified in SP/B5.

A change will be accepted only if to a valid basis and the Revenue would expect the opening valuation to be adjusted to a consistent basis, but would not normally seek to revise the valuations of earlier years unless the difference is substantial. A mere refinement in the basis of valuation is unlikely to require any adjustment to the opening stock values.

Retention of title

6–012   Goods acquired in the normal course of trade are regarded as the property of the purchaser for tax purposes, even though title in the goods is retained by the vendor until they are paid for. It follows that such goods held at the end of the accounting period are treated as stock and valued at the lower of cost or market value irrespective of the legal ownership (SP/B6).

CHAPTER 7

Herd Basis

Definition

Animals and other living creatures kept by a farmer for the purposes of **7–001**
his farming are in the first instance treated as trading stock: I.C.T.A 1988,
s.97 and Sched. 5, para. 1. If, however, the animals form part of a production herd and an election for the herd basis is made, the animals are effectively treated as a fixed asset and not as trading stock: Sched. 5, para. 1(2),
(3). References in this chapter are to I.C.T.A. 1988, Sched. 5, unless
otherwise indicated. The Revenue published booklet I.R.9 on the herd
basis in 1984.

A production herd is defined as "a herd of animals of the same species
(irrespective of breed) kept by [the farmer] wholly or mainly for the sake of
the products which they produce for him to sell, being products obtainable
from the living animal": para. 8(5).

Products obtainable from the living animal means (a) the young of the **7–002**
animal, or (b) any other product obtainable from the animal, not being a
product obtainable only by slaughtering the animal itself: para. 8(5)(a) and
(b). The term "herd" includes a flock, and any other collection of animals
however named: para. 8(1). Animals kept wholly or mainly for the work
they do in connection with the carrying on of the farming are specifically
excluded from the herd basis: para. 7.

Immature animals are normally excluded from the herd: para. 8(2). For
this purpose female animals are treated as becoming mature when they
produce their first young: para. 8(4) and laying birds become mature when
they first lay: para. 9(3). Immature animals can be included as part of the
herd where the land on which the herd is kept is such that the animals
which die cannot be replaced except by animals bred and reared on that
land, and the immature animals in question are bred in the herd, are maintained for the purpose of replacement and are no more in number than is
necessary to prevent a fall in the numbers of the herd: para. 8(2) and (3).
This would apply, *e.g.* to hill sheep and hill cattle which have adapted to
their environment and where an animal brought in from outside would be
unlikely to prosper.

There is no limit to the number of animals in a herd and the herd basis can be applied to a single animal: para. 9(4). This would most commonly be applied in the case of, *e.g.* a champion bull. Interestingly it could be argued that stud fees received for the services of the bull would not be part of the ordinary farming activities but would constitute a separate trade: *Malcolm* v. *Lockhart* (1919) 7 T.C. 99, *Marshall* v. *Tweedy* (1926) 11 T.C. 524. The herd basis would, however, still apply because it can apply to trades other than farming: para. 9(1). Animals kept primarily for exhibition, racing or other competitive purposes may not be included within the herd basis: para. 9(5).

7–003   Production herds are defined as being of the same class where the herds are of the same species (irrespective of breed) and the products produced by the herd are of the same kind: para. 8(6). If a herd basis election is made it must cover all herds of a particular class and includes therefore herds of that class which the claimant has ceased to keep before making the election, or first begins to keep after making the election: para. 2(1). This means, *e.g.* that if a farmer kept a herd of Jersey cattle and a herd of Guernseys and both were kept for their milk, the two herds would be regarded as of the same class even though they were physically separated and maintained as pedigree herds quite distinct from each other. The herd basis would have to be made for both herds or neither. If the farmer also had an Aberdeen Angus herd kept for the sale of the young for beef production, this would be a herd of a different class, although of the same species, and a herd basis election could be made independently of the two dairy herds.

A flock of sheep kept for the production of wool would be a separate herd from the cattle, being of a separate species, but separate flocks of different varieties of sheep would be regarded as of the same class unless, exceptionally, a flock were kept exclusively for the production of milk in which case it would constitute a separate class and a herd basis election could be made independently.

Flying flocks, *i.e.* animals held for a short period with a view to fattening for sale, are not production herds and do not qualify for the herd basis.

Election

7–004   An election for the herd basis must be made in writing to the inspector and must specify the class of herds to which it relates, *e.g.* dairy cattle: para. 2(2). An election once made is irrevocable and applies for the first chargeable period to which it relates and all subsequent chargeable periods: para. 2(4).

Normally the time-limit for an election for the herd basis is two years after the end of the first chargeable period for which the farmer is chargeable under Schedule D, Case I in respect of his farming activities and in which he maintains a production herd of the class in question: para. 2(3). For example, if a farmer makes up his accounts, as many do, to September 30, and if he kept a production herd for the first time during the year ended September 30, 1988, (having previously been purely an arable farmer) that would form the basis of assessment for 1989–90 and the election would have been due by April 5, 1992. In the case of a company the election would have been due by September 30, 1990, two years after the end of the accounting period.

If, however, there is a loss set off against other income for income tax purposes under I.C.T.A. 1988, ss.380, 382 a loss in the year to September 30, 1988, could be relieved against other income in 1988–89 and 1989–90 and the election would have to be made by April 5, 1991, two years after the first tax year, the profits of which are computed by reference to the profits of a period, that is the year ended September 30, 1988, during some part of which the farmer kept a production herd. In the case of a company if the loss for the year ended September 30, 1988, were carried back against farming profits for the year ended September 30, 1987, under I.C.T.A. 1988, s.393(2) the herd basis election would have to be made by September 30, 1989.

It may be possible to engineer an opportunity for a herd basis election by taking in a partner, or having a change in partners; this does not involve a disposal of the herd for stock valuation purposes if a continuation election for income tax purposes is made. A transfer to a company which elects for the herd basis could be a problem in that the herd would be deemed to be sold at market value, not cost, as the herd would be a fixed asset of the company, not stock: I.C.T.A. 1988, s.100(1)(b), which could increase substantially the profit in the transferor's business.

If a herd basis election affects a previously final assessment it must be re- **7–005**
opened and the appropriate adjustments made: para. 11.

On a change of partners in a farming partnership there is a change in the farmer keeping the animals for the purposes of farming and therefore a new opportunity for making a herd basis election. Indeed, even where an election is in force before the change, it is necessary for the new partnership to claim the herd basis, a point that is often overlooked in practice. This applies whether or not an election for continuation is made under I.C.T.A. 1988, s.113. If a new herd basis election is not made the herd would normally be treated as if sold in the open market (under para. 5) as a sale to connected parties, which would amount to the disposal of a herd and treated accordingly, see para. 7–011.

A farmer must keep, and if requested by a notice from the inspector produce to the Revenue, the records necessary for the calculation of the herd basis computations: para. 10.

Computations

The herd basis computation is spelt out in considerable detail in order to **7–006**
cover the consequences of the treatment as a fixed asset instead of trading stock in various circumstances: para. 3(1). The value of the herd is not credited to the profit and loss account because it is treated as a fixed asset and consequently the initial cost and the cost of any additions to the herd are not allowed as expenses: para. 3(2). If an animal is transferred from trading stock to the herd as an addition to the herd it is necessary to bring in as a credit to the profit and loss account the cost of the animal to the farmer and any further costs incurred in rearing it to maturity: para. 3(3)(b).

In the case of an animal bred by the farmer the cost of breeding and rearing it to maturity is credited to the profit and loss account and debited to the asset account: para. 3(3)(a). There is often a problem in ascertaining with any degree of accuracy the cost of an animal and in the absence of more precise information the Revenue will normally accept that the cost of rearing to maturity is for cattle 60 per cent. of their market value and for sheep and pigs 75 per cent. of their market value, following an agreement with the National Farmers' Union. If the farmer's own costing records are sufficiently sophisticated to show the actual cost and this is below the appropriate percentage of the market value it is clearly beneficial to adopt the actual figures.

7–007   If an animal dies or ceases to be part of the herd, *e.g.* on a sale, the sale proceeds of the live animal, its carcass or any part thereof, or any insurance proceeds: para. 3(12), are included as a trading receipt: para. 3(4)(a).

The cost of the replacement animal is allowed as a trading expense: para. 3(4)(b). If, however, the replacement animal is of superior quality to that disposed of, only the appropriate proportion of the cost applicable to an animal of the same quality is allowed as an expense: para. 3(5). If the replacement animal is of worse quality, only the actual cost is allowed, but if the animal which is replaced was slaughtered compulsorily the amount to be included as a trading receipt is limited to the amount spent on the lower quality replacement animal: para. 3(6).

EXAMPLE

Mr. Long sold three Friesians, Buttercup, Daisy and Effie for £400, £500 and £600 respectively, from his herd of 100 cattle, which had a book cost of £10,000.

He transferred from his breeding stock to the herd four Friesian cows which had just reached maturity and which had a market value of £500 each. He also purchased a mature cow, Fanny, to replace Effie at a cost of £900 although an animal of similar quality would cost £700.

	£
Sale proceeds	1,500
Cost of Fanny—limited to equivalent animal	700
Cost of two home bred replacements already allowed so	Nil
Addition to profits	£800

Addition to trading profits for two additional cows added to the herd.
2 × £500 × 60%   £600

This will be the deemed cost of these animals on a subsequent sale. This may be reflected in the herd, for tax purposes, as follows:

	£
Herd at start of year	
100 cows @ £100	10,000
Transactions during year	
$\frac{3}{97}$ Disposals @ £100 (cost)	300
	9,700
2 Replacements of equal value	200
1 Replacement of enhanced value £100 + £200	300
$\frac{2}{102}$ Additions @ £300 (60% of cost £500)	600
	£10,800

	£
Trading Account	
Asset enhanced by	800
Cash enhanced by	
(Sales £1,500 less purchase £900)	600
Trading account credit	£1,400

At the end of the day there are NET 2 additions.

Herd replacement

Where the whole herd is sold but another production herd of the same **7–008** class is acquired this is merely treated as a sale and replacement of the lesser of the numbers in the original herd and in the replacement herd. If the numbers in the replacement herd are greater than in the original herd the additional number would be treated as additions to the herd and the cost disallowed. If the number in the replacement herd is lower than in the original herd certain of the animals in the original herd will be treated as having been sold without being replaced: para. 3(7).

Sales

If a grown animal is sold without being replaced only the profit or loss on **7–009** the disposal of that particular animal is brought into the computation. This is arrived at by comparing the proceeds of sale of the animal with its cost of breeding and rearing to maturity in the case of a home-bred animal: para. 3(10)(a), or the cost of purchasing the animal, together with any costs incurred in rearing it to maturity: para. 3(10)(b). If the animal was acquired other than for valuable consideration, *e.g.* as a gift, the market value at acquisition would be taken into account instead of cost, together with any costs of rearing to maturity: para. 3(10)(b). It is the cost of the actual animal sold that is compared with the sale proceeds, which is why the farmer must keep detailed records.

Because the proceeds of sale of an animal that is replaced are credited in the herd basis computation and the cost of the replacement debited, it is sometimes the practice in preparing farm accounts to pass these transactions through the trading account instead of through the fixed asset herd account, leaving the cost in the herd account unaffected. Only when an animal is sold without replacement is part of the herd value removed but this may well result in the cost of the wrong animal being compared with the sale proceeds and does not give the correct answer required by the legislation. The correct treatment is to pass all purchases and sales of herd animals through the herd account in the same way as for any other fixed asset, so that the total of the herd account represents the total cost of the animals which at any time constitutes a herd.

Strictly speaking a replacement animal is only a replacement if acquired in the same accounting period. However, in practice, a replacement animal acquired within 12 months of the disposal is regarded as a replacement and the previous year's computation is adjusted accordingly unless the proceeds of sale have been held over in suspense pending the acquisition of the replacement animal (I.R. 9, para. 7). It appears that the Special Commissioners have allowed a replacement 22 months after the disposal: Accountancy Age, September 5, 1985.

Disposal of herd

7–010   A major advantage of the herd basis applies where there is a disposal within a 12-month period of the entire herd without its replacement by a new herd of the same class: para. 3(8)(a), or there is a substantial reduction made in the number of animals in the herd: para. 3(8)(b). In these cases any profit or loss on the disposal of the herd is excluded from the tax computation for income tax, corporation tax and C.G.T. purposes (para. 3(6); C.G.T.A. 1979, ss.37(1), 127(1)).

If, however, within five years of the sale of the herd or a substantial proportion of it, the seller acquires or begins to acquire a production herd of the same class, the sale proceeds are taxable and the cost of the replacements are allowable in the same way as a replacement of an individual animal. There is a limitation that the sale proceeds are not to exceed the replacement cost of lower quality animals if the sale was forced upon the seller by causes wholly beyond his control: para. 3(9)(a). The sale proceeds are taxed not when received but when the appropriate replacements take place within the five-year period: para. 3(9)(b).

7–011   On a partnership change or dissolution with no herd basis election by the new partners or continuing sole trader there would be a deemed sale at market value of the herd, under para. 5, giving rise to a tax-free profit under para. 3. The continuing partners or sole trader would not be regarded as sellers who have replaced a herd within five years because the seller was the old partnership, a different farmer from the new partnership or sole trader.

The definition of a substantial reduction in the number of animals in the herd is not further defined but the Revenue states:

> "It will depend on the facts of the case whether a particular reduction in number is substantial, but the Board of Inland Revenue will normally be prepared to consider a reduction of 20 per cent or more as substantial." (I.R. 9, para. 7).

Where there is a five-year interval between disposing of one production herd and acquiring a further herd of the same class, a herd basis election may be made for the new herd as if there had never previously been a production herd of this class: para. 4.

Connected parties

If any animals constituting part of a production herd are sold or trans- **7–012** ferred other than at a full arm's length price and the transferor is a body of persons over whom the transferee has control or vice versa, or both transferor and transferee are bodies of persons over whom some other person has control, open market value is substituted for the actual proceeds or transfer value: para. 5(1)(a). For this purpose, body of persons includes a partnership, and control means voting control in accordance with I.C.T.A. 1988, s.840: para. 5(2). The substitution of market value also applies where there is a transfer in which the sole or main benefit, or one of the main benefits, which might have expected to accrue to the parties or any of them was a benefit from obtaining a right to elect for the herd basis to apply or from varying the effect of such an election: para. 5(1)(b).

Compulsory slaughter

Where the whole or a substantial part of a production herd is compulsor- **7–013** ily slaughtered, a late election for the herd basis may be made: para. 6(1). Such an election must be made within two years of the end of the chargeable period which is affected by the compensation payment; that is normally two years from either the end of the company's accounting period in which the compensation is received or the end of the tax year in the basis period of which the compensation is received, unless the period is brought forward by reason of set-off of losses in the same way as for an original election under para. 2(3). If an election is subsequently made to put a second year of assessment onto an actual basis under I.C.T.A. 1988, s.62(2), (3), the herd basis election is effective if given not later than the s.62(2) election; para. 6(3).

If a late election is given as a result of compulsory slaughter it applies only for the period to which it first relates and subsequent periods: para. 6(4). This includes the period affected by loss where there is a carry back under I.C.T.A. 1988, s.380: para. 6(4) exception. This does not apply for corporation tax purposes so the first period affected is that in which the compensation is received. Compensation is received in an accounting period for this purpose if the event giving rise to the compensation took place within the accounting period, even though the cash is received in a later period: in other words, if under normal accounting practice it would be shown as a debtor in the accounts at the year end: para. 6(5).

Ownership

It is worth noting that the herd basis applies to a farmer *keeping* farm **7–014** animals, and ownership as such is not essential. It would appear therefore that the rules will be applied for animals bought on hire-purchase or deferred terms where the vendor retains title until the full purchase price has been paid. Where there is an interest element in the cost of animals acquired on deferred terms this would be treated as a normal trading expense and only the proportion of the cost relating to the actual purchase would be the cost of the animal for the herd basis calculations.

Share farming

7–015   It is by no means clear how the herd basis would be applied to share farming but it is thought in practice that the Revenue would accept a combined herd basis computation that was then apportioned to the sharers in accordance with their rights.

Planning aspects

7–016   In the majority of cases a farmer with a production herd will wish to elect for the herd basis so that the profit on sale of the herd or a substantial proportion of it will be tax free, even though such sale may be unlikely to occur for some considerable period. It is, of course, possible that the disposal of the herd could result in a loss which on a herd basis election would be unrelievable. However, in terms of generally rising prices it would seem probable that the likelihood of a profit on eventual disposal is greater than the likelihood of a loss.

One of the disadvantages of the herd basis is the degree of record keeping needed, but an efficient farmer would probably keep much of this information for the purpose of running the farm properly in any event.

Obviously the herd basis is most advantageous where the tax free profit on sale is likely to be the largest and this is where the animals are home-bred or at least home-reared to maturity, rather than bought in as mature animals. As a general plan it is preferable to buy in replacements but to breed additions to the herd. The lower the cost of rearing that can be substantiated, the greater the benefit. A substantial reduction in a herd is more likely if there are several smallish herds rather than a single large herd, but whether this can be achieved given the classification of what constitutes a separate herd depends on the individual circumstances of each case. If possible the herd should be expanded when prices are low to keep down the base value of the herd animals.

7–017   If a number of sales are contemplated these should, if possible, be concentrated within a 12-month period in order to achieve a substantial reduction. It might be worth selling marginal animals rather than retaining them, in order to obtain the benefit of the tax-free profit on the substantial reduction. Where the herd is increased in size in a period in which there are also sales, the animals with the highest costs should be chosen as replacements (on which the cost is deductible) and those with the lowest costs chosen as additions (on which there is no allowance). It is necessary to plan whether to have either (a) a sale followed by a replacement, where both would be reflected in the trading account, or (b) an increase in the herd followed by a subsequent reduction, where the cost of the increase would not be allowed but only the profit on the disposal without replacement would be taken into account. This obviously depends on the respective values of the various animals.

7–018   Animals on which a loss on sale would arise should be sold outside the 12-month period in which there is a substantial reduction in the herd in order to obtain relief for the loss. Similarly, any replacements of worn out animals should be made outside such a period so that the allowable cost as a replacement exceeds the sale proceeds. Again it might be possible to choose which animals are to be regarded as additions and which as replacements in order to maximise the available relief.

CHAPTER 8

Grants and Compensation

Grants and compensation receipts

8–001   Grants in respect of trading activities will normally be taxed as trading income or deducted from the allowable expenditure claimed as a trading expense which in both cases makes the receipt fully taxable. (*I.R.C.* v. *Falkirk Ice Rink Ltd.* [1975] S.T.C. 434, *Ryan* v. *Crabtree Denins Ltd.* [1987] S.T.C. 402). Grants and compensation receipts so treated have included ploughing grants under the Agricultural Development Act 1939 (*Higgs* v. *Wrightson* (1944) 26 T.C. 73), reimbursement of development expenditure under the Agriculture and Horticulture Act 1964, s.2 and Small Horticultural Production Business Scheme 1964 (S.I. 1964 No. 963), grants under the Dairy Herd (Alternative Enterprise) Scheme, grants under the Hill Livestock Compensatory Allowance, Sheep Annual Premium Scheme, Suckler Cow Premium Scheme, Agricultural Training Board grants, grants under the EC Farm and Horticulture Development Scheme of 1974, compensation for loss of profits, *Burmah Steam Ship Co. Ltd.* v. *I.R.C.* (1930) 16 T.C. 67 and disturbance, *Stoke-on-Trent City Council* v. *Wood Mitchell & Co. Ltd.* [1979] S.T.C. 197, SP 8/79. Other relevant cases include *Lang* v. *Rice* [1984] S.T.C. 172 (compensation for loss of profits), *Donald Fisher (Ealing) Ltd.* v. *Spencer* [1987] S.T.C. 423 (damages from estate agent for negligence), *London & Thames Haven Oil Wharves Ltd.* v. *Attwooll* (1967) 43 T.C. 491 (compensation for loss of use of damaged jetty), *Gray* v. *Lord Penrhyn* (1937) 21 T.C. 252 (defalcations made good by auditors) and *Rolfe* v. *Nagel* [1982] S.T.C. 53 (compensation for loss of prospective customer).

Compensation for a capital loss or sterilisation of a capital asset is a capital receipt, *Glenboig Union Fireclay Co. Ltd.* v. *I.R.C.* (1922) 12 T.C. 427, *Van den Berghs Ltd.* v. *Clark* (1935) 19 T.C. 390.

8–002   Flood rehabilitation grants under the Coastal Flooding (Emergency Powers) Act 1953 were held to be entirely capital in the case of *Watson* v. *Samson Brothers* (1959) 38 T.C. 346, as the sums were paid to restore a capital asset to production; however it is possible that only the excess over the actual rehabilitation costs should have been treated as capital, but this argument was advanced by the Revenue only in the High Court instead of before the Commissioners and was rejected as inadmissible.

Compensation for reorganisation or compulsory acquisition under the Agriculture (Miscellaneous Provisions) Act 1968, ss.9 or 12 is regarded as a tax free capital receipt for up to four times the rent with any excess taxable as income.

Cereal deficiency payments in respect of home grown cereals are treated as trading income in the accounting year in which the payments are notified, although in strictness they should be credited by reference to the dates of harvesting the grain. This concession under Extra-Statutory Concession B6 (1988) does not apply where the commencement or cessation rules are in force. Similar provisions are applied in practice to lime subsidies.

Compensation for the compulsory slaughter of animals is treated as income. However it is possible in such circumstances to make a late election for the herd basis to apply and this may be made under I.C.T.A. 1988, Sched. 5, para. 6 in which case the herd basis rules would apply as explained in Chapter 7. If the herd basis does not apply the compensation may be spread equally over three years beginning in the accounting period following that in which the compulsory slaughter took place. (Extra-Statutory Concession B11 (1988)). **8–003**

Under the Milk Supplementary Levy (Outgoers) Scheme 1984 compensation may be taken either in a capital form, subject to C.G.T. for the surrender of milk quota or as compensation for milk quota loss of profits, in which case the compensation is spread equally over seven years and taxed as income when received. In many cases the value of the quota exceeds the compensation available under the outgoers scheme and this is currently little used in practice.

Receipts under the Dairy Herd Conversion Scheme were held to be capital in *White* v. *G & M Davis* [1979] S.T.C. 415 and *I.R.C.* v. *W. Andrew Biggar* (a firm) [1982] S.T.C. 677.

Capital expenditure grants

In some cases grants are made in respect of capital expenditure which would qualify for capital allowances and as such has to be deducted from the expenditure before the allowances are claimed, unless it is a grant made under the provisions of Part II of the Industrial Development Act 1982 or Part I of the Industry Act 1982, or similar grants in Northern Ireland: C.A.A. 1968, s.84. Grants deducted before capital allowances would include those under the EC Farm and Horticultural Development Schemes 1974 and 1980, the Agriculture Improvement Scheme (EC), the Agriculture Improvement Scheme (National), and those under the Agricultural Act 1967, ss.31, 32 and 33 and those under the Agriculture and Horticulture Grant Scheme 1980, the Agriculture and Horticulture Grant (Variation) Scheme 1981 and the Regulations made thereunder. Where such grants relate to reimbursement of revenue expenditure, it would be reduced by the grant before claiming the net cost as a trading expense. Other grants reducing capital allowances would be those under the EC Marketing and Processing Regulations. **8–004**

Statutory compensation received under the Agricultural Holdings Act 1948, s.34 by an outgoing tenant of an agricultural property following notice to quit is totally tax free, following *Davis* v. *Powell* [1977] S.T.C. 32, as is statutory compensation to a tenant paid under the Landlord and Tenant Act 1954, s.37 (*Drummond* v. *Austin Brown* [1984] S.T.C. 321). If, **8–005** however, the tail end of a lease is surrendered for a cash sum which takes into account the statutory compensation but without the service of a notice to quit there would be a normal chargeable disposal. Statutory compensation to tenants in respect of milk quota under the Agriculture Act 1986 and S.I. 1987, No. 626 is not, in the Revenue view, similar to that payable under I.C.T.A. 1954, s.37 and would be subject to C.G.T. (Farm Tax Brief Vol. 2, No. 7, p. 58).

A grant for giving up an uncommercial agricultural unit under the Agriculture Act 1967, s.27 is also entirely tax free (C.G.T.A. 1979, s.112). Compensation for compulsory acquisition however is subject to C.G.T. under C.G.T.A. 1979, ss.110 to 111B (see Chapter 15).

Capital sums under the Farm Structure (Payments to Outgoers) (Variation) Scheme 1981 on giving up or changing the use of non-economic land is a non-taxable capital receipt, although an annuity under the same scheme payable to an individual aged 65 years or more is taxable as income.

8–006 Compensation for the surrender of milk quota under the Milk Supplementary Levy (Outgoers) Scheme 1984, which may be taken instead of the compensation for milk quota loss of profits which is taxable as income, is subject to C.G.T. and dealt with in Chapter 15. The compensation is now payable over seven years at 27.489 pence per litre.

Local Authority Housing Grants for the improvement of farm properties would be deducted from the allowable capital expenditure subject to agricultural buildings allowances.

Planting and restocking grants for commercial woodlands are dealt with under Chapter 12.

Countryside Commission grants for improvement to the environment such as the construction of nature trails, picnic areas, etc., would not be taxable except to the extent that if the expenditure is allowable revenue expenditure the amount claimed would have to be reduced by the grant received.

CHAPTER 9

Rents

In these days of mechanisation many farmers find that they have what **9–001** used to be farmworkers' cottages or other properties which are surplus to immediate requirements and are let.

The tax treatment on income from property depends on the terms under which the property is let. If property is let with services, including the provision of meals, broadly in line with a guest house or hotel, as is often done by country farmers in holiday areas, the profits would be assessed as earned income under Schedule D, Case I and any losses available would be computed under Case I rules and be available in the same way as any other trading losses. The distinction between earned and unearned income is not currently important except when computing pension contributions.

Furnished holiday lettings

In *Griffiths* v. *Jackson, Griffiths* v. *Pearman* [1983] S.T.C. 184 and *Gittos* **9–002** v. *Barclay* [1982] S.T.C. 390 it was held that furnished property lets to short-term occupiers does not amount to a trade, despite the provision of some services, and the lettings income therefrom is assessable under Schedule D, Case VI as investment income. The Government therefore changed the treatment of short-term holiday lets to enable them to count as earned income and as a business for C.G.T. purposes, enabling roll-over relief and retirement relief to be claimed. The distinction between a hotelier and the owner of property who lets furnished rooms is a fine one considering the trend towards self-service hotels and motels, as was pointed out by Vinelott J. in the *Jackson* and *Pearman* cases.

As the trend for self-catering holidays developed the owners of small guest houses and others found it more difficult to compete. Consequently some were forced into the expensive process of converting their properties into self-catering holiday flats only to find the taxation treatment very restrictive.

The Government therefore introduced what is now I.C.T.A. 1988, ss.503, 504 to alleviate their plight providing a considerable array of reliefs.

The holiday accommodation, which must be in the U.K., has to be let furnished on a commercial basis for short holiday lets. It must be available for commercial lettings for not less than 140 days in a qualifying reference period during which it must be let for a minimum period of 70 days. Relief is not available where lettings are to persons in the same occupation for a continuous period in excess of 31 days.

9–003 Where the letting conditions are satisfied the tax benefits to the proprietor are as follows:

1. Although taxed under Schedule D, Case VI the profits are treated as earned income. Accordingly tax is paid in two instalments as for any trader;
2. Where the accommodation is owned or part-owned by the wife, a claim for the separate taxation of wife's earnings may be made up to 1989–90; from 1990–91 wives are taxed independently from their husbands;
3. Capital allowances may be claimed;
4. Relief for losses is available for set-off against other income, etc.;
5. Profits constitute "relevant earnings" enabling retirement annuity or personal pension premiums to be paid to provide for a pension;
6. All lettings made by a person, partnership or "body of persons" are treated as one trader similar to the treatment of farming income enabling profits and losses to be pooled;
7. Certain C.G.T. relieving provisions apply, notably retirement relief, roll-over relief and the provisions relating to gifts of business assets.

The business property relief which ordinarily applies to the transfer of a business for inheritance tax purposes has not been extended to the letting of furnished holiday accommodation.

Where the business of organising and running the lettings is conducted from the home address of the owner it should be possible to claim the cost of travelling to and from the property as a tax deductible expense, together with all other costs of accountancy, secretarial, postage, stationery, agent's fees, repairs and telephone expenses.

Furnished property lettings

9–004 Where a property is let furnished other than on a holiday let, the entire rents are normally assessed as unearned income under Schedule D, Case VI (*Shop Investments Ltd.* v. *Sweet* (1940) 23 T.C. 38), subject to the taxpayer's right to elect for the unfurnished element of the rents to be assessed under Schedule A, leaving the proportion relating to the provision of furnishings to be taxed under Schedule D, Case VI: I.C.T.A. 1988, s.15(1)(4).

In most cases it pays to have the assessment entirely under Schedule D, Case VI as the Revenue is prepared to allow an annual flat depreciation allowance of 10 per cent. based on the rent received, less rates or other tenants' expenses, in accordance with the Inland Revenue Press Release of October 13, 1977, (SP/A19) in addition to expenses incurred in the repair, maintenance, insurance and management of the property. Where the 10 per cent. deduction is allowed relief cannot also be claimed for the cost of renewing items of furnishing or furniture. It would also be desirable to claim Schedule D, Case VI treatment for furnished lettings where there is any likelihood of losses arising from other Schedule D, Case VI activities as such losses can only be set against other Schedule D, Case VI sources of income.

Unfurnished property lettings

Property which is let unfurnished is assessable under Schedule A: **9–005**
I.C.T.A. 1988, s.15.

Rents from income overseas are assessed under Schedule D, Case V: I.C.T.A. 1988, s.18(3). The rents charged are those due to be received in the tax year whether payable in arrears or in advance, and whether paid late or on time (*Strick* v. *Longsdon* (1953) 34 T.C. 528). Relief for rents not paid at all is given under I.C.T.A. 1988, s.41.

The definition of rents has been considered in *Martin* v. *Routh* (1964) 42 T.C. 106 and *Jeffries* v. *Stevens* [1982] S.T.C. 639. Exceptions include annual interest and rents from mines and quarries (including gravel pits, sand pits and brick-fields), iron works, gas works, sale springs or works, alum mines or works, and water works and streams of water, canals, inland navigations, docks and drains or levels, fishings, rights of markets and fairs, tolls, bridges and ferries, railways and other ways, other concerns of a like nature and tied premises. Rental income received by such bodies or from such sources are taxed under Schedule D and not Schedule A.

Receipts from the sale of turf can be either Schedule A or Schedule D, Case I (*Lowe* v. *J.W. Ashmore Ltd.* (1970) 46 T.C. 597) as can grazing (*Bennion* v. *Roper* (1969) 46 T.C. 613) but the sale of colliery dross bings was held to be capital in *Roberts* v. *Lord Belhaven's Executors* (1925) 9 T.C. 501.

Relief was given against the Schedule A assessment in respect of any income from easements chargeable under Schedule B until its abolition in 1988–89. The Schedule A assessment was limited to any excess over the assessable value of the occupation under Schedule B: I.C.T.A. 1988, s.16.

By concession, a trader who lets part of his building in which he carries **9–006**
on his business can treat the rent as part of his trading receipts chargeable under Schedule D, Case I. It does not apply where the business consists of rendering services to tenants (Revenue leaflet I.R. 27, paras. 18, 107). Where farm cottages are not required for farm workers and are let, but there is a possibility that they will be used by a farm worker in the future, it is desirable to try and retain the income within the Schedule D assessment, both to obtain roll-over relief and so that the profit is available for pension contributions. Income from caravan sites where a trade is carried on may include rental income (Extra-Statutory Concession B29 (1988)).

Letting property can constitute a business (*American Leaf Blending Co. Sdn. Bhd.* v. *Director-General of Inland Revenue* [1978] S.T.C. 561, P.C.).

The person entitled to the rents is the person chargeable under Schedule A: I.C.T.A. 1988, s.21. The Schedule A assessment on a company is part of the income of the company under I.C.T.A. 1988, s.6.

9–007 Where, however, the landlord is not resident in the U.K. the Schedule A assessment can be made on an agent under I.C.T.A. 1988, s.43 or the tax can be collected from the tenant or agent under I.C.T.A. 1988, s.23.

It is usual to have one Schedule A assessment raised by the tax office to which tax returns are sent, but provision is made for separate assessments to be raised for each property or group of properties and this procedure is usually applied where property is dealt with by different agents in different parts of the country. The total assessable income under Schedule A is not affected by the number of assessments: I.C.T.A. 1988, s.22.

The assessment is on a current year basis which means that, *e.g.* the Schedule A assessment for 1989–90 is based on the income less expenses for the year ended April 5, 1990.

If the landlord prepares proper accounts including a balance sheet in respect of his rental income the receipts and expenses can be treated on a normal accruals basis (Revenue leaflet I.R. 27, para. 102).

The assessment will, however, normally be raised some months earlier so that the tax which is due on January 1, in the year of assessment, *i.e.* January 1, 1990, in this example, can be paid on the due date. The assessment is therefore based in the first instance on the net rents assessed for the preceding year. When the correct figures are known the assessment is automatically adjusted to the actual net rents for the tax year, and it is not necessary to appeal against the original assessments.

The landlord can appeal against the assessment based on the preceding year's rents if he can show that the aggregate gross rents he is due to receive in the current tax year are less than for the preceding year, because, *e.g.* one of the properties has been sold. The landlord cannot appeal against the original assessment on the grounds that the expenditure is in excess of the preceding year, although in cases of hardship the Revenue will usually hold over part of the tax (Revenue leaflet I.R. 27, para. 29).

If Schedule A is not paid by the landlord the collector may serve a notice on the tenant, sub-tenant or agent requiring them to pay the tax direct to the Revenue and deduct the amount paid over from the amount they are due to pay the landlord. The amount demanded cannot exceed the amount due to be paid to the landlord: I.C.T.A. 1988, s.23.

Categories of tenancy

9–008 It is necessary to divide properties under Schedule A into three categories:

1. those let at full rents, not on tenant's repairing leases, which are pooled so far as income and expenditure are concerned;
2. those let at full rents, on tenant's repairing leases, which are not pooled but dealt with on a property-by-property basis; and

3. those let at less than a full rent which are also dealt with on a property-by-property basis.

Rent includes a tenant's contribution towards repairs.

A tenant's repairing lease is one where the lessee has to maintain and repair substantially the whole of the premises: I.C.T.A. 1988, s.24(6). If the tenant is not liable for fair wear and tear, a tenant's repairing lease does not exist. A full rent is not necessarily a rack rental for full market value but is merely a rent sufficient to enable the landlord to break even taking one year with another: I.C.T.A. 1988, s.24(7). Any mortgage interest payable on the property would be ignored in ascertaining whether or not a property was let at a full rent (Revenue leaflet I.R. 27, para. 72).

Expenses

Revenue expenses relating to the property, *i.e.* maintenance, repairs, **9–009** insurance and management are allowable deductions from rent: I.C.T.A. 1988, s.25(2). The deduction is made in the tax year in which the payment is made, but if relief cannot be given in that year a deduction may be claimed for the year in which the expenditure was incurred. If relief still cannot be given the unrelieved expenditure may be carried forward: Revenue leaflet I.R. 27, para. 69 and I.C.T.A. 1988, s.31, subject to the accounts basis being followed (Revenue leaflet I.R. 27, para. 102).

For a property let at full rent and not a tenant's repairing lease, it is permissible to claim repair expenditure relating to dilapidations arising during the currency of the lease, or a preceding lease or a void period, I.C.T.A. 1988, s.25(2), in between such leases, I.C.T.A. 1988, s.25(2)–(5). The income and expenditure in relation to properties let on full rents are pooled and a loss on one property can be set against a profit on another lease which is not a tenant's repairing lease: I.C.T.A. 1988, s.25(6). An excess of expenditure over income can be carried forward to future years. It should be noted that expenditure before a property is let at a full rent can only be claimed so far as it relates to dilapidations arising from any period of ownership before this first letting during which the property was empty or was let at a full rent. Relief cannot therefore be claimed for putting into good lettable order a property acquired in a dilapidated condition.

A tenant's repairing lease often yields only a ground rent and the only **9–010** expense incurred is likely to be the cost of collecting the rent. If there is any exceptional repair expenditure, *e.g.* because a landlord is liable to make good structural faults, any excess expenditure over the income from that lease can be set against income from property let at full rents (Revenue leaflet I.R. 27, para. 77). Otherwise expenditure can be deducted only from the rent from the particular lease concerned. Any expenditure relating to a previous lease or a void between leases cannot be set against the income from any other tenant's repairing lease.

EXAMPLE

A landlord received rents and paid expenses on let properties as follows:

	1. (leases at full rent)		3. (tenant's repairing lease at full rent)	4. (lease not at full rent)
	£	£	£	£
Rents received	1,000	900	500	100
Expenses	500	1,600	40	150
Profit (loss)	500	(700)	460	(50)
Loss set off against property let at full rent not being a tenant's repairing lease	500	500	—	—
Assessable	Nil	Nil	£460	Nil
Losses c/fwd		(200)		(50)
		(against future income on Nos. 1 or 2)		(against future income on No. 4 under same lease)

Expenditure on property not let at a full rent can only be set against the income from that property under the current lease.

9–011 The Revenue allows expenses in respect of dilapidations before the ownership of the landlord where the previous owner was the spouse or trustee for the spouse (Extra-Statutory Concession A21 (1988)). Repairs obviated by improvements may be claimed under Extra-Statutory Concession B4 (1988).

Legal expenses relating to a lease have been disallowed (*I.R.C.* v. *Wilson's Executors* (1934) 18 T.C. 465) as have road charges (*Davidson* v. *Deeks* (1956) 37 T.C. 32) and rents misappropriated by an agent (*Pyne* v. *Stallard-Penoyre* (1964) 42 T.C. 183).

Collection charges payable under a will were annual payments not expenses in *Clapham's Trustees* v. *Belton* (1956) 37 T.C. 26.

Where sums other than rent under a lease are received such as rent charges, feu duties and payments for easements, sporting rights or rights of way, any revenue expenditure incurred in relation to the land as a result of such receipt may be deducted. For example, the costs of maintaining a river bank which is let out for fishing would be allowable: I.C.T.A. 1988, s.28, see Chapter 10.

Interest

9–012 The deduction of interest as a trading expense is dealt with in Chapter 3 and generally in Chapter 21. Interest paid on a loan taken out to purchase or improve an estate or interest in land, or buildings on the land, or repay such a loan, is allowed as a charge on income, *i.e.* as a deduction from total income not specifically from rents received, but only if certain conditions are met: I.C.T.A. 1988, ss.353, 354, 367, see Chapter 21.

No relief is available for overdraft interest: I.C.T.A. 1988, s.353(3)(a), *Walcot-Bather* v. *Golding* [1979] S.T.C. 707 or at more than a reasonable commercial rate: I.C.T.A. 1988, s.353(3)(b), I.C.T.A. 1988, ss.354, 355, 367.

Eligible loans for interest purposes are further restricted to (a) the main residence of the owner (*Frost* v. *Feltham* [1981] S.T.C. 115) or of a separated spouse or of a dependent relative if let rent free and without any other consideration or (b) property let at a commercial rent for 26 weeks out of 52 and available for letting or under reconstruction or repair for the remainder of the time: I.C.T.A. 1988, s.355(1).

Interest paid in respect of property let is only available against rental income on that or any other land and cannot be relieved against income other than from property but can be carried forward against future property income: I.C.T.A. 1988, s.355(4).

It should be noted that the property has to be let, a mere licence is not sufficient.

The land has to be in the U.K. or the Republic of Ireland, I.C.T.A. 1988, s.354(1); no relief is available for interest on non-trading loans to acquire other properties even if fully let (*Ockenden* v. *Mackley* [1982] S.T.C. 513).

Land managed as one estate

An election for land to be treated as managed as one estate can be made where an election had been made in respect of land managed as one estate in 1962–63, and the appropriate election is made by the end of the year of assessment after that in which the land is acquired: I.C.T.A. 1988, s.26.   **9–013**

Where a valid election was made it is possible to aggregate expenditure property let at a full rent and property occupied by the owner or let at less than the full rent, (but not under a tenant's repairing lease) provided that the property is managed as one estate and provided that the gross rateable value of properties not included at a full rent is included in the computation at notional income.

This treatment is advantageous where the expenditure on maintaining the properties not let on a full rent exceeds the gross rateable value, but if the appropriate election was not made by the then owner (in 1962–63) no future claim for treatment of property as one estate will be considered.

Miscellaneous

Capital expenditure on a sea wall to protect existing property may be treated as repair expenditure on the property equally over 21 years: I.C.T.A. 1988, s.30. If a property is sold the allowance passes to the purchaser (*Hesketh* v. *Bray* (1888) 2 T.C. 380).   **9–014**

Allowable deductions for Schedule A must be relieved as soon as possible: I.C.T.A. 1988, s.31.

Where there is an excess of expenditure over income the expenditure may be allocated to the taxpayer's best advantage. Deductions cannot be claimed for expenses borne by someone else such as an insurance company or by a person other than a company, under deduction of tax. No double allowance is permitted.

Capital allowances

9–015   An election can be made to claim capital allowances in respect of plant and machinery on various classes of plant and machinery used in the maintenance, repair or management of property: I.C.T.A. 1988, s.32. Such allowances are added to allowable expenditure for Schedule A purposes and not set against general income. Balancing charges in excess of capital allowances are allowable expenditure for a year are charged under Schedule D, Case VI.

Agricultural land

9–016   An excess of expenditure (including capital allowances) over income in relation to agricultural land which is let may be set against agricultural income for the year of assessment or accounting period in which the expenditure was incurred or on election against any other income for that year of assessment or accounting period: I.C.T.A. 1988, s.33. If an election is not made or an unrelieved balance remains, the excess can be carried forward indefinitely against future income from farming or letting agricultural land. This claim can be extended to maintenance costs of uncommercial farms under Extra-Statutory Concession B5 (1988).

Tax treatment of receipts and outgoings on sale of land

9–017   On a sale of an interest in property, rents and related outgoings are apportioned in accordance with the agreement: I.C.T.A. 1988, s.40. The purchase is charged to tax on the net amount received less that which is apportioned to the vendor on the basis of the actual receipt date. The amount apportioned to the vendor is deemed to be received by him immediately before the date up to which it is apportioned. For example, a sale might take place on the basis that the vendor would be entitled to rents up to say, March 31, 1989, with subsequent rents going to the purchaser. The quarter's rent paid on June 24 would be charged on the purchaser in 1989–90, less six days' rent apportioned to the vendor which would be treated as having been received by him on March 31, 1989, and therefore assessable in 1988–89.

Relief for rent, etc., not paid

9–018   Rents are taxed on the landlord on the amount due to be received, not the amount actually received: I.C.T.A. 1988, s.41. Bad debts which remain after reasonable steps to enforce payment or as a result of rent being waived in order to avoid hardship to the tenant may be treated as if they were never due and any appropriate adjustments to the assessments have to be made. The relief has to be claimed and applies to rents and premiums taxable under Schedule D, Case IV as well as under Schedule A. If, however, the lost rent is subsequently received the Revenue must be informed within six months and the appropriate assessments will be made. A gratuitous waiver of rent (under seal) before the due date of payment is accepted by the Revenue as an alteration in the rent due and therefore becomes an effective waiver (Inland Revenue letter to the authors, January 22, 1979).

Non-residents

If a tenant pays rent or a premium or disguised premium direct to a non-resident landlord he must deduct tax at the basic rate and pay the amount so deducted over to the Revenue under I.C.T.A. 1988, ss.349, 350, 43. In such circumstances the non-resident would not be charged in the name of any agent in the U.K. If the non-resident pays excessive tax as a result of deduction at source compared with the assessment on the rents received less allowable expenses, he may make the appropriate repayment claim. **9–019**

Non-residents are liable to tax at higher rates, if appropriate, in respect of rents arising from property in the U.K., although the Revenue does not attempt to pursue such claims. If, however, rents are received through an agent in the U.K., T.M.A. 1970, s.78 would apply and the U.K. agent would be assessed to both the basic and higher rate taxes as appropriate and would have the right to withhold any payments made under T.M.A. 1970, s.83. The CCAB Press Release, February 19, 1974, confirms that payment to a U.K. account in ignorance of the landlord's place of abode exonerates the lessee.

Premiums

Where property is let and a premium (or its Scottish equivalent) charged, it is necessary to consider the taxability of the premium. A premium charged on the assignment of a lease will give rise to a C.G.T. assessment on the assignor. As far as the assignee is concerned, the premium paid represents a capital payment of a depreciating asset. No tax relief for the amortisation of the lease is allowed to the lessee. **9–020**

If a premium is charged on the grant of a new lease for less than 50 years, a part of the premium is charged to tax under Schedule A as rental income: I.C.T.A. 1988, s.34(1). The part so taxed is the total premium less two per cent. a year for each year of the lease other than the first, so that for a 10-year lease all but 18 per cent. of the premium would be charged to tax under Schedule A. The premium is assessable in the year in which the lease is granted. As far as the tenant paying the premium is concerned, if he is a trader he may claim as a trading expense the annual equivalent of the proportion of the premium taxed on the landlord under Schedule A: I.C.T.A. 1988, s.87. In the example just quoted 8.2 per cent. (i.e. $(100 - 18) \div 10$) of the premium would be allowed as a trading expense of the trader for each of the 10 years of the life of the lease.

Certain items, although not directly paid as premiums, may nevertheless be taxable as deemed premiums. A deemed premium can arise in a number of situations. Those most frequently met in practice are:

1. where the terms of the lease oblige the tenant to bear the cost of the capital improvements to the fabric of the property;

2. where under the terms of a lease a sum becomes payable by the tenant in lieu of rent, or in consideration for the surrender of the lease;

3. where in consideration for the variation or waiver of any of the terms of a lease a sum becomes payable by the tenant (*Banning* v. *Wright* (1972) 48 T.C. 421).

Such premiums paid in kind would be taxed on the lessor in the same way as if they had been paid in cash: I.C.T.A. 1988, s.34(2)–(5). It should be noted, however, that while a trader may obtain relief for premiums paid under I.C.T.A. 1988, s.34 and 35, relief is not available on a disguised premium falling under 1. above where the capital works qualify for capital allowances: I.C.T.A. 1988, ss.34(2) and 87(8).

A premium or other capital payment made to a person connected with the landlord is taxable on the recipient under Schedule D, Case VI.

Where a premium is payable by instalments the recipient may pay the tax by instalments over the period of the instalment receipts or eight years, whichever is shorter, provided that he can convince the Revenue that he would otherwise suffer undue hardship, see Hansard, June 22, 1972.

Top-slicing relief under I.C.T.A. 1970, Sched. 3, was available only until 1987–88.

Rules for ascertaining duration of leases

9–021   A lease is deemed to terminate at the date when, under the terms of the lease, it is most likely to terminate so that if there are break or renewal clauses which are likely to be taken advantage of, the length of the lease takes these into account: I.C.T.A. 1988, s.38. This is an anti-avoidance section with the usual provisions relating to connected persons so that if they become entitled to an extension of the lease it is treated as if the tenant were so entitled.

The Revenue has power to demand information from any person in connection with the duration of a lease, except that a solicitor may claim privilege and merely divulge his client's name and address.

Schedule D charge on assignment of lease granted at undervalue

9–022   If a lease is granted at an undervalue such that a premium could have been demanded, the premium forgone may be assessed on the assignee or a subsequent assignee under Schedule D, Case VI if the assignee subsequently assigns the lease at a premium: I.C.T.A. 1988, s.35. The assessment would not be in excess of the assignee's premium received. This is an anti-avoidance provision to prevent a landlord granting a lease to, *e.g.* a relation enabling the relation to receive a capital profit on reassigning at a premium.

EXAMPLE

Assignment of lease granted at undervalue

A Ltd. granted a lease to A for 21 years without taking a premium, although a premium of £20,000 could have been charged:

	£	£
Notional premium (amount forgone)		20,000
Less: 2% a year for 20 years (21–1)		8,000
Notional Schedule A charge		£12,000

A assigns the lease to B Ltd. for £15,000
Schedule D, Case VI charge on A:

$$£15,000 \times \frac{12,000}{20,000}$$

(Capital gain: £15,000 − £9,000 = £6,000)

£9,000

B Ltd. assigns to C Ltd. for £10,000
(no Schedule D, Case VI charge as the consideration does not exceed that paid by B Ltd.)

C Ltd. assigns to D Ltd. for £20,000

Excess consideration received by C Ltd.	10,000	
Amount forgone by A Ltd.	20,000	
Less: Excess charged on A's assignment	15,000	
	£5,000	

Schedule D, Case VI charge on C Ltd. $\times \frac{12,000}{20,000}$ 3,000

Total assessed   £12,000

Schedule D charge on sale of land with right to reconveyance

A sale of land with a right to repurchase is taxed under Schedule D, Case **9–023** VI on the excess of the sale price over the repurchase price, subject to a deduction of two per cent. a year for each complete year other than the first between the date of sale and the earliest date of repurchase: I.C.T.A. 1988, s.36. The transaction is also caught if the repurchase is not by the vendor but by somebody connected with him. The grant of a lease to the vendor or connected person under the terms of the sale is treated as a reconveyance unless the lease is granted and begins to run within one month after the sale.

Premiums paid, etc.: deduction from premiums and rents received

An intermediate landlord who pays a premium to a head landlord may **9–024** deduct the chargeable amount of the premium spread over the period of the lease as if it were an expense of the property: I.C.T.A. 1988, s.37.

Relief is available in respect of an actual or deemed premium paid within I.C.T.A. 1988, ss.34 and 35 even if the recipient is exempt from tax for some other reason, *e.g.* because of charitable status.

If the intermediate landlord himself charges a premium on granting a sub-lease he may deduct from the chargeable proportion of this further premium a proportion of the chargeable premium paid for the lead lease. The proportion is that which the duration of the sub-lease bears to the duration of the lead lease.

Rent, etc., payable in connection with mines, quarries and similar concerns

9–025　Rent of land used by mines and quarries, mine-works, gas-works, water-works, canals, docks, fisheries, market rights, tolls, railway and similar organisations is paid under deduction of tax as if it were patent royalties and the recipient is taxed under Schedule D under I.C.T.A. 1988, s.55. This does not apply where the rent is paid in kind, in which case the assessment is under Schedule D, Case III on the recipient, on the value of the goods (*I.R.C.* v. *Baillie* (1936) 20 T.C. 187, *I.R.C.* v. *New Sharlston Collieries Co. Ltd.* (1939) 21 T.C. 69, *I.R.C.* v. *Hope* (1937) 21 T.C. 116, *Earl Fitzwilliam's Collieries Co.* v. *Phillips* (1943) 25 T.C. 430, *Stratford* v. *Mole & Lea* and *Old Silkstone Collieries Ltd.* v. *Marsh* (1941) 24 T.C. 20, *Craigenlow Quarries Ltd.* v. *I.R.C.* (1951) 32 T.C. 326, *Rogers* v. *Longsdon* (1966) 43 T.C. 231, *T & E Homes Ltd.* v. *Robinson* [1979] S.T.C. 351) (I.C.T.A. 1988, s.119).

Rent, etc., payable in respect of electric line wayleaves

9–026　Wayleaves payable in respect of pylons and telegraph poles, etc., are paid under deduction of tax if they exceed £2.50 a year and the recipient is assessed under Schedule D, Case III under I.C.T.A. 1988, s.120.

Investment companies

9–027　It used to be popular to hold investment properties through a private property investment company in order to withdraw the rents in the guise of directors' remuneration taxed as earned income. This is normally no longer advantageous as the level of remuneration that would be allowable, either as a property outgoing or management expense, is extremely limited, *L.G. Berry Investments Ltd.* v. *Attwooll* (1964) 41 T.C. 547, and on the sale of a property there would be a charge to corporation tax on the capital gains in the company. The disposal would be reflected in the value of the company's shares which would be expected to increase in proportion to the net amount of the chargeable gain. This means that before any surplus can be paid out to the shareholders, *e.g.* on a liquidation of the company or the sale of the shares, there would be a further C.G.T. liability on the individual's own shareholding. Trading premises should not normally be owned through an investment company because of the denial of C.G.T. retirement relief on retirement and realisation of the shares.

9–028　One of the problems with a closely controlled property investment company is that it is likely to be a close investment holding company which for periods beginning after March 31, 1989, is not eligible for the small companies' rate of corporation tax of 25 per cent. but has to pay the full rate of 35 per cent.

CHAPTER 10

Sporting Rights

Stud fees

Some sporting activities are regarded as a trade under Schedule D, Case **10–001**
I. These can be regarded as a separate trade where, *e.g.* stallion fees are
obtained for serving mares outside the farm as in *Malcolm* v. *Lockhart*
(1919) 7 T.C. 99, *McLaughlin* v. *Bailey* (1920) 7 T.C. 508 and *Cloghran
Stud Farm* v. *Birch* [1936] I.R. 1. If, however, the serving is confined to
animals on the farm the breeding activities could be part of the farming
trade (*Lord Glanely* v. *Wightman* (1933) 17 T.C. 634).

Stallion fees unconnected with a trading activity would be assessed under
Schedule D, Case VI (*Earl of Jersey's Executor's* v. *Basson* (1926) 10 T.C.
357, *Benson* v. *Counsell* (1942) 24 T.C. 178). Stallions used for breeding
purposes are not plant for the purposes of capital allowances (*Earl of
Derby* v. *Aylmer* (1915) 6 T.C. 665), although a working horse was held to
be plant in the workman's compensation case of *Yarmouth* v. *France*
(1887) 19 Q.B.D. 647.

Where there is both a trade of providing the services of stallions and private breeding for personal purposes only the service fees for privately
owned mares must be charged at the full market value (*Sharkey* v. *Wernher*
(1955) 36 T.C 275). The serving fees would then be taxed as those of a separate trade (*Wernher* v. *I.R.C* (1942) 29 T.C. 20).

Rights over land

Where sporting rights over land are let the receipts would be assessable **10–002**
to Schedule A under I.C.T.A. 1988, s.15(1) para. 1(c). It is possible to
deduct expenses of maintenance, repairs, insurance or management,
I.C.T.A 1988, s.28(a), any rent paid for the premises: I.C.T.A. 1988,
s.28(b), and other non-capital expenses of obtaining the income: I.C.T.A.
1988, s.28(c) and to carry forward any excess: I.C.T.A. 1988, s.28(d).

If in any year the sporting rights are not granted the expenses remain
deductible as if the land had been let at a full rent: I.C.T.A. 1988, s.29(1).
Where, however, the sporting rights are exercised by the owner or tenant
or by a guest of his, or by a director or participator of a close company,
owner or tenant, the expenses claimable for the maintenance of the sport-

ing right has to be reduced by the full market value of the rights enjoyed by the proprietor, his guests and directors or participators: I.C.T.A. 1988, s.29(2).

10–003   Where, as would normally be the case, the sporting rights to a director of a company are assessed as a benefit in kind under I.C.T.A. 1988, s.154, they would not also be deducted as from the expenses as notional income: I.C.T.A. 1988, s.29(3). Where the owner or tenant entitled to the sporting rights is a company, the allowance for expenses under I.C.T.A. 1988, s.29 is calculated by reference to years of assessment apportioned over the company's accounting periods as necessary: I.C.T.A 1988, s.29(4). Sporting rights for these purposes means rights of fowling, shooting or fishing or of taking or killing game, deer or rabbits: I.C.T.A. 1988, s.29(5).

Rates

10–004   Until March 31, 1990 (March 31, 1989, for Scotland) sporting rights are not exempt from rates even though the agricultural land itself would be exempt under the General Rate Act 1967, s.26, see Chapter 20. Sporting rights for rating purposes are defined by G.R.A. 1967, s.16 as any right of fowling, shooting or taking or killing game or rabbits or of fishing. There are a number of alternatives in connection with the liability to rates. If the occupier of agricultural land lets the land to a tenant for farming but retains the sporting rights for himself he is liable for the rates on the sporting rights under G.R.A. 1967, s.16(e) and 29(e) (*Lord Hastings* v. *Walsingham Revenue Officer* [1930] 2 K.B. 278, *Cleobury Mortimer R.D.C.* v. *Childe* [1933] 2 K.B. 368).

If the occupier of non-agricultural land lets the land to a tenant but retains the sporting rights for himself the tenant is liable for the rates of the land which is not exempt as agricultural land and this includes the rates for the sporting rights, although he can recover the rates in respect of the sporting rights from the occupier under G.R.A. 1967, s.29(1) and ss.65 and 66 (*Rogers* v. *St. Germans Union* (1976) 35 L.T. 332).

10–005   If the occupier of land, whether agricultural or not, lets the sporting rights separately to a tenant by deed but retains the other rights for himself rates may be collected either from the occupier or the tenant under G.R.A. 1967, s.29(2) (*Kennick* v. *Guilsfield Overseas* (1879) 5 C.P.D. 41). If, however, the ownership of the sporting rights is assigned absolutely the rates would be collected either from the tenant or from the assignee of the sporting rights (G.R.A. 1967, ss.29(3) and (4), *Swayne* v. *Howells* [1927] 1 K.B. 385). If the sporting rights are granted to a tenant other than by deed where the owner retains the other rights over the land himself it is the tenant who would be liable for the rates (G.R.A. 1967, s.16(e), *Cleobury Mortimer R.D.C.* v. *Childe* [1933] 2 K.B. 368).

From April 1, 1990 (April 1, 1989 for Scotland) agricultural land is exempt from non-domestic rating under L.G.F.A. 1985, Sched. 5, paras. 1–8, but agricultural land does not include land used mainly or exclusively for purposes of sport or recreation: L.G.F.A. 1988, Sched. 5, para. 2(2)(d): see Chapter 20. Certain rights of fishing are exempt under L.G.F.A. 1988, Sched. 5, para. 10.

Casual sporting rights over agricultural land where the rights are not severed from the land do not give rise to any charge to rates so long as the primary use remains agricultural because the land retains the exemption (*Swayne* v. *Howells* [1927] 1 K.B. 385, *Towler* v. *Thetford Rural Council* (1929) 99 L.J. K.B. 258). If the occupier of non-agricultural land grants sporting rights on a casual basis without severing the rights from the land he will be liable to rates on the basis of the land value, there being no agricultural exemption, and in valuing the land for rating purposes the sporting rights would be taken into consideration (*R.* v. *Williams* (1854) 23 L.T.O.S. 76).

A material alteration to a sporting rights agreement can cause the severance from the land to cease (*Burnell* v. *Dawnay* (1952) 45 R. & I.T. 790, *Sherborne* v. *Windrush Estates Ltd.* v. *Dyer* (1957) 50 R. & I.T. 378).

CHAPTER 11

Minerals

11–001   Farmers and landowners are naturally interested in what is under the soil as well as what is on the surface. The extraction of oil and natural gas is an unlikely pastime for the normal landowner and involves a specialist area of taxation, including petroleum revenue tax, under the Oil Taxation Act 1975, as amended. This topic is outside the scope of this book. However, the extraction of sand and gravel and the abstraction and sale of water are by no means uncommon activities of farmers and landowners and such activities are not included as part of the farming trade unless purely incidental to the farming activities. They are taxed as a deemed trade under Schedule D, Case I whether or not normal trading activities are carried out.

Rents and other income from property accruing to mining and other concerns are taxed under Schedule D, Case I I.C.T.A. 1988, ss.55(1), 15(1)(3)(*b*) (*Humber Conservancy Board* v. *Bater* (1914) 6 T.C. 555, *Port of London Authority* v. *I.R.C.* (1920) 12 T.C. 122, *Mersey Docks and Harbour Board* v. *Lucas* (1883) 1 T.C. 385, 2 T.C. 25)).

Such concerns are defined by I.C.T.A. 1988, s.55(2) as:

(a) mines and quarries (including gravel pits, sand pits and brickfields);
(b) ironworks, gasworks, salt springs or works, alum mines or works (not being mines falling within the preceding paragraph) and waterworks and streams of water;
(c) canals, inland navigation, docks and drains or levels;
(d) fishings;
(e) rights of markets and fairs, tolls, bridges and ferries;
(f) railway and other ways; and
(g) other concerns of the like nature as any of the concerns specified in paragraphs (b) to (e) above.

Tax is deducted at source under I.C.T.A. 1988, s.119 from rents paid to such concerns. If there is no concern tax is not deducted (*Duke of Fife's Trustees* v. *George Wimpey & Co. Ltd.* (1943) 22 A.T.C. 275).

Mineral royalties

11–002   Mineral royalties received by a U.K. resident after April 5, 1970, are split into two, one-half being taxed as income and one-half being taxed as a capital gain: I.C.T.A. 1988, s.122. Half the management expenses relating to the mineral royalties may be deducted from the income element but there is no such deduction from the capital element.

Mineral royalties are defined very widely for this section and although they would usually be subject to deduction of tax at source under I.C.T.A. 1988, s.119 could include sums chargeable by direct assessment under Schedule A or D. The payer of the royalty, if I.C.T.A. 1988, s.119 applies, will deduct tax from the whole payment in the normal way. The income element of the royalty is taxed on the recipient to corporation tax or basic and higher rate income tax as appropriate and the capital element to C.G.T. or corporation tax on chargeable gains as if it were the amount of the chargeable gain. If too much tax has been deducted under I.C.T.A. 1988, s.119, after offsetting any C.G.T. due, any balance is repayable. Terminal loss relief is available in respect of capital losses during the currency of the agreement in accordance with F.A. 1970, Sched. 6.

The Mineral Royalties (Tax) Regulations 1971, S.I. 1971 No. 1035 set out the rules for apportionment between mineral and non-mineral content of royalties.

Terminal loss relief

F.A. 1970, Sched. 6 provides a C.G.T. terminal loss relief if the tax- **11–003** payer entitled to receive mineral royalties disposes of, or under C.G.T. rules is deemed to have disposed of, his interest in the land, or if the mineral royalty agreement terminates.

On the termination of the royalty agreement the taxpayer may claim that the interest in land is deemed to have been disposed of and immediately reacquired at market value provided that at that date there is not a deemed or actual disposal under the normal C.G.T. rules and provided that a capital loss would result. Such a loss is known as the terminal loss (F.A. 1970, Sched. 6, para. 4).

The terminal loss may be claimed in the year of assessment or accounting period in which the royalty agreement expires (F.A. 1970, Sched. 6, para. 5). Alternatively, it may be claimed under F.A. 1970, Sched. 6, para. 7 below.

If there is a deemed or actual disposal for C.G.T. purposes the taxpayer may elect for the loss to be dealt with in accordance with F.A. 1970, Sched. 6, para. 7 below. otherwise normal C.G.T. rules will apply in determining in which period any loss accrues (F.A. 1970, Sched. 6, para. 6).

On a claim under F.A. 1970, Sched. 6, paras. 4 to 6 the terminal loss may be set against capital gains chargeable within the prior 15 years ending with the date of disposal or deemed disposal (F.A. 1970, Sched. 6, para. 7).

The terminal loss may however be claimed only against a capital element **11–004** of the mineral royalty chargeable to C.G.T. or corporation tax on the chargeable gain under I.C.T.A. 1988, s.122. The terminal loss must be relieved in a later year rather than an earlier year, but may be carried back for the full 15 years. If the terminal loss exceeds the chargeable gains from mineral royalties the balance is allowed as a capital loss in the year of disposal or deemed disposal.

A claim must be made within six years of the disposal or deemed disposal (F.A. 1970, Sched. 6, para. 8).

These provisions are to be construed as one with the C.G.T. provisions in C.G.T.A. 1979 (F.A. 1970, Sched. 6, para. 9).

11–005   The purchase by a mine-working or gravel-extracting business of mineral-bearing land with an agreement for resale after reinstatement to the vendor or the payment of a lump sum for a licence to enter onto the land and extract minerals would be a capital transaction and not the acquisition of trading stock for the purchaser (*Knight* v. *Calder Grove Estates* (1954) 35 T.C. 447, *H.J. Rorke* v. *I.R.C.* (1960) 39 T.C. 194). Conversely it would be a capital disposal of the vendor and would be dealt with under the C.G.T. rules as explained in Chapter 15. A lease of the site to the gravel extractor would give rise to the normal Schedule A assessment on the recipient as explained in Chapter 9 and premiums under I.C.T.A. 1970, s.80, and deemed premiums on the sale of land with an option to repurchase under I.C.T.A. 1970, s.82, would give rise to Schedule A charges on the recipients also, as explained in Chapter 9.

Expenditure on obtaining, or attempting to obtain, planning permission is capital (*ECC Quarries Ltd.* v. *Watkis* [1975] S.T.C. 578, SP4/78 November 6, 1978).

De-watering a mine shaft is capital expenditure (*United Collieries Ltd.* v. *I.R.C.* (1930) 12 T.C. 1248) but the cost of keeping it free of water is a proper revenue expense (*Naval Colliery Co. Ltd.* v. *I.R.C.* (1928) 12 T.C. 1017, *Thomas Merthyr Colliery Co. Ltd.* v. *Davis* (1933) 17 T.C. 519). The cost of removing top soil from a quarry was regarded as a revenue expense in *Milverton Quarries Ltd.* v. *Revenue Commissioners* (1960) I.R. 224.

Capital expenditure on the extraction of minerals depends on whether or not the expenditure was originally incurred on or after April 1, 1986, or before that date, subject to transitional provisions. It is therefore necessary to consider both the old and the new allowances. It should be noted that there is no equivalent of the mineral depletion allowance under C.A.A. 1968, s.60, after March 31, 1986.

Old code of allowances

11–006   "Qualifying expenditure" means capital expenditure on testing for minerals and extracting them and the construction of works likely to be of little value when the deposit has been fully exploited (C.A.A. 1968, s.51). The mines, oil wells, etc., allowance did not extend to refineries and buildings which could possibly qualify for an industrial buildings allowance.

Machinery and plant used for exploration

11–007   Mines, oil wells, etc., allowances did not usually apply to machinery and plant, except for that used for mineral exploration (C.A.A. 1968, s.52).

Exploration machinery not disposed of when the source is worked out is effectively valued, and only the fall in value while being used for a qualifying purpose is eligible for mines, oil wells, etc., relief (C.A.A. 1968, Sched. 5, para. 1).

There are provisions to ensure that if the allowable qualifying expenditure is limited to the diminution in value of the exploration equipment, any balancing allowance on its subsequent sale is limited to the excess of the sale proceeds over the expenditure, less the allowances given (C.A.A. 1968, Sched. 5, para. 2).

Overseas mineral rights

Capital expenditure on working overseas mineral deposits within **11–008** C.A.A. 1968, Sched. 5, was qualifying expenditure, apart from land covered by C.A.A. 1968, s.54, plant or machinery, buildings or structures. Capital expenditure includes minimum royalties. The balancing charge is again limited to allowances actually given (C.A.A. 1968, s.53).

The mines, oil wells, etc., allowances available in respect of expenditure on the acquisition of overseas mineral rights after April 6, 1949, acquired from a U.K. owner, were limited to the expenditure previously incurred by the vendor; unless the transaction is a sale of an overseas source as a going concern. There are provisions for splitting multiple acquisitions (C.A.A. 1968, Sched. 5, para. 3).

If the source was worked by the vendor his capital expenditure on extraction rights is reduced by applying the fraction $\dfrac{A}{A + B}$ where:

A = the total potential future output from the source
B = the vendor's output

It is only the vendor's residue of expenditure so calculated that counted as qualifying expenditure in the hands of the purchaser (C.A.A. 1968, Sched. 5, para. 4).

Similar provisions applied to acquisitions of overseas mineral bearing land (C.A.A. 1968, Sched.5, para. 5).

Acquisition of land outside the U.K.

Capital expenditure on mineral bearing land outside the U.K. within **11–009** C.A.A. 1968, Sched. 5, was qualifying expenditure, unless it is expenditure on machinery or plant, buildings or structures. Any balancing charge is limited to the allowances actually given (C.A.A. 1968, s.54).

Demolition costs

Costs of demolition were added to the written-down value of the asset. **11–010** The net cost of demolition is the total cost less any sale proceeds (C.A.A. 1968, s.55).

Initial allowances

Before April 1, 1986, an initial allowance of 40 per cent. was given in **11–011** respect of qualifying expenditure on mines and oil wells. F.A. 1971, s.52 increased the initial allowance to 100 per cent. in development areas and Northern Ireland and provided that an initial allowance could have been disclaimed in whole or in part (C.A.A. 1968, s.56).

Rates were as follows:

	Initial Alone %	With investment allowance Initial %	Investment %
6/4/44–5/4/52	10	—	—
6/4/52–14/4/53	—	—	—
15/4/53–6/4/54	40	—	—
7/4/54–17/2/56	40	—	20
18/2/56–7/4/59	40	—	—
8/4/59–5/11/62	40	20	20
6/11/62–16/1/66	20	20	30
17/1/66–26/10/70	40	—	—
27/10/70–31/3/86	40(a)	—	—

Note (a) 100 per cent. in development areas and Northern Ireland.

Writing-down allowances

11–012   A writing-down allowances was available for each year of the greater of five per cent. of the residue of qualifying expenditure or a proportion based on output in the period and potential future output. The residue of qualifying expenditure was that actual expenditure less any initial allowance and any previous writing-down allowances (C.A.A. 1968, s.57).

Mining, etc., allowances were to be deducted in the first instance from the trading profits. If they turned a profit into a loss this could have been utilised in the normal way (C.A.A. 1968, s.66).

Sale of source

11–013   If qualifying assets were sold there was a balancing charge in respect of any excess of the sale proceeds over the residue of expenditure, and a balancing allowance in respect of any deficiency. There are provisions which might require apportionment where the qualifying expenditure was made before April 6, 1952.

The balancing charge cannot exceed the written-down value. The purchaser can base his allowances on the lower of the price attributable to the qualifying assets acquired or the residue of expenditure on those assets immediately after the sale (C.A.A. 1968, s.58).

Expenditure incurred by persons not engaged in the trade of mining, etc.

11–014   A non-trader who incurred qualifying expenditure obtained no mine, oil well, etc., allowance, but if he sold the assets the purchaser was entitled to the allowance provided he was carrying on a suitable trade, based on the lower of the vendor's qualifying expenditure or the purchase price (C.A.A. 1968, s.59).

Mineral depletion allowances

11–015   A trader who works a mine, oil well, etc., for example, a gravel extractor was entitled to an annual allowance in respect of capital expenditure before April 1, 1986, on mineral bearing deposits of a wasting nature, such as gravel-bearing land. The allowance was 50 per cent. of the royalty value of the output in the basis period in which the expenditure was incurred and for the next nine years and thereafter the allowance was 25 per cent. of the royalty value for the next 10 years and thereafter was 10 per cent. of the royalty value of the output.

The actual writing-down allowance together with the notional allowances, which would have been granted for the years preceding 1963–64, when mineral depletion allowances were first introduced, cannot exceed the actual capital expenditure incurred. If the minerals ceased to be worked there was a balancing charge if the writing-down allowances actually granted exceed the net cost of acquiring the *post* April 3, 1963, output. There was a balancing allowance if the allowances fell short of this cost.

The net cost of acquiring the output from the source was the cost of the minerals, for example the land less its market value on cessation, and less any capital receipts which might have been received in the meantime.

If part of the expenditure was incurred before April 6, 1963, apportionment was necessary on the basis of the mineral royalty value.

Reinstatement expenditure could only be allowed if taken into account in computing the balancing charge or allowance. A temporary interruption in working the source was ignored. A transfer to another party under the same control could not increase the allowable capital expenditure. Further capital expenditure on the same source was related back to the initial expenditure on that source, but this did not affect the pre-April 4, 1963, position.

Qualifying expenditure for depreciation of mineral-bearing land, etc., **11–016** under this section did not include qualifying expenditure for mines, oil-wells, etc., allowances or expenditure on plant or buildings. If relief against trading profits had been granted in respect of a proportion of the premium payable to an original landlord on the granting of a lease, the expenditure qualifying for mineral depletion allowances was reduced accordingly.

The practical problem of fixing the royalty value was usually referred by the taxpayer's inspector to the mineral valuer attached to the local District Valuer's office.

Allowances under these provisions ceased to be available for expenditure incurred on or after April 1, 1986, or April 1, 1987, if the appropriate election was made under F.A. 1986, Sched. 14, para. 2, although the outstanding balance, *i.e.* the amount on which a writing-down allowance would have been made in the year beginning April 1, 1986 (or 1987), which is the expenditure incurred on the mineral assets qualifying for relief, less the notional and actual relief given under C.A.A. 1968, s.60, qualifies for relief under the new provisions as expenditure on the acquisition of a mineral asset (F.A. 1986, Sched. 13, para. 4(1)(*b*)).

Contributions

Expenditure on buildings for occupation by employees or the supply of **11–017** services thereto such as water, gas or electricity, or welfare services which were likely to be of little value when an overseas mineral source was depleted qualified for a writing-down allowance of 10 per cent. a year on a straight-line basis (C.A.A. 1968, s.61). There were no balancing allowances or charges in connection with this relief which was closely comparable with the old agricultural buildings allowances under C.A.A. 1968, s.68.

Abortive exploration

11–018   Expenditure by a person carrying on the trade of mineral extractor on abortive expenditure which would have been qualifying expenditure under C.A.A. 1968, s.51, had it led to the working of a productive source was allowed as a trade expense unless already relieved under some other section (C.A.A. 1968, s.62).

For abortive expenditure on planning permission see *ECC Quarries Ltd.* v. *Watkis* [1975] S.T.C. 578, and SP4/78, November 6, 1978.

Expenditure before commencement of trade

11–019   Eligible expenditure in connection with mining, etc., incurred before the commencement of trade was allowable as if incurred on the first day of trade. This did not apply to mineral depletion allowance.

New code of allowances

11–020   The old mineral and oil well allowances cease to have effect on March 31, 1986, subject to the transitional provisions introduced by F.A. 1986, Sched. 14 (F.A. 1986 s.55).

Expenditure from that date qualifies for relief as specified in F.A. 1986, Sched. 13.

For the purpose of capital allowances a company's accounting period which straddles March 31, 1986, is divided into two periods, the second of which commences on April 1, 1986. Similarly for income tax purposes a new basis period begins on April 1, 1986.

For the purposes of capital allowances on plant and for short-life assets where new expenditure, *i.e.* after March 31, 1986, is incurred before the commencement of a trade of mineral extraction, the plant is deemed to have been sold and immediately reacquired on the day of actual commencement at the original cost. Where a person is carrying on a trade of mineral extraction, any expenditure in connection with the provision of machinery or plant for mineral exploration and access is deemed to be wholly and exclusively for the purposes of the trade. Where the machinery or plant is abandoned its disposal value, if any, is brought into account. The previous provisions for capital allowances for plant and expenditure before the commencement of a trade in F.A. 1971, Sched. 8, para. 7(2) are repealed.

Application of Capital Allowances Act 1968, etc.

11–021   The provisions of the C.A.A. 1968, ss.70–89 on the making of allowances, meaning of basis period, etc., are applied to the new code, mines and oil wells allowances as they apply to the old code (F.A. 1986, Sched. 13, para. 2).

Time when expenditure is incurred

Pre-trading expenditure is deemed to be incurred on the first day of **11–022** carrying on the trade, although pre-trading expenditure on machinery or plant or on exploration is treated as incurred on the first day of carrying on a trade of mineral extraction (F.A. 1986, Sched. 13, para. 3).

Qualifying expenditure—general provisions

The capital expenditure qualifying for allowances is that on mineral **11–023** exploration and access, on the acquisition of a mineral asset, on the construction of any works used for the working of a source of mineral deposits which are likely to be of little or no value, to the person working it, immediately before cessation of working and, in the case of a foreign concession, expenditure on the construction of works likely to be valueless when the concession comes to an end. Assets claimed as plant and machinery for capital allowances purposes are excluded from the mines and oil wells allowances.

Qualifying expenditure does not include expenditure on the site itself or rights in or over any such site. Expenditure on works constructed wholly or mainly for subjecting the raw products to any process other than preparing it for use may well qualify for industrial buildings allowances. Expenditure on buildings for the welfare of workers, offices or buildings containing offices amounting to more than 10 per cent. of the total cost is excluded. Expenditure on abortive planning permission is deemed to be on mineral exploration and access (F.A. 1986, Sched. 13, para. 4).

Pre-trading expenditure on machinery or plant which is sold, etc.

If capital expenditure is incurred on the provision of machinery or plant **11–024** used for mineral exploration and access which is sold, demolished, destroyed or abandoned before the commencement of a trade, the loss on disposal is deemed to have been incurred on the first day of carrying on a trade of mineral extraction provided that the trade commenced within six years (F.A. 1986, Sched. 13, para. 5).

Pre-trading exploration expenditure

Pre-trading expenditure, other than on plant, on mineral exploration **11–025** and access at any source may be treated as incurred on the first day of trading if within six years of a trade of mineral extraction at that source (F.A. 1986, Sched. 13, para. 6).

Contributions

Capital expenditure on contributions towards the cost of buildings or **11–026** supplying gas, electricity or other services for mineral extraction outside the U.K. which are likely to be of little or no value once the source ceases to be worked is qualifying expenditure (F.A. 1986, Sched. 13, para. 7).

Restoration expenditure

11–027   Qualifying expenditure extends to expenditure on restoration of land after the trade of mineral extraction has ceased, which has been incurred within three years of cessation. Restoration includes landscaping and, in the case of land in the U.K., works required by the grant of planning permission. Such expenditure is treated as qualifying expenditure incurred on the last day of trading. The allowable net cost is the cost of restoration which is not allowed as a trading expense, less the receipts from sale of top soil, tipping rights, etc., which are exempt from tax as income to the extent to which they reduce the qualifying expenditure (F.A. 1986, Sched. 13, para. 8).

Writing-down and balancing allowances

11–028   A writing-down allowance is to be made on a pool basis on the excess of the expenditure over the disposal receipts in the accounting period. The allowance is 25 per cent., except in the case of pre-trading expenditure on plant and machinery sold before trading commenced, pre-trading exploration expenditure and the expenditure on the acquisition of a mineral asset, when the rate of allowance is 10 per cent. The allowance is calculated on the written-down value. There is no initial allowance (F.A. 1986, Sched. 13, para. 9).

Manner of making allowances and charges

11–029   Allowances are given in the first instance against the profits of the trade of mineral extraction (F.A. 1986, Sched. 13, para. 15).

Disposal receipts

11–030   When an asset on which qualifying expenditure has been incurred ceases permanently to be used for the purpose of the trade of mineral extraction by the person who incurred the expenditure, the disposal value has to be brought into account (F.A. 1986, Sched. 13, para. 10). The disposal value, as for plant and machinery, is the open market value, or if it is sold to a person who carries on a similar trade, the sale proceeds. In the case of land the disposal value is restricted under F.A. 1986, Sched. 13, para. 18.

Balancing charges

11–031   The excess of the disposal receipts and allowances already given over the initial qualifying expenditure is recovered by means of a balancing charge which is limited to the allowances actually given (F.A. 1986, Sched. 13, para. 11).

Balancing allowances

11–032   A balancing allowance is given in a period in which the trade of mineral extraction is permanently discontinued or where particular mineral deposits cease to be worked without the trade ceasing.

The balancing allowance is the excess of the qualifying expenditure over the disposal receipts and any allowances already given. A balancing allowance is also to be made where an asset permanently ceases to be used and the disposal value is brought into account if the expenditure is in excess of the disposal value and allowances given. Balancing allowances may also be given on pre-trading expenditure on machinery or plant or exploration and for abortive expenditure in searching, exploration or obtaining planning permission. There may also be a balancing allowance where possession of the asset is lost or it ceases to exist, or begins to be used for a purpose other than for a trade of mineral extraction (F.A. 1986, Sched. 13, para. 12).

Treatment of qualifying expenditure

Expenditure in connection with a trade is treated as being for the pur- **11–033** poses of a trade of mineral extraction (F.A. 1986, Sched. 13, para. 13).

Demolition costs

The cost of demolition of an asset is treated as qualifying expenditure **11–034** (F.A. 1986, Sched. 13, para. 14).

Restriction of qualifying expenditure

Expenditure representing the undeveloped market value of an interest in **11–035** land does not qualify for allowances, *i.e.* as if there was no source of mineral deposits on or in the land, and it was unlawful to carry out development other than that in respect of which planning permission or general consents already exist. Land outside the U.K. is deemed to be situated in England and Wales for this purpose. As the undeveloped market value includes the value of any buildings or structures on the land if these permanently cease to be used when the land is acquired for mineral extraction, and there is unrelieved value in the buildings, this qualifies for writing-down allowances at 10 per cent. per annum on the reducing balance. The unrelieved value is the value of the building excluding the land when the interest is acquired, less any allowances as reduced by balancing charges given in respect of the buildings (F.A. 1986, Sched. 13, para. 16).

Where the acquisition of the interest in land represents a premium on a lease for which relief would be allowable under I.C.T.A. 1970, s.134, the cost is treated as reduced by the amount qualifying for such relief (F.A. 1986, Sched. 13, para. 17).

Restriction of disposal receipts

Where a disposal receipt is brought into account on the disposal of a **11–036** mineral asset only the excess over the undeveloped market value is taken into account in calculating the balancing allowance or charge (F.A. 1986, Sched. 13, para. 18).

Assets formerly owned by traders

Where expenditure has been incurred on an asset after March 31, 1986, **11–037** and that asset is acquired by a purchaser, the allowances are limited to the previous trader's qualifying expenditure unless the purchaser's capital expenditure is incurred under a contract entered into before July 16, 1985 (F.A. 1986, Sched. 13, para. 19).

Limitation of expenditure

11–038   Where the previous trader's qualifying expenditure qualified for allow-ances the purchaser's relief is limited to the previous trader's residue of expenditure in the same way as for industrial buildings allowances (F.A. 1986, Sched. 13, para. 20).

Mineral exploration

11–039   Where the purchased asset is a mineral asset and part of its value is attributable to the previous trader's expenditure on mineral exploration and access, a just and reasonable proportion of the buyer's expenditure is similarly so attributed and therefore qualifies for relief at 25 per cent. a year on the reducing balance and the balance is attributed to the acquisition of a mineral asset written down at 10 per cent. a year on the reducing balance (F.A. 1986, Sched. 13, para. 21).

Transfer of mineral assets within a group

11–040   Where assets are transferred between companies under common control the transferee company's expenditure for the purposes of allowances is limited to the transferor company's qualifying expenditure, unless an election is made to take over the written-down value under C.A.A. 1968, Sched. 7, para. 4 or the asset is an oil licence (F.A. 1986, Sched. 13, para. 23).

Assets formerly owned by non-traders

11–041   Where assets representing expenditure on mineral exploration and access by a non-trader are sold to a trader his allowances are restricted to the vendor's qualifying expenditure (F.A. 1986, Sched. 13, para. 24).

Mineral extraction: old expenditure

11–042   Various terms are defined by F.A. 1986, Sched. 14, para. 1, in particular outstanding balance, *i.e.* the unrelieved expenditure under the old code provisions.

Where the old expenditure fell within C.A.A. 1968, s.57, the outstanding balance is the residue of expenditure under the old code for the accounting period or basis period beginning on April 1, 1986. Where the old expenditure related to mineral depletion allowances under C.A.A. 1968, s.60, the outstanding balance is the amount on which a writing-down allowance would have been made for the accounting period or basis period beginning on April 1, 1986. Where the old expenditure was a contribution to public service outside the U.K. within C.A.A. 1968, s.61, the outstanding balance is the excess of the expenditure over writing-down allowances received for the accounting period or basis period beginning on April 1, 1986.

An amount equal to the outstanding balance is treated for the new code scheme as expenditure incurred on April 1, 1986 (or 1987 if an election is made) (F.A. 1986, Sched. 14, para. 3).

Where there is no outstanding balance on April 1, 1986 (or 1987), the expenditure is deemed to have been incurred on April 1, 1986 (or 1987), and therefore any disposal receipts would be dealt with under the new code provisions (F.A. 1986, Sched. 14, para. 4).

Election

Expenditure incurred before April 1, 1987, under a contract entered into **11–043** before July 16, 1985, may, on election, be treated as qualifying for the old code allowances in which case the outstanding balance is the figure at April 1, 1987 (F.A. 1986, Sched. 14, para. 2).

Unrelieved expenditure

Any unrelieved expenditure on mineral exploration and access at April **11–044** 1, 1986 (or 1987), is deemed to be qualifying expenditure under the new code provisions provided that the person who incurred the expenditure began to carry on a trade of mineral extraction before April 1, 1986 (or 1987), or after such dates before mineral exploration and access ceases at the source in question (F.A. 1986, Sched. 14, para. 5).

If no allowances have been given under the old code the expenditure is treated as incurred on April 1, 1986 (or 1987), and given under the new code (F.A. 1986, Sched. 14, para. 6).

Old expenditure on construction

Old code expenditure for which an allowance has been given in respect **11–045** of works in connection with mineral deposits being of little value when the source ceases to be worked are deemed to be incurred on April 1, 1986 (or 1987), and qualify for new code allowances (F.A. 1986, Sched. 14, para. 7).

Balancing charges

Where a balancing charge is made under the new code in respect of old **11–046** code expenditure any allowances under the old code have to be included in calculating the balancing allowance or charge (F.A. 1986, Sched. 14, para. 8).

The result of mineral extraction is very often a large hole in the ground which under the terms of the original planning permission for the extraction of minerals required to be filled and the land passed back to agricultural use. A common method of filling the hole is to make it available for the tipping of waste products.

Waste deposits

11–047   After gravel or other minerals have been extracted there often remains a large hole in the ground which is much in demand for the deposit of refuse and other waste products, until the hole is filled and the land reinstated for agricultural relief.

Under I.C.T.A. 1988, s.15, annual profits or gains in respect of receipts from the ownership of an interest in land are taxable under Schedule A. A lump sum received on the granting of a licence to deposit waste is a receipt from the ownership of land, but it is not an annual profit or gain because it is a receipt of a capital nature (*Nethersole* v. *Withers* 28 T.C. 501). Mining royalties and rents charged by I.C.T.A. 1988, s.119, are excluded from Schedule A, but section 119 would not apply to the lump sum because it is not within the definition of "rent" in I.C.T.A. 1988, s.119(3).

Under I.C.T.A. 1988, s.34, a percentage of the lump sum would be taxable as income if it were a premium for a lease with less than 50 years to run. A licence is not a lease for this purpose. Care has therefore to be taken to draft documentation so that it will operate as a licence. It is not possible to ensure that a transaction is not a lease merely by the label attached to it (*Addiscombe Gardens Estates* v. *Crabbe* [1958] 1 Q.B. 513). An important case in this whole area of whether or not the documentation could create anything other than the licence is *Shellmex* v. *Manchester Garages* [1971] 1 W.L.R. 612.

On the assumption that there is no right of reconveyance I.C.T.A. 1988, s.36 should not apply.

Provided that the lump sum is of a capital nature it ought not to be treated as profits or gains of a deemed trade under I.C.T.A. s.53(3) which relates to the occupation of land on a commercial basis.

11–048   I.C.T.A. 1988, s.776 provides that certain gains of a capital nature relating to land are to be treated as income assessable under Schedule D, Case VI. In particular I.C.T.A. 1988, s.776(2) provides that the section applies wherever:

(a) land, or any property deriving its value from land, is acquired with the sole or main object of realising a gain from disposing of the land, or

(b) land is held as trading stock, or

(c) land is developed with the sole or main object of realising a gain from disposing of the land when developed.

It is considered that regard being had to I.C.T.A. 1988, s.776(13) the "tipping rights" will not be an interest in land for the purposes of the section.

Should the licence constitute an interest in land then it would be necessary to establish that the transaction is not covered by I.C.T.A. 1988, s.776(2)(a), (b) or (c). The position of the vendor would be strengthened if he had held the land for a significant period of time, had purchased the land merely to exploit the mineral content or had owned it previously and if the possibility of granting the licence to tip was not originally envisaged when purchasing the land.

The sale and lease-back rules concerning the taxation of consideration received, in I.C.T.A. 1988, s.780, merit consideration. However in the circumstances envisaged that section should not apply because there is no surrender or assignment of a lease.

Trading income

Naturally, if the lump sum could be shown to be a revenue receipt from **11–049** the mineral or waste disposal trade of the company it would become income assessable under Schedule D, Case I. However, it is difficult to see how the sum could be regarded as arising from the mineral trade because it relates to the exploitation of a fixed asset subsequent to the completion of mineral extraction. Also as the vendor is prevented from using the asset for trading purposes there is a case for regarding it as a transaction of a non-trading nature in that there has been a sterilisation of assets. An important case is *Glenboig Union Fire Clay Co. Ltd.* v. *I.R.C.* 12 T.C. 427.

In that case the taxpayer owned leasehold rights in fireclay seams with the right to extract minerals. The seam went under a railway track. The railway company eventually paid a large sum for not exploiting the fireclay. The House of Lords held that the sum was capital in the following oft-quoted words:

> "In truth the sum of money is the sum paid to prevent the Fireclay Company obtaining the full benefit of the capital value of that part of the mines which they are prevented from working by the railway company. It appears to me to make no difference whether it be regarded as the sale of the asset out and out, or whether it be treated merely as a means of preventing the acquisition of profit which would otherwise be gained. In either case the capital asset of the company to that extent has been sterilised and destroyed, and it is in respect of that action that the sum . . . was paid. . . . It is now well settled that the compensation payable in such circumstances is the full value of the minerals that are left unworked, less the cost of working, and that is of course the profit that would have been obtained were they in fact worked. But there is no relation between the measure that is used for the purpose of calculating a particular result and the quality of the figure that is arrived at by means of the application of that test."

If it could be established that the consideration arose from the trade then **11–050** it will be necessary to argue that it is of a capital nature having regard to such matters as the period of the licence agreement. Also it will greatly assist arguments if similar transactions have not occurred in the past nor are likely to occur in the immediate future.

There is a fine dividing line between a receipt falling to be assessed under Schedule D, Case I as being of a revenue nature or alternatively falling to be regarded as a capital sum. Indeed, the judiciary has even indicated that the matter could possibly be determined by the toss of a coin.

In the event that the vendor carries on a waste disposal business it will be of assistance if that trade were transferred to another company in a 75 per cent. group before the receipt of the lump sum relating to the granting of the licence. Clearly, the greater the interval between the date of transfer and the receipt the better.

The transfer should be free of adverse taxation consequences for the fol- **11–051** lowing reasons:

(a) capital gains will not arise (I.C.T.A. 1970, s.273).

(b) the indexation for capital gains is preserved by F.A. 1982, Sched. 13, paras. 2 and 3.

(c) balancing charges in respect of (i) plant and (ii) industrial buildings ought not to arise (I.C.T.A. 1988, s.343; C.A.A. 1968, Sched. 7, para. 4).

(d) trading losses of the transferor can be carried across to the transferee (I.C.T.A. 1988, s.343).

Also, if the opportunity were to present itself to the vendor that he could acquire further land where the value of the actual or potential tipping right was a major commercial attraction and a further capital sum was likely to be sought for the tipping rights then that land should be acquired by a separate company thereby frustrating suggestions of contamination.

Capital gains—roll-over relief

11–052   On the assumption that the lump sum falls to be assessed as a capital gain it should be possible to obtain roll-over relief under C.G.T.A. 1979, s.115, on the acquisition of appropriate new assets purchased in the period one year before and three years after the date of granting the licence. It is important that such new assets are taken into use at the time of purchase to comply with the requirements of the section. Where the land has been used and occupied for the purposes of the mineral trade there should be little difficulty in obtaining the relief. One matter that not infrequently arises is that mineral businesses often arrange for third parties to extract minerals from the land under some royalty arrangement and if this is the case then a proportion of the gain could not be rolled over because for the duration of that arrangement the Revenue would probably regard the land as not having been used for trading purposes. C.G.T.A. 1979, s.115(6) provides for apportionment "having regard to the time and extent to which it was, and was not, used" for the purposes of the trade. Whilst there may be a certain measure of uncertainty as to what is meant by "extent" the Revenue normally works on the basis of a straight line apportionment between the qualifying and non-qualifying periods in computing the roll-over relief.

Value added tax

11–053   A licence agreement for waste disposal should provide for the payment of V.A.T. on the lump sum. The reasons for this are:

(i) V.A.T. is properly chargeable unless the grant of the tipping rights is exempt under V.A.T.A. 1983, Sched. 6, Gp. 1. The practice is stated on page 6 of V.A.T. News No. 12 as follows:

"It is now clear that making a charge for the right to tip refuse or industrial waste is not the granting of a right over land, and from December 1, 1976, such charges are liable to V.A.T. at the standard rate.
However, where a landowner grants the right to occupy and use land for the purposes of operating a tip, the charge for such a grant remains exempt, since the occupier has a more extensive right than simply a right to tip."

From August 1, 1989, however, the landlord must charge V.A.T. if he has made an election under F.A. 1989, Sched. 3, para. 2.

A claim for exemption on the grounds that the third party has a more extensive right than simply a right to tip could well prejudice the tax position, *e.g.* whether there is a lease or licence agreement.

(ii) Such V.A.T. exemption would not be beneficial to the recipient of the lump sum because it would restrict the recovery of input tax, unless a supply fell to be ignored under the V.A.T. (General) Regulations 1985, S.I. 1985 No. 886, reg. 30(3)(b)(i).

(iii) The third party paying the V.A.T. should not be disadvantaged.

Stamp duty

A non-exclusive licence should not subject to *ad valorem* duty.   **11–054**

Balancing charges/allowances

A disposal before April 1, 1986, may given rise to a balancing charge   **11–055** with regard to mineral depletion allowances. However, such a charge might well have arisen on an earlier occasion if the source of deposits had ceased to be worked (C.A.A. 1968, s.60(4), (5)).

Woodlands

Introduction

12–001   Governments have provided tax concessions to encourage investment in woodlands and the Forestry Commission gives grants towards the cost of planting new woodlands, restocking existing woodlands and rehabilitating existing unproductive woodlands by selective planting and natural regeneration. Radical changes were announced by the Chancellor of the Exchequer in his Budget Statement on March 15, 1988; from that date many of the tax advantages of investing in forestry were withdrawn, but on the other hand the grants available were substantially increased.

Grants

12–002   Details of the Woodland Grant Scheme which came into effect on March 15, 1988, are given in the table below. The Conifer Planting and Broadleaved Planting Grants are payable in three instalments, 70 per cent. on completion of planting, 20 per cent. after five years and 10 per cent. in the tenth year.

Woodland Grant Scheme—Planting Grants

Area of Wood	Conifers per hectare	Broadleaves per hectare
0.25 ha–0.9 ha ($\frac{1}{2}$–$2\frac{1}{2}$ acres)	£1,005	£1,575
1.0 ha–2.9 ha ($2\frac{1}{2}$–7 acres)	£880	£1,375
3.0 ha–9.9 ha (7–25 acres)	£795	£1,175
10.0 ha and over (25 acres and over)	£615	£975

Natural Regeneration Grants apply at the same rates, but only 50 per cent. is payable after completion of approved work, with 30 per cent. when adequate stocking has been achieved and 20 per cent. five years later.

Neglected woods in need of refurbishment which are under 20 years old, not previously grant aided and adequately stocked with suitable species, are eligible for half planting grant levels, payable as to 60 per cent. on completion of approved operations with the remaining 40 per cent. five years later.

New planting on arable or improved grassland is eligible for a supplement on top of the indicated grant levels of £200 per hectare (£80 per acre) which is payable with the first instalment of the planting grant.

Grants may also be available for tree replacement under the Storm Damage Recovery Scheme run by the Ministry of Agriculture (Hansard November 30, 1987, Vol. 123, col. 141).

Economic arguments

12–003 The fact that there are strong economic arguments for increasing this country's wood producing capability is important in establishing confidence in an investment which may take 50 years or longer to mature.

The economic arguments include the following:

(a) experts forecast a world shortage of timber by the end of the century;

(b) total imports of wood and wood products into the U.K. are about £4½ billion a year and represent 90 per cent. of the requirements of the country on an annual basis;

(c) large areas of the U.K., especially Scotland, are extremely suitable for growing trees, and the loss in agricultural production would be very small indeed;

(d) timber-exporting countries are supplying less of the basic raw materials and more of the processed and semi-processed products; their own requirements are increasing and the cost of transport is rising rapidly;

(e) substitute materials for wood such as steel, aluminium, cement and bricks are energy-intensive and are therefore liable to become more expensive in the long run;

(f) there are a number of conservation reasons in favour of forestry. Although generally these are not of a commercial nature, such considerations as rental income from shooting rights may be of interest.

12–004 Forestry investment does not rely for its existence either on loopholes in the tax law or on clever manipulation of funds. The tax advantages have always been well known and support for woodlands has in the past been forthcoming from governments of all persuasions.

Returns from forestry in the U.K. have traditionally been very low, and are comparable with those associated with tenanted farm land. This is partly because the increase in capital value reduces the current income yield as a percentage. The intending forestry investor should consider the projected yield to redemption taking into account any tax implications.

Assessments of future returns from forestry investments are based upon the present value of the cash flow at current prices from timber sales throughout the crop rotation. Yields of between 4 and 5 per cent. a year are common, ignoring any increase in timber values.

Taxation treatment

Income tax before March 15, 1988

12–005   The normal basis of tax assessment of commercially managed woodlands in the U.K. was under Schedule B until its abolition from 1988–89: I.C.T.A. 1988, s.16. The occupier, *i.e.* the person having the use of the land, I.C.T.A. 1988, s.16(4), was deemed to receive income of one-third of the annual value of the land in its unimproved state, I.C.T.A. 1988, s.16(3)(*a*), *I.R.C.* v. *Lord Trent's Executors* (1959) 38 T.C. 326), *Lady Miller* v. *I.R.C.* (1930) 15 T.C. 25, *Joly* v. *Pinhoe Nurseries Ltd.* (1936) 20 T.C. 271, *Tennant* v. *Smith* (1892) 3 T.C. 158, *Dawson* v. *Counsell* (1938) 22 T.C. 149, 158); see also the husbandry cases referred to in Chapter 3.

The annual value was the gross rateable value, I.C.T.A. 1988, ss.16(3)(*b*), 837, ignoring the trees growing on the land and pending any revaluation. The assessment was constant, being commonly less than £1 per acre making the Schedule B assessment about 30p per acre. The cost of assessment and collection might well have exceeded the tax payable and for this reason the assessment was often not raised by the inspector. Under Schedule B there was no tax liability on income from sales of thinnings, mature timber and forestry grants. However, no expenses could be deducted from the assessment, nor were capital allowances available. The occupation of commercial woodlands was not itself a trade, and the growing timber was not stock, *Coates* v. *Halker Estates Company* (1961) 40 T.C. 75.

Sales of timber included sales from a saw-mill connected with the woodlands at least up until the rough planking stage, *Christie* v. *Davies* (1945) 26 T.C. 398, *I.R.C.* v. *Williamson Bros.* (1950) 31 T.C. 370, but not where the timber was, say, made up into boxes, *Collins* v. *Fraser* (1969) 46 T.C. 143. C.G.T. was not applicable to the timber element in respect of a sale of commercial woodlands in view of C.G.T.A. 1979, s.34(2) as the profits would have been covered by the Schedule B assessment under I.C.T.A. 1988, s.16 or the Schedule D assessment under I.C.T.A. 1988, s.54.

12–006   From March 13, 1984, timber merchants who occupied woodlands as part of their normal trading operation would be taxed under Schedule D, Case I and not Schedule B, negating the case of *Russell* v. *Hird & Mercer* (1983) S.T.C 541 and *McLellan Rawson & Co. Ltd.* v. *Newall* (1955) 36 T.C. 117, I.C.T.A. 1988, s.16(5). It had been held in some cases that the purchase, and sale, of standing timber by a timber merchant was part of his trade, *Murray* v. *I.R.C.* (1951) 32 T.C. 238, *Hopwood* v. *C.N. Spencer Ltd.* (1964) 42 T.C. 169.

A dealer in land who bought and sold woodlands left out of his trading account the value of the trees and underwood to prevent his selling the trees under Schedule B and claiming a trading loss, I.C.T.A. 1988, s.99(1).

Schedule B taxation treatment was normally desirable only where woodlands investment was generating substantial income in relation to modest expenditure; this typically applied to plantations over 10 years of age, at the least.

Schedule D election

Where it was expected that expenditure was to be substantial and **12–007** income minimal or non-existent, the occupier of commercially managed woodlands could elect to be assessed under Schedule D: I.C.T.A. 1988, s.54. The election had to be made in writing to the inspector of taxes not later than two years after the end of the chargeable period in respect of which it was to apply: I.C.T.A. 1988, s.54(2). Where the election was made it did not have to cover the whole area of the woodland estate as an election could be made for all or part of the young plantation under 10 years old to be treated as a separate woodland estate, leaving the remainder of the woodland to be assessed under the original Schedule B basis: I.C.T.A. 1988, s.54(3).

Once the election was made, provided that the woodlands were occupied on a commcial basis, the profits and losses arising from the occupation of the woodlands with a view to realisation of ultimate profits could be treated in the same way as those deriving from a trade. The election was irrevocable except on a change of occupancy: I.C.T.A. 1988, s.54(4).

It follows, therefore, that for higher rate taxpayers the plan of action was to start with bare land, arrange for a planned planting programme, and elect for assessment on a Schedule D basis. The losses incurred could be set against the investor's other sources of income under I.C.T.A. 1988, s.380(4), but not under I.C.T.A. 1988, s.381 (early years of trade) (see Chapter 4). Exceptionally large amounts of income could be balanced by additional expenditure on the woodlands, as could income subject to top slicing, such as single premium bonds.

Once the planting programme had been completed (usually after five years) consideration was given to passing the woodlands to a family trust, or younger members of the family, thus automatically returning one particular woodlands to the Schedule B basis of assessment. There had to be a change of occupier as well as a change of owner and the Revenue usually argued that a transfer to a spouse would not change the *de facto* occupier, and the onus was on the taxpayer to prove the point, if necessary.

The land itself could be leased in which case the rent payable was deductible in the Schedule D computation.

Income tax after March 14, 1988

F.A. 1988 abolished Schedule B with effect from 1988–89 and the right **12–008** to elect for commercially managed woodlands to be assessed under Schedule D was withdrawn from March 15, 1988, subject to certain transitional provisions. Consequently, unless the transitional provisions apply, any losses arising after March 14, 1988, are not available for relief by being set off against the taxpayer's other income, nor are profits taxable.

Transitional provisions apply if before March 15, 1988, the taxpayer either occupied the woodlands, or had contracted or arranged either to occupy them or to plant new trees, or had applied for a Forestry Commission grant (or Northern Ireland equivalent). In these circumstances an election for the woodlands to be assessed under Schedule D may be made or, if already made, will continue to be effective until 1992–93.

Capital allowances

12–009 Where the owner or tenant of any agricultural land, or forestry land assessed under Schedule D (C.A.A. 1968, s.69—definition of forestry land), incurred any capital expenditure on forestry land, *e.g.* on fences, roads, etc., the expenditure attracted agricultural buildings allowances as described in Chapter 5 (F.A. 1986, Sched. 15).

Where allowances were given under C.A.A. 1968 or F.A. 1986, Sched. 15 the balance of allowances was available to a new occupier whether or not he elected for Schedule D treatment: C.A.A. 1968, s.68(4); F.A. 1986, Sched. 15, para. 4.

Capital gains tax

12–010 C.G.T. is payable on disposals of forestry property, but only on the increase in the value of the land, not on the growing timber (C.G.T.A. 1979, s.113). It is necessary therefore when considering the disposal of woodlands to divide the proceeds between the underlying land and the timber crop to determine the potential C.G.T liability. The availability of indexation relief will go some way to mitigating the liability which might otherwise arise on a disposal of the land.

The disposal of Schedule D assessable commercial woodland could qualify for C.G.T. "roll-over relief" (C.G.T.A. 1979, ss.115–121) in respect of the taxable proceeds of the forestry land (C.G.T.A. 1979, s.121(1)(b)). The replacement assets to which the relief applies need not be used in the same trade as those formerly owned. It is apparent that this relief may also be claimed following the chargeable disposal of a business, the proceeds of which are reinvested in forestry land. In addition to roll-over relief, gains on the gifting of a commercial woodland can be postponed until the sale of the property by the donee (F.A. 1980, s.79) as can a transfer of a forestry business to a company under C.G.T.A. 1979, s.123, (see Chapter 16).

Inheritance tax

12–011 The transfer of woodland property during lifetime or on death is treated in a similar manner as transfer of any business property (I.H.T.A. 1984, s.105(1)(a)). A lifetime transfer may escape tax as a potentially exempt transfer (I.H.T.A. 1987, s.3A) see Chapter 18. Once a two–year ownership period has been established, forest land and timber qualifies for business property relief. Where the relief is available, the value of the woodland property is reduced by 50 per cent. for all I.H.T. purposes (I.H.T.A. 1984, s.104(1)(a)). It qualifies as property on which the tax is payable by instalments over 10 years under I.H.T.A. 1984, s.229.

Relief for woodlands is available if claimed by notice in writing within two years of death, or such longer time as the Board may allow, and if certain conditions are met (I.H.T.A. 1984, s.125). Relief is only available if the deceased held the land beneficially for the five years preceding his death or acquired it otherwise than for money or money's worth. The land must be situated in the U.K. which excludes the Channel Islands and the Isle of Man, unlike agricultural property. The relief only applies on death.

It does not apply where the woodlands are included as agricultural property under I.H.T.A. 1984, ss.115–124B. The value of the trees or underwood is left out of account at death, but I.H.T. will probably become chargeable at a later date under I.H.T.A. 1984, s.126. In practice it is often preferable not to claim this relief and to pay I.H.T. on the value at death.

Charge to tax on disposal of trees or underwood

I.H.T. is payable on a disposal in relation to the last death on which the **12–012** trees or underwood passed (I.H.T.A. 1984, s.126). A person entitled to the sale proceeds (or who would be so entitled if the disposal was a sale) is liable to the tax. An inter-spouse disposal is ignored. I.H.T. is charged only on the first disposal following the death, at the death rates in force at the time of the disposal, and it is not charged again in relation to the same death on a subsequent disposal of the same trees or underwood.

Amount subject to charge

I.H.T. is calculated on the net sale proceeds on a sale for full consider- **12–013** ation in money or money's worth and on the net value at the date of disposal in other cases (I.H.T.A. 1984, s.127). The amount of tax is calculated as if the chargeable amount were added to the estate of the latest person on whose death the property passed, and the amount payable is the additional tax so calculated.

It should be noted that the amount left out of account on a death is the value of the trees and underwood at that time, but the amount subsequently charged on a disposal is calculated by adding in the value at the date of the disposal at which time the value may be considerably in excess of the value at death.

Provided that at the date of death the trees or underwood would have qualified as relevant business property for business relief under I.H.T.A. 1984, ss.103–114, a 50 per cent. deduction is allowed from the disposal proceeds when computing the tax liability. This applies to disposals after October 26, 1977, even if the death occurred before that time.

EXAMPLE

Disposal of trees or underwood: basis and rate of tax **12–014**
Roland died on September 30, 1989, and left the following estate, there having been no lifetime transfers within the seven years before death and ignoring business relief.

	£
Sundry assets	150,000
Value of trees and underwood	60,000
	£210,000
I.H.T. payable at death: on £150,000	£12,800

181

On January 1, 1998 the trees and underwood were sold
after allowing for selling and replanting expenses for: 120,000
Other assets at death 150,000

£270,000

I.H.T. payable
First £118,000 Nil
Next 152,000 @ 40% 60,800

£270,000 60,800

Less: Paid on death 12,800

Payable on sale of timber £48,000

If no claim were made, inheritance tax on death would have been:

First £118,000 Nil
Next 92,000 @ 40% 36,800

£210,000 £36,800

However the additional £24,000 (£36,800–£12,800) which would have
been payable on the death would have been expended nine years
earlier and at a time when the cash may not have been available even
taking into account the instalments payments option (I.H.T.A. 1984,
s.229).

Rate of charge

12–015   The charge under I.H.T.A. 1984, s.126 above in respect of a sum determined under I.H.T.A. 1984, s.127 above is computed at the rate in force at
the date of death, the amount being so charged being the highest part of
the estate (I.H.T.A. 1984, s.128).

Credit for tax charged

12–016   If the later disposal is itself a chargeable transfer the value of the trees or
underwood for the later transfer is regarded as diminished by the tax
chargeable under I.H.T.A. 1984, s.126 (I.H.T.A. 1984, s.129). See also
I.H.T.A. 1984, s.114(2) for avoidance of double relief.
    Net values are after selling expenses and expenses of replanting within
three years of such longer time as the Board may allow (I.H.T.A. 1984,
s.130).

Potentially exempt transfers

Where estate duty was postponed on death it was cancelled by a first sub- **12–017**
sequent transfer under F.A. 1975, s.49(4). If that transfer is a gift it cannot
be potentially exempt, F.A. 1986, Sched. 19, para. 46. This could extend
to the whole of the property transferred if the woodland consisted only of a
part.

Value added tax

Customs and Excise normally permits voluntary registration by a person **12–018**
undertaking commercially the planting and maintenance of trees. There
may well be no possibility of income for 20 years but Customs and Excise
requires a continuing intention to make taxable supplies in due course
before input V.A.T. is reclaimable (V.A.T.A. 1983, Sched. 1,
para. 11(1)(*b*).

In all cases professional forestry advice should be sought before embark-
ing on any forestry enterprise.

Heritage Property

13–001   The Government has consistently endeavoured to mitigate the effects of death duties and I.H.T. on national treasures retained in kind, including stately homes and supporting land.

Inheritance tax provisions

Conditionally exempt transfers

13–002   Lifetime transfers of historic houses and land of outstanding scenic, historic or scientific interest, works of art, etc., may be conditionally exempt from I.H.T. provided that the transferor or his spouse had between them been beneficially entitled to the property throughout the six years ended with the transfer, or the asset was acquired on death when the asset was conditionally exempt or left out of account under the relevant estate duty rules (I.H.T.A. 1984, s.30). Whether or not a transfer will qualify as a potentially exempt transfer made after March 17, 1986, will be decided without reference to these provisions, probably on the grounds that under I.H.T.A. 1984, s.3A(1)(b) a transfer already exempt cannot qualify as being potentially exempt. It will be noted that a claim for conditional exemption will not be allowed until after the transferor's death and that conditional exemption will not be available to the extent that property comprised in a potentially exempt transfer has been sold by the transferee before the transferor's death. I.H.T.A. 1984, s.30 applies to transfers after April 6, 1976. I.H.T.A. 1984, ss.30–35 inclusive provide the detailed rules applicable to such assets, and are explained in some length in I.R. 67 "Capital Taxation and the National Heritage."

The relief also applies on death without the necessity of having owned the asset for six years or having inherited it.

Designation and undertakings

13–003   The objects which the Treasury has power to designate as qualifying works of art, etc., are defined and include pictures, books, works of art, scientific collections, land or buildings, amenity land and historic objects (I.H.T.A. 1984, s.31). Land essential for the protection of the character and amenities of a building of outstanding historic or architectural interest no longer has to adjoin the building for exemption to apply. The Treasury may require separate undertakings where different people are entitled to different properties. The owner has to give an undertaking that the property will be kept permanently in the U.K. and that reasonable steps will be taken for its preservation and for securing reasonable access to the public. Where a claim for designation is made in respect of a transfer of value occurring after March 17, 1986, which qualified as a potentially exempt transfer under I.H.T.A. 1984, s.3A but which becomes chargeable (because of the donor's death within seven years of the gift), a decision as to whether the property is suitable for designation will be based on the circumstances existing following the transferor's death. Ancient semi-natural woodlands may be eligible for exemption.

Chargeable events

A conditionally exempt transfer becomes liable to I.H.T. on the happening of a chargeable event at the rates in force at that time unless the transfer qualified as a potentially exempt transfer made after March 17, 1986 (see I.H.T.A. 1984, s.3A) when tax will be chargeable only if the transferor dies within seven years of making the gift (I.H.T.A. 1984, s.32). This can happen when an undertaking has not been observed in a material respect on death, sale, gift or other disposal, unless the disposal is a sale to an approved institution or the asset is transferred to the Board in satisfaction of I.H.T. However, a death or gift is not a chargeable event if the transfer is itself a conditionally exempt transfer with the legatee or donee giving the appropriate undertakings to the Treasury. See H.C. Written Answer, June 25, 1980, Vol. 987, col. 202. **13–004**

Associated properties

As various people may be required to give undertakings, conditional exemption is provided for associated properties, that is an historic building, its amenity land and objects historically associated with it (I.H.T.A. 1984, s.32A). The death of the owner or the disposal of any of the associated properties is a chargeable event for I.H.T., unless the personal representatives or trustees dispose of the property by private treaty to a museum or similar body or the property is transferred *in specie* in settlement of tax or the disposal qualifies as a potentially exempt transfer made after March 17, 1986 (see I.H.T.A. 1984, s.3A). If there is a part disposal, the exemption does not extend to the remainder unless the appropriate undertaking is given over the remaining property. If the disposal occurs on death and the appropriate undertaking is given, the transfer may again be conditionally exempt. If there is a partial disposal which is not itself exempt, it is regarded as a disposal of the whole property unless the appropriate undertaking is given for the remainder and it still qualifies under the heritage property rules. **13–005**

Amount of charge

I.H.T. is payable on a chargeable event on a transfer of value equal to the value of the property at the time of the chargeable event and at one half the death rate from March 18, 1986, under I.H.T.A. 1984, s.7, if what is **13–006** termed the relevant person is alive, and if he is dead, at the appropriate table rate as if it were the highest part of his estate (I.H.T.A. 1984, s.33). As a transitional measure following the introduction of I.H.T., from March 18, 1986, when calculating tax on a chargeable event on or after that date by reference to a death occurring before that date, it is to be assumed that the amendments to I.H.T.A. 1984, s.7 (rates) made by F.A. 1986 were in force at the time of the earlier death—thus securing the benefit of the shorter seven year cumulation period and higher tax threshold.

Under I.H.T.A. 1984, s.33(2A) the death of the relevant person after a lifetime chargeable event does not result in a recomputation of the tax rate on that earlier event; this should exclude the possibility of potentially exempt transfers made after March 17, 1986, but before the chargeable event, entering into the tax calculation at that time, should they become chargeable on the donor's death within seven years: see I.H.T.A. 1984, s.3A.

A sale at an arm's length value not intended to confer any gratuitous benefit is deemed to take place at the sale price. See *Tyser* v. *Att.-Gen* [1938] Ch. 426 where, in a similarly-worded estate duty provision, "proceeds of sale" was interpreted as meaning the net proceeds of sale after deducting the expenses of sale. Under I.H.T.A. 1984, s.78, in certain circumstances the rates of tax under I.H.T.A. 1984, s.33(1)(b) are modified for discretionary trusts.

For the purposes of computing the I.H.T. payable on a chargeable event (other than one within I.H.T.A. 1984, s.79) it is necessary to identify the "relevant person" who will be deemed to have made the transfer. However, this section is concerned only with chargeable events following transactions which have either been conditionally exempt transfers, or have occurred on "conditionally exempt occasions," *i.e.* generally transfers by trustees which are not chargeable under I.H.T.A. 1984, ss.58–85 because of I.H.T.A. 1984, s.78. Where the "last transaction" (see below) was a conditionally exempt transfer the relevant person is the original transferor. However, where the last transaction was a "conditionally exempt occasion," the relevant person is the settlor (or if there is more than one, such person as the Board may select).

Where there have been two or more transactions the "last transaction" is the latest transaction within the period of 30 years before the chargeable event. Where there has been more than one transaction within the 30-year period the Board may select the "last transaction". I.H.T. on the chargeable event is calculated as though the relevant person had made an additional chargeable transfer. If the relevant person is alive the amount is added to his cumulative gifts total and computed at one half the table rate, whereas if he has died it is added to the amount of his estate.

Reinstatement of transferor's cumulative total

13–007   It is provided that a chargeable event is treated as part of the cumulative total of the person who made the last conditionally exempt transfer before the chargeable event (I.H.T.A. 1984, s.34) although he may not have been designated the relevant person for I.H.T.A. 1984, s.33. If he is alive this affects subsequent chargeable transfers by him. If he is dead the amount on which the tax becomes payable on the chargeable event is added to the value of his estate at his death. It is also provided that if the property has been comprised in a settlement made less than 30 years before the chargeable event a settlor who has made a conditionally exempt transfer of the property can be placed in the position of a person who made the last conditionally exempt transfer before the chargeable event.

Maintenance funds for historic buildings, etc.

The Treasury will approve a maintenance fund provided that the trustees approved by the Treasury include a trust corporation, solicitor or accountant and are resident in the U.K., *i.e.* the general administration is carried on, and a majority of the trustees are resident, in the U.K. (I.H.T.A. 1984, s.77, Sched. 4, paras. 1 and 2). **13–008**

"Accountant" means a member of an incorporated society of accountants, and "trust corporation" means one for the purposes of the Law of Property Act 1925, or for the purposes of the Administration of Estates (Northern Ireland) Order 1979, Article 9.

The requirements to be met are:

(a) That within the period of six years of becoming an approved trust its funds must not be capable of being used for any purpose other than for the maintenance, repair or preservation of the historic building, etc., (for the definition of qualifying property see below), or for the maintenance, repair, preservation or improvement of the trust's own property.

(b) That should any of the property cease to be held on the trusts within the six-year period, or before the settlor's death (within that time), it can pass only to a national body (museums, etc.) or a qualifying charity existing for the preservation of maintaining historic buildings, etc.

(c) That any income arising from trust property at any time and not applied for the maintenance, repair, etc., of the building must be applied for the benefit of national bodies or qualifying charities as (c) above (I.H.T.A. 1984, Sched. 4, para. 4).

After the six-year period, income remains subject to the same restrictions, although capital is not, albeit the latter is subject to the consequences contained in I.H.T.A. 1984, Sched. 4, para. 8 on property leaving the maintenance fund.

Under the old rules (see F.A. 1980, s.89(4)(d)) and under the revised provisions property may be transferred between approved maintenance funds within the prescribed time scales without crystallising a tax charge. In such cases the six-year period during which the property of the fund is restricted will run from the date of approval of the transferor fund. **13–009**

Qualifying property is defined to include land which in the opinion of the Treasury is of outstanding scenic or historic or scientific interest: buildings of outstanding historic or architectural interest and ancillary land and objects historically associated with such buildings. Property will qualify only where the requisite undertaking has been given to the Treasury (as to the preservation of the property, facilities for public access, etc.) and provided that such undertaking has not been breached.

Where property is comprised in a settlement exempt under the old provisions it will be deemed to be approved under these provisions (I.H.T.A. 1984, Sched. 4, para. 4).

13–010 The Treasury may withdraw approval of a fund if it considers that its property or its administration cease to warrant its continuance (I.H.T.A. 1984, Sched. 4, para. 5).

13–011 The Treasury has power to request the trustees to provide it with such accounts and other information as it may require relating to the fund and also have power to approve funds in advance of transfers of property to them (I.H.T.A. 1984, Sched. 4, para. 6).

13–012 The Treasury has the right and powers of a beneficiary as respects the appointment, removal and retirement of trustees where a direction has been issued (I.H.T.A. 1984, Sched. 4, para. 7).

Property leaving maintenance funds

13–013 Property held in an approved maintenance fund for historic buildings, etc., is not relevant property and as such is not subject to exit and 10 yearly charges under the normal discretionary trust rules. However, where property ceases to be so held, or is applied other than for the prescribed purposes of the fund either directly or by a depreciatory transaction, or where there is an omission to exercise a right (unless without gratuitous intend) there is a tax charge under I.H.T.A. 1984, Sched. 4, para. 8. Dispositions which would be exempt under I.H.T.A. 1984, s.16 (grant of tenancies of agricultural property) are excluded.

The tax charge is calculated on the diminution in the value of the trust property as a result of the disposition and this is grosses up where the tax is paid out of the trust funds. The rate of tax is as prescribed in paras. 11–15 below. There are anti-avoidance provisions where an interest under the settlement has been acquired by a charity for a consideration in money or money's worth.

13–014 Where property leaves an approved maintenance fund the charge may be avoided if the property is resettled within 30 days on another approved maintenance fund by an exempt transfer under I.H.T.A. 1984, s.27. The 30-day period is increased to two years where the occasion of charge is the death of the settlor. There are similar anti-avoidance provisions and rules as to calculation of any excess which may be chargeable as under I.H.T.A. 1984, Sched. 4, paras. 16–18 (I.H.T.A. 1984, Sched. 4, para. 9).

Whilst the revertor to settlor provisions have generally been abolished they still apply to approved maintenance funds and there will be no charge under I.H.T.A. 1984, Sched. 4, para. 8 on property leaving such a fund and reverting to the settlor or his spouse, or where the settlor has died and the property reverts to the widow or widower within a two-year period from the death. There are similar anti-avoidance provisions and calculation of a tax charge on any excess transfer as under I.H.T.A. 1984, Sched. 4, paras. 16–18, below.

13–015 There are a number of relatively complex calculations as to the rate of tax applicable on property leaving an approved maintenance fund. The rate prescribed in I.H.T.A. 1984, Sched. 4, para. 11 will apply to most situations where the property in the maintenance fund has been derived from a previous discretionary trust, or where discretionary trust property has been distributed to an individual who has made an exempt transfer under I.H.T.A. 1984, s.27 to a maintenance fund.

The rate of tax under para. 11 depends upon the relevant period which is defined as commencing at the latest of:

(a) the date of the last 10-year anniversary of the settlement in which the property was comprised before it last ceased to be relevant property (in most cases on transfer to the maintenance fund, or to an individual who makes the transfer);

(b) the date the property became (or last became) relevant property before ceasing to be as such and

(c) March 13, 1975.

The relevant period ends on the day before the event giving rise to the charge.

The rate of tax is the aggregate of the following percentages:

0.25 per cent. for each of the first 40 complete successive quarters in the relevant period.
0.20 per cent. for each of the next 40.
0.15 per cent. for each of the next 40.
0.10 per cent. for each of the next 40.
0.05 per cent. for each of the next 40.

The aggregate percentage (the maximum being 30 per cent.) is applied to the amount of the transfer calculated under I.H.T.A. 1984, Sched. 4, para. 8.

Where tax is chargeable under para. 8 and the tax rate under para. 11 **13–016** does not apply, the rate of tax is the greater of the first rate and the second rate described below.

The first rate is the aggregate of the same percentages as defined (I.H.T.A. 1984, Sched. 4, para. 11) above although the relevant period is redefined and commences with the date that the property became held by an approved maintenance fund and ends on the day before the occasion of charge. No period before a previous occasion of charge under para. 8 in respect of the same property is included in the relevant period (I.H.T.A. 1984, Sched. 4, para. 13).

If the settlor is alive the second rate is the effective rate which would be **13–017** charged on a notional lifetime transfer by the settlor on the date (and on the amount under I.H.T.A. 1984, para. 8) of the occasion of charge (I.H.T.A. 1984, Sched. 4, para. 14).

The subsequent death of the settlor is not to affect the rate of tax on any earlier lifetime charge—this should exclude the possibility of any previous potentially exempt transfers made by the settlor after March 17, 1986 (see I.H.T.A. 1984, s.3A) from entering into the calculations should they become chargeable on his death within seven years. As a transitional measure following the introduction of I.H.T. rules from March 18, 1986, when calculating tax on a chargeable event on or after that day by reference to a pre-March 18, 1986, death, it is to be assumed that the amendments to I.H.T.A. 1984, s.7 (rates) made by F.A. 1986 were then in force—thus securing the benefit of the shorter seven-year cumulation period and higher tax threshold.

13–018   If the settlor is dead, the second rate is the effective rate arrived at by including the amount under I.H.T.A. 1984, Sched. 4, para. 8 as the top slice of his estate, see I.H.T.A. 1984, Sched. 2, para. 6 where there has been a reduction in rates of I.H.T. However, if the settlor died before March 13, 1975, the second rate is found by including the amount as the top slice of the estate subject to estate duty and recomputing the tax at the death rates at the date of the occasion of charge.

There are detailed provisions to cover the position where the property has been comprised in more than one settlement; these enable the Board to select the settlor for the purposes of this paragraph. The effective rate is defined as the fraction found by expressing the tax chargeable as a percentage of the amount on which it is charged.

Where property became comprised in a maintenance fund under the old rules it will be deemed to have been so held under the revised rules in I.H.T.A. 1984, Sched. 4, para. 8 above for the purposes of paras. 11–14 above (I.H.T.A. 1984, Sched. 4, para. 15).

Property becoming comprised in maintenance funds

13–019   There is no exit charge under I.H.T.A. 1984, s.65 where property, on ceasing to be relevant property, becomes comprised in a maintenance fund for historic buildings, etc., approved under I.H.T.A. 1984, Sched. 4. The relief is not available where an interest under the settlement has previously been acquired by the trustees for a consideration in money or money's worth or if they became entitled to the interest as a result of transactions (either directly or indirectly) including a disposition for such consideration, *i.e.* the associated operations provisions contained in I.H.T.A. 1984, s.268 may be appropriate in view of the definition of disposition. There will, however, be a charge where the value of the transfer computed on the consequential loss basis (without grossing up) but ignoring business and agricultural property relief exceeds the value of the property received by the maintenance fund (less any consideration given by them).

Where an individual becomes entitled to trust property (which thus ceases to be relevant property) the exit charge under I.H.T.A. 1984, s.65 does not apply provided that the individual resettles the property within 30 days on an approved maintenance trust by an exempt transfer under I.H.T.A. 1984, Sched. 4. The limit is extended to two years where the person became entitled to the property on death. Similar rules apply to individuals as for the trustees where the individual has acquired an interest in the property for a consideration in money or money's worth, or where an excess transfer is chargeable to tax as discussed above (I.H.T.A. 1984, Sched. 4, paras. 16–18).

Capital gains tax relief

13–020   Gifts of historic buildings, land of scenic, historic or scientific interest, works of art, etc., and gifts to heritage maintenance funds qualify for C.G.T. relief if they attract exemption from I.H.T.: C.G.T.A. 1979, ss. 147, 147A. Tax on any gain accrued up to the date of the gift can be deferred until the donee disposes of the asset.

Income tax relief

Maintenance funds for historic buildings

A settlement covering a maintenance fund for historic buildings and **13–021** other qualifying property is almost impossible to draft without retaining an interest for the settlor which would mean that the trust income was still deemed to be the settlor's under I.C.T.A. 1988, ss.673–677. The trustees can elect in writing within two years of the end of a year of assessment to have the trust income taxed separately and not as if it were the settlor's: I.C.T.A. 1988, s.691.

Maintenance funds: charge of income tax

Any accumulated income which is applied other than for the purposes of **13–022** the fund or is transferred other than to a qualifying charity or museum is subject to additional income tax at 5 per cent. It is exempt if it is already taxed as the income of the settlor under some other provision: I.C.T.A. 1988, s.694. Where the transfer to a qualifying charity or museum is made for a consideration in money's worth the transferor will lose the income tax exemption although the transferee will not. Tax exemption is not lost where property is transferred free of I.H.T. to another settlement.

Maintenance funds: one estate elections

The advantages for estates (including historic buildings, etc.) of the "one **13–023** estate" election under I.C.T.A. 1988, s.26, where an approved fund has been set up by the transfer to it of part of an estate subject to such an election are preserved by I.C.T.A. 1988, s.27. This applies also where ownership of the historic building, etc. (comprised in the estate) changes provided that a I.C.T.A. 1988, s.26 election was previously in force, or where the property is transferred within 30 days to a further approved settlement. The trustees are allowed to deduct from rents proper revenue expenses relating to any part of the estate not held by the trust and similar expenses paid by the owner not absorbed by current rents. The owner is also entitled to relief of the trustees' excess maintenance expenditure on trust property under I.C.T.A. 1988, s.33.

It is made clear that notwithstanding these provisions the trust income is not to be treated as being that of the owner of the remaining estate.

Maintenance funds: reimbursement of settlor

Where the income of a maintenance fund is treated as the income of the **13–024** settlor for 1982–83 onwards and he is carrying on a trade of running a stately home, any reimbursement of maintenance expenditure by the trustees is not be regarded as income of the trade. This enables him still to claim the maintenance costs as a trading expense: I.C.T.A. 1988, s.692.

Maintenance funds: severance of non-exempt part

13–025   Where a fund is partially exempt as a maintenance fund or ceases to be exempt the non-exempt proportion will be subject to tax as a normal trust: I.C.T.A. 1988, s.693.

Grounds open to the public

13–026   Where an estate is so organised that part of it is open to the public, this will normally constitute a separate trade distinct from any farming activities on the estate. Whether it amounts to a single comprehensive trade of running a stately home, museum, fun-fair, wildlife park, etc., as a single global activity, or as a number of separate activities, will depend on the way in which the business is organised. The profits will be calculated in the same way as for any other trade as described in Chapter 3, although obviously the special reliefs relating to farming would be inapplicable.

In certain circumstances there may be a farm which is organised so that public visitors contribute directly to the running of the farm. This is particularly common where rare breeds of animals are being bred which would otherwise be uneconomic to preserve in modern conditions. It also applies where activities on the farm such as the making of butter or cheese are sufficiently interesting to enable the proprietors to charge the public to visit the premises. This in turn means that the provision of a cafeteria and shop selling the farm produce may well be set up.

Whether these activities are part of the overall farming trade or represent a separate trade depends upon the circumstances, normally on whether the bulk of the produce sold is from the farm and ancillary activities, or whether it is bought in. It is by no means uncommon for shops on farms to provide a wide range of eggs, fruit and vegetables, little of which emanates from the farm itself. This is in reality a retail enterprise quite distinct from the farm and may be taxed as a separate trade, although in practice the Revenue will often treat the whole of the activities as part of the single trade of farming if there is unlikely to be any significant tax difference at the end of the day.

In a number of cases with a stately home or estate open to the public, some or all of the activities will be hived off to third parties, for example, the catering facilities or animal park. In such cases the proprietor may receive a rent which in many cases will be linked to the turnover of the tenant. The premises however will be either let furnished, in which case the rents will be assessable under Schedule D, Case VI, or let unfurnished; in which case the assessment will be under Schedule A, as explained in Chapter 9. Alternatively the arrangement may amount to a joint venture or partnership and be taxed accordingly.

It is most unlikely that entrance fees from members of the public would be regarded as casual receipts under Schedule D, Case VI rather than a trade under Schedule D, Case I.

Sites of special scientific interest

13–027   Under the Wildlife and Countryside Act 1981 the Nature Conservancy Council, which is appointed by the Environment Secretary and funded by the Department of the Environment, may declare an area of land as being a site of special scientific interest (SSSI) for the purpose of conserving wild-life or plants or for its geological or physiological features. See Appendix 2. At the present time some 4,000 sites have been so designated. The sites themselves may be large or small.

If a site has been listed, the landowner is prevented from carrying out various operations on the site such as drainage of marsh lands. The Nature Conservancy Council usually enters into a management agreement with the owner of the SSSI under which the use of the land is regulated and con-fined to those activities which would not harm the land for the purpose for which it was designated. Under such a management agreement the Nature Conservancy Council may require work to be done on the site and it is empowered to make grants, lump sum payments or loans to compensate for the loss of profit from inhibiting the owner's use of the land.

Compensation for loss of profit of this nature is taxable as income or as a deduction from expenditure incurred. Lump sum compensation for dimi-nution in capital value will be subject to C.G.T. on a partial disposal. The designation of an area of land as an SSSI could reduce its capital value which would be reflected in the valuation for I.H.T. and C.G.T. purposes. If the Nature Conservancy Council requires or approves of capital work on an SSSI it may make a capital grant towards the costs and this would be deducted from the amount of the cost in computing agricultural buildings allowances on the works.

If the landowner objects to the designation of a site as an SSSI or wishes **13–028** to carry out listed works contrary to that approved by the Nature Conser-vancy Council he must give the appropriate notice to the Council. If agree-ment is ultimately not reached it is likely that a nature conservation order will be made over the land under which the owner would be entitled to compensation or the land may be subject to a compulsory purchase order. In each case the compensation would be capital and subject to C.G.T. as a part disposal.

Instead of an SSSI, land may be designated as a nature reserve, a national park, area of outstanding natural beauty, limestone pavement, ancient monument, archaeological area or conservation area. Where land is so affected it may merely mean that planning permission is more difficult or is required whereas in the case of agricultural buildings planning per-mission would not normally be needed, or it may directly affect the value of the landowner's property for example, where he has a field containing an ancient monument, and in that case the landowner is likely to receive compensation for the loss of profits or diminution in value caused by the prohibition against disturbing the monument. Compensation for loss of profits is taxable as a trading receipt, compensation for a capital loss as a capital gain.

Charities

The whole area of charities and taxation is one of considerable complica- **13–029** tion and is outside the scope of this book. It is, however, worth mentioning that landed estates are not infrequently transferred to charities such as the National Trust. The Trust will not take a property which is incapable of generating sufficient income to maintain itself unless there is additional income-bearing land or some other form of endowment transferred at the same time. On the other hand the National Trust will be prepared in many cases to grant long leases to the donor and his family for occupation at a nominal rent with limited access for the public to parts of the property. This can be an effective way of keeping the use of the property for members of the family in a tax efficient manner.

The transfer to a charity would be exempt from C.G.T. under C.G.T.A. 1979, s.146 and I.H.T. under I.H.T.A. 1984, s.23. The income on investments or rents from surrounding land transferred to the charity would be tax free under I.C.T.A. 1970, s.360, but income from a trading activity such as farming by a charity would be taxable at the basic rate. In such circumstances it is normal for the trading to be carried out by a company owned by the charity and for the whole of the profits to be paid up by way of covenant or dividend during the year so that in the case of a covenant the gross payment is a deduction for tax purposes from the company's profits and the tax deducted on payment is recoverable in the charity, or in the case of a dividend the corporation tax is effectively covered by the advance corporation tax on the dividend which is in turn recoverable as a tax credit in the hands of a charity.

13–030   Where the property is not suitable for transfer to the National Trust or it is desired not to make such a transfer it may be possible to set up a private charity. The concept of a charity comprises four principal divisions:

Trusts for the relief of poverty.
Trusts for the advancement of education.
Trusts for the advancement of religion.
Trusts beneficial to the community and not falling under any of the other preceding heads (*Special Commissioners* v. *Pemsel* (1891) 3 T.C. 53).

It is not by any means impossible for a farm to qualify as a charity for the benefit to the community, particularly where it includes for example an SSSI or is in a national park or area of outstanding natural beauty.

CHAPTER 14

Domicile, Residence and Farming Overseas

Introduction

A person's liability to U.K. tax depends on his domicile and whether he **14–001** is resident or ordinarily resident in any tax year. Nationality and citizenship are not relevant. In the case of a company its country of incorporation must also be considered.

Changes to the law of domicile were suggested in "Private International Law, The Law of Domicile," Cm 200 published by the Law Commission in 1987, but that report did not address the tax implications of any changes. A Consultative Document entitled entitled "Residence in the United Kingdom: The Scope of U.K. Taxation for Individuals" was published by the Board of Inland Revenue in July 1988 but on 15 March 1989 the Government stated that it did not intend to bring forward any proposals at that time.

Domicile

A domicile of choice is basically the country in which an individual **14–002** intends to make his permanent home, but it is not normally possible to claim a domicile of choice without actually living in the country claimed (*I.R.C.* v. *Duchess of Portland* [1982] S.T.C. 149). A child acquires as a domicile of origin the domicile of his father at the time of birth. A posthumous or illegitimate child takes his mother's domicile at his birth as his domicile of origin. In England and Wales the domicile of a child under 16 is a dependent one and would usually follow any change in the father's domicile. A woman already married on January 1, 1974, would have acquired her husband's domicile on marriage as a domicile of dependence, now regarded as one of choice, but under the Domicile and Matrimonial Proceedings Act 1973 a woman who marries after that date retains an independent domicile (1979 B.T.R. 398, December 31, 1979). A married woman with a dependent domicile can by her actions now establish an independent domicile of choice, but her domicile of origin does not automatically revert, even on divorce (*Faye* v. *I.R.C.* (1961) 40 T.C. 103), or on her husband's death (*In re Wallach dec'd Weinschenk* v. *Treasury Solicitor* (1949) 28 A.T.C. 486, *I.R.C.* v. *Duchess of Portland* [1982] S.T.C. 149). A minor may acquire an independent domicile of choice at the age of 16 or earlier marriage (except in Scotland).

As domicile is very much a matter of fact and of intention (*Iveagh Lord*

v. *Revenue Comrs.* [1930] I.R. 431), mere length of stay in the U.K. is not sufficient to establish a U.K. domicile. In (*Winans* v. *Att.-Gen.* [1904] A.C. 287 and *Bowie (or Ramsay)* v. *Liverpool Royal Infirmary* [1930] A.C. 588 residence for nearly 40 years did not establish a change of domicile. Similarly in the more recent cases of *Buswell* v. *I.R.C.* [1974] S.T.C. 266 and *I.R.C.* v. *Bullock* [1976] S.T.C. 409 a foreign domicile of origin was not overruled by residence in the U.K. for a considerable period in the face of the taxpayer's avowed intention not to change his domicile. In *Steiner* v. *I.R.C.* [1973] S.T.C. 547, *Re Lawton* (1958) 37 A.T.C. 216, and *Re Furse (dec'd) Furse* v. *I.R.C.* [1980] S.T.C. 596, however, a change of domicile was on the facts of each case upheld. It is difficult to throw off a U.K. domicile or origin (*I.R.C.* v. *Cohen* (1937) 21 T.C. 301, *Plummer* v. *I.R.C.* [1987] S.T.C. 698) and if a foreign domicile of choice is obtained a return to live in the U.K. would resurrect the domicile of origin (*Fielden* v. *I.R.C.* (1965) 42 T.C. 501).

Individuals treated as domiciled in U.K. for inheritance tax

14–003 An individual not domiciled in the U.K. at the time of the transfer is treated as if he were domiciled in the U.K. if:

(a) he was domiciled under general law in the U.K. on or after December 10, 1974, and within the three years immediately preceding the transfer, or
(b) he was resident in the U.K. on or after December 10, 1974, and in at least 17 out of the previous 20 years (I.H.T.A. 1984, s.267).

For this purpose residence is to be determined in accordance with the usual income tax rules except that any house available for use in the U.K. is ignored, I.C.T.A. 1988, ss.334–336, *re Clore (deceased)* (No. 2), *Official Solicitors* v. *Clore and Others* [1984] S.T.C. 609, the deceased failed to lose his English domicile.

It should be noted that the deemed domicile provisions do not apply to certain excluded property (I.H.T.A. 1984, ss.6, 48).

Residence

14–004 Temporary absence from the U.K. does not make a British subject non-resident for tax purposes (I.C.T.A. 1970, s.49, *Reed* v. *Clark* [1985] S.T.C. 323).

A person with a full-time employment outside the U.K. has his residence for tax purposes determined without regard to whether he maintains a place of abode in the U.K. for his use. Similar provisions apply if a trade, profession or vocation is carried on entirely outside the U.K. Incidental duties in the U.K. are ignored (I.C.T.A. 1970, s.50, *Robson* v. *Dixon* (1972) 48 T.C. 527, SP/A10, October 28, 1985).

A temporary visitor to the U.K. is not liable to tax under Schedule D, Cases IV or V on remittances unless he has spent six months in the U.K. in any year of assessment: I.C.T.A. 1988, s.336, C.G.T.A. 1979, s.18(3).

Similarly for Schedule E purposes a visitor will be treated as a non-resident unless the length of stay is six months or more.

A house available for use in the U.K. implies residence (*Lloyd* v. *Sulley* (1884) 2 T.C. 37) unless there is complete absence from the U.K. (*Turnbull* v. *Foster* (1904) 6 T.C. 206). Two months spent in the U.K. in each year for three years was sufficient to give rise to a tax charge as resident in the case of *Cooper* v. *Cadwalader* (1904) 5 T.C. 101), but staying in hotels implies non-residence (*I.R.C.* v. *Zorab* (1926) 11 T.C. 289) and *I.R.C.* v. *Brown* (1926) 11 T.C. 292). Living in a yacht anchored in the U.K. (*Bayard Brown* v. *Burt* (1911) 5 T.C. 667) and use of a hunting lodge (*Loewenstein* v. *De Salis* (1926) 10 T.C. 424) has resulted in residence. Temporary visits abroad do not cause U.K. residence to be lost (*I.R.C.* v. *Combe* (1932) 17 T.C. 405) nor do temporary visits to the U.K. give rise to residence (*Withers* v. *Wynyard* (1938) 21 T.C. 724).

Ordinary residence

Residence and ordinary residence are not defined by the Taxes Acts. In **14–005** *Levene* v. *I.R.C.* (1928) 13 T.C. 486 the dictionary definition of place of usual abode was followed. Ordinary residence has been contrasted with casual or occasional residence (*Lysaght* v. *I.R.C.* (1928) 13 T.C. 511, *Thomson* v. *Minister of National Revenue (Canada)* [1946] S.C.R. 209) and may be treated as meaning in the normal course of his life the taxpayer has resided in the U.K. Actual cases relating to residence show that where there is some business or permanent tie in the U.K. the taxpayer tends to be treated as resident subject to the specific exemptions contained in I.C.T.A. 1988, ss.334–336, although the purpose of the visit will not outweigh the time spent (*Lord Inchiquin* v. *I.R.C.* (1948) 31 T.C. 125).

Revenue practice

The Revenue practice relating to residence is set out in Revenue booklet **14–006** I.R. 20 and SP3/81. If no place of abode is maintained in the U.K. the taxpayer is resident for any year of assessment in which visits to the U.K. total 183 days or if visits are habitual and substantial, *i.e.* on average three months a year for four years. If the visits are habitual and substantial ordinary residence is also claimed. Ordinary residence and residence will still be claimed where the absence abroad amounts to less than three years and the taxpayer visits the U.K. for three months in a tax year. The three-year period is reduced where there is a full-time occupation overseas and the taxpayer has been abroad for a complete tax year (subject to permitted visits to the U.K.).

If a place of abode is maintained in the U.K., which is not to be ignored under I.C.T.A. 1988, s.335 (because of a full-time employment overseas or being engaged full-time in a trade or profession outside the U.K.), the taxpayer who has previously been resident in the U.K. is resident for any year in which he is physically present in the U.K. for however short a length of time and ordinarily resident if the visits are habitual and substantial.

Under Extra-statutory Concessions A11 and D2 (1988) a taxpayer coming to the U.K. is normally treated as resident in the U.K. in the year of arrival from the date on which he arrives, and a taxpayer leaving the U.K. is treated as resident in the year of emigration only up until the date of his departure, see *R.* v. *I.R.C., ex parte Fulford-Dobson* [1987] S.T.C. 344. However, a full year's personal allowances is given in each case against the income for the resident period, which is calculated on the preceding year basis if appropriate (*Fry* v. *Burma Corporation Ltd.* (1930) 15 T.C. 113).

A non-resident is liable to tax on U.K. income at basic and higher rates and to the investment income surcharge (*Brooke* v. *I.R.C.* (1917) 7 T.C. 261) provided that the Revenue can collect the tax, but see Extra-statutory Concession B13 (1988) which effectively exempts U.K. bank deposit on building society interest.

14–007 As with residence, there is no statutory definition of ordinary residence and it was held in the case of *Lysaght* v. *I.R.C.* (1928) 13 T.C. 511 to be the converse of casual or occasional residence. It may therefore be treated as meaning in the normal course of events the taxpayer is resident for substantial periods of time in the U.K. For example, in *Kinloch* v. *I.R.C.* (1929) 14 T.C. 736 it was held that a widow who lived mainly in hotels and had spent substantial periods in the U.K. to visit her son at school was ordinarily resident. See also *Peel* v. *I.R.C.* (1927) 13 T.C. 443, *Miesegaes* v. *I.R.C.* (1957) 37 T.C. 493, *Reid* v. *I.R.C.* (1926) 10 T.C. 673 and *Elmhirst* v. *I.R.C.* (1937) 21 T.C. 381; also *R.* v. *Barnet London Borough Council, ex parte Shah* (1980) The Times, July 21), *Circutti* v. *Suffolk County Council* (1980) The Times, July 30.

The six-month period in I.C.T.A. 1988, s.336 is calculated exactly (*Wilkie* v. *I.R.C.* (1951) 32 T.C. 495). Schedule E appeals as to domicile or ordinary residence are dealt with under I.C.T.A. 1988, s.207.

Dual residence

14–008 It should be emphasised that it is perfectly possible for an individual to have more than one country of residence for the tax purposes of each country. Although an individual may be resident and ordinarily resident in the U.K. he may be treated as being resident in the other country and not in the U.K. under the "tiebreaker" provisions of a double taxation relief agreement. In such circumstances the double taxation relief agreement overrules the U.K. legislation in view of I.C.T.A. 1988, s.788.

Husband and wife

14–009 It should be noted that the residence of a husband and wife are considered independently. If one spouse is regarded as resident in the U.K., but the other is non-resident, under I.C.T.A. 1988, s.282 they would be taxed in the U.K. as if they were permanently separated, although this provision is not to increase the tax payable (*Gubay* v. *Kington* [1984] S.T.C. 99).

Foreign possessions

14–010 Farming overseas is not part of a U.K. trade of farming under I.C.T.A. 1988, s.53 (*Sargent* v. *Eayrs* [1973] S.T.C. 50). It is likely that for practical reasons the management of the farm would be outside the U.K. and a U.K. resident farmer would therefore be assessable on the income as income from a foreign possession taxable under Schedule D, Case V under I.C.T.A. 1988, ss.15 and 65–67. The farm may well be liable to overseas tax, being a permanent establishment in the overseas country. Non U.K. tax liabilities are outside the scope of this book.

Credit should be available for any overseas tax paid against any U.K. liability either under a double taxation treaty or as unilateral relief under I.C.T.A. 1988, s.790. If it is advantageous the U.K. taxpayer may alternatively claim the overseas tax as an expense as he would do, for example, if there were a loss for U.K. tax purposes, which would enable the losses carried forward to be increased, as a credit would be worthless if there is no U.K. liability: I.C.T.A. 1988, s.811.

The profits are calculated under Schedule D, Case V as if there were a U.K. resident trade although the special provisions such as farm averaging relating to a U.K. farm would not be available. Losses would be calculated in a similar manner: I.C.T.A. 1988, s.391(1). Capital allowances are also available as if the trade were in the U.K.: I.C.T.A. 1988, s.65(3).

The reliefs which used to exist for foreign employment earnings and **14–011** income from foreign trade has largely been removed by F.A. 1984, s.30. However, if a U.K. resident farmer has a non U.K. domicile for tax purposes, or if he is a British subject or citizen of the Republic of Ireland not ordinarily resident in the U.K., I.C.T.A. 1988, s.65(5)(a), he will be assessed only on remittances of profits from abroad to the U.K.

It is worth noting that roll-over relief for C.G.T. purposes is available, even if the reinvestment is outside the U.K. (C.G.T.A. 1979, ss.115 to 121) as is the relief for the transfer of a business to a company which is available even if the company is non-resident (C.G.T.A. 1979, s.123) (see Chapter 15).

Agricultural relief for I.H.T. purposes is restricted to agricultural property in the U.K., the Channel Islands or the Isle of Man (C.T.T.A. 1984, s.115(5)). However, property situated outside the U.K. is excluded property if the person beneficially entitled to it is an individual domiciled outside the U.K. (I.H.T.A. 1984, s.6(1)). There is an extended meaning to the term domicile for I.H.T. purposes (I.H.T.A. 1984, s.267) which is explained in Chapter 18.

Company residence

With effect from March 15, 1988, a company incorporated in the U.K. is **14–012** resident in the U.K. A company incorporated overseas but with its central management and control in the U.K. is also resident in the U.K. There are transitional provisions covering the five years to March 14, 1993, for certain companies which had emigrated before March 15, 1988.

Until March 14, 1988, a company's residence was determined by the place of its central management and control (*Calcutta Jute Mills Co. Ltd.* v. *Nicholson* (1876) 1 T.C. 83, *Cesena Sulphur Co. Ltd.* v. *Nicholson* (1876) 1 T.C. 88, *Imperial Continental Gas Assoc.* v. *Nicholson* (1877) 1 T.C. 138). The criterion still obtains for companies not incorporated in the U.K. It applies irrespective of where the head office or shareholders' general meetings are held (*De Beers Consolidated Mines Ltd.* v. *Howe* (1906) 5 T.C. 198).

14–013   Before March 15, 1988, a U.K. incorporated company could have a foreign residence (*Todd* v. *Egyptian Delta Land and Investment Co. Ltd.* (1928) 14 T.C. 119) but see *Eccott* v. *Aramayo Francke Mines Ltd.* (1925) 9 T.C. 445, *Egyptian Hotels Ltd.* v. *Mitchell* (1915) 6 T.C. 542, *Noble (BW), Ltd.* v. *Mitchell* (1926) 11 T.C. 372, *American Thread Co.* v. *Joyce* (1913) 6 T.C. 163 and SP6/83.

A company incorporated outside the U.K. can have a U.K. residence (*New Zealand Shipping Co. Ltd.* v. *Stephens* (1907) 5 T.C. 553, *Bullock* v. *The Unit Construction Co. Ltd.* (1959) 38 T.C. 712). Such a company could be resident both in the country of incorporation and in the U.K. if its central management and control is in the U.K.: see *Swedish Central Railway Co. Ltd.* v. *Thompson* (1925) 9 T.C. 342, *John Hood & Co. Ltd.* v. *Magee* (1918) 7 T.C. 327 and *Union Corp. Ltd.* v. *I.R.C.* (1953) T.C. 207.

Companies not resident in the U.K.

14–014   A non-resident company is chargeable to corporation tax only on the worldwide trading profits and other U.K. source income of a branch or agency in the U.K. and on chargeable gains on the disposal of assets situated in the U.K. and used by the branch or agency for the purpose of the trade: I.C.T.A. 1988, s.11. Other overseas income of the company is not chargeable to U.K. tax. Other U.K. source income is subject to basic rate income tax.

A U.K. resident company with overseas branches is usually assessed under Schedule D, Case I on its worldwide profits (*London Bank of Mexico and South America* v. *Apthorpe* (1891) 3 T.C. 143, *San Paulo (Brazilian) Railway Co.* v. *Carter* (1895) 3 T.C. 344, 407, *Denver Hotel Co.* v. *Andrews* (1895) 3 T.C. 356, *Grove* v. *Elliots & Parkinson* (1896) 3 T.C. 481, *Apthorpe* v. *Peter Schoenhofen Brewing Co. Ltd.* (1899) 4 T.C. 41). Overseas subsidiary companies may be genuinely non-resident (*Kodak Ltd.* v. *Clark* (1903) 4 T.C. 549, *Stanley* v. *Gramophone & Typewriter Ltd.* (1908) 5 T.C. 358).

Anti-avoidance

14–015   There are a number of anti-avoidance measures aimed at overseas activities of U.K. residents which are outside the scope of this book.

CHAPTER 15

Capital Gains and Losses

General principles

Capital gains and losses of individuals, partnerships and companies are **15–001** broadly calculated on the same basis. The assessable gain is the gain arising since acquisition or, if later, March 31, 1982. The net sale proceeds, after deducting allowable selling expenses, are compared with the base value which is normally the cost or, if held then, the market value at March 31, 1982, plus allowable enhancement expenditure to produce a capital gain or loss subject to indexation. The gain or loss adjusted for indexation, based on the increase in the retail prices index since March 1982, and the result-ant gain or loss is subject to C.G.T. at marginal income tax rates (25 per cent. or 40 per cent. or a combination of the two) in the case of an individ-ual or partnership and to corporation tax (25 per cent. or 35 per cent. or marginal rate of 37.5 per cent.) in the case of a company.

Before the new rules were introduced with effect from 1988–89, capital gains were calculated using the base date of April 6, 1965, so that gains since then, apportioned if the asset was held before that date, were subject to tax. For disposals after April 5, 1988, the base date is March 31, 1982, but the gains are calculated under the old rules if those produce a lesser gain or a smaller loss; and if one method produces a gain and the other a loss, it is treated as giving rise to neither gain nor loss.

Chargeable gains accruing to a company, other than a life assurance company, in respect of disposals after March 16, 1987, are subject to cor-poration tax at the full rate of 35 per cent. or the small companies' rate of

25 per cent. (for the year commencing April 1, 1989), or at an average rate between those two rates if the company's profits for the year are between £150,000 and £750,000 (which figures are divided by the total number of companies under common control if appropriate). Advance corporation tax, including surplus A.C.T. carried forward, can be set off against corporation tax on chargeable gains.

15–002 One important difference between losses in a company and losses by an individual or partnership is that a company's chargeable gains are part of its total profits for the year: I.C.T.A. 1988, ss.8 and 345(1)(*a*). This means that the corporation tax assessment is based on the total income from various sources plus chargeable gains, I.C.T.A. 1988, s.9(3), and means that it is possible to set against a capital profit:

(a) a trading loss for the year or the following year under I.C.T.A. 1988, s.393(2);

(b) charges on income: I.C.T.A. 1988, s.338;

(c) management expenses: I.C.T.A. 1988, s.75;

(d) advance corporation tax: I.C.T.A. 1988, s.239.

In the case of an individual or partnership there are no provisions to enable a trading loss to be set against a capital profit.

C.G.T. is levied on persons resident or ordinarily resident in the U.K. (C.G.T.A. 1979, s.2(1)). By concession, an individual (but not a company or a trust) who becomes resident in the U.K. is charged only on disposals made after his arrival, and an individual who leaves the U.K. and becomes not resident and not ordinarily resident is not charged on disposals after the date of his departure unless they relate to assets used in a branch or agency in the U.K. (Extra-statutory Concession D2 (1988)). Extra-statutory Concessions cannot be used for tax avoidance: *R.* v. *I.R.C. ex p. Fulford-Dobson* [1987] S.T.C. 344. Non-residents are subject to C.G.T. only on assets used in a branch or agency in the U.K. (C.G.T.A. 1979, s.2(2) and 12).

C.G.T. is payable on December 1, following the end of the tax year in which the gain is realised in the case of an individual or trust (C.G.T.A. 1979, s.7). In the case of a company the tax is payable on the normal due date for the payment of corporation tax: I.C.T.A. 1988, s.345 and Sched. 30, para. 1 and F.A. 1987, s.36.

Gains arising from the sale of assets outside the U.K. are taxable in the case of a U.K. resident with double taxation relief for any C.G.T. paid overseas (C.G.T.A. 1979, ss.10, 11). There is relief available if it is impossible to remit the gain because of foreign exchange control provisions (C.G.T.A. 1979, s.13).

15–003 Disposal proceeds include insurance claims and compensation (C.G.T.A. 1979, ss.20(1), 21, Extra-statutory Concession D19 (1988)), except for the lessee of a short lease reinstating under the terms of the lease (Extra Statutory Concession D1 (1988)). There is a deemed disposal on the value of an asset becoming negligible (C.G.T.A. 1979, s.22) and for this purpose it is possible to separate the building from the land (*Williams* v. *Bullivant* [1983] S.T.C. 107, *Cleveleys Investment Trust Co.* v. *I.R.C.* (No. 2) [1975] S.T.C. 457).

The date of disposal for C.G.T. purposes is the date of the contract and not the date of completion (C.G.T.A. 1979, s.27(1), *Magnavox Electronics Co. Ltd. (in liquidation)* v. *Hall* [1985] S.T.C. 260), or, in the case of a conditional contract, when it becomes unconditional, or, in the case of a compulsory purchase acquisition, when the compensation is agreed or the land occupied (C.G.T.A. 1979, s.27(2)) (*Stanton* v. *Drayton Commercial Investment Co. Ltd.* [1981] S.T.C. 525; C.G.T.A. 1979, s.111).

The market value is substituted for the sale proceeds or cost as appropriate if the transaction is other than at arm's length, *e.g.*, as a gift or transfer into settlement (C.G.T.A. 1979, s.29A(1), *Aspden* v. *Hildesley* [1981] S.T.C. 206, *Berry* v. *Warnett* [1982] S.T.C. 396), or where there is a disposal for consideration which cannot be valued unless there is no corresponding disposal (C.G.T.A. 1979, s.29A(2)).

There is a general exclusion from C.G.T. for any gain subject to tax as income (C.G.T.A. 1979, s.31) apart from the capitalised value of rent charges (C.G.T.A. 1979, s.31(3)).

Base value

The base value for C.G.T. is limited under C.G.T.A. 1979, s.32(1) to: **15–004**

(a) the cost;
(b) enhancement expenditure reflected in the state or nature of the asset (*Aberdeen Construction Group Ltd.* v. *I.R.C.* [1978] S.T.C. 127, *Emmerson* v. *Computer Time International Ltd.* [1977] S.T.C. 170, *Allison* v. *Murray* [1975] S.T.C. 524, *Oram* v. *Johnson* (1980) S.T.C. 222, *Drummond* v. *Austin Brown* [1984] S.T.C. 321), *Chaney* v. *Watkis* [1987] S.T.C. 89;
(c) the costs of defending title;
(d) the actual incidental costs of disposal (*I.R.C.* v. *Richards Executors* (1971) 46 T.C. 626, *I.R.C.* v. *Chubbs' Trustees* (1971) 47 T.C. 353).

There is normally no deduction for interest except, exceptionally, capitalised interest in a company (C.G.T.A. 1979, s.32(3), I.C.T.A. 1970, s.269) (see Chapter 20). Nor is there any deduction for expenses allowable against income (C.G.T.A. 1979, s.33), or where capital allowances are available (C.G.T.A. 1979, s.34). The cost of an asset includes the cost of an asset derived from another asset (C.G.T.A. 1979, s.36, *Aberdeen Construction Group Ltd.* v. *I.R.C.*, *Bayley* v. *Rogers* [1980] S.T.C. 544). For example, the cost of a freehold might be made up of the original cost of a lease, duly depreciated to the date of the acquisition of the freehold reversion, plus the cost of the freehold reversion.

Government or local authority grants are excluded from the base value for C.G.T. purposes (C.G.T.A. 1979, s.42) except grants for relinquishing occupation of uncommercial agricultural units under the Agriculture Act 1967, s.27, which are tax free under C.G.T.A. 1979, s.112.

It is permissible to make any appropriate apportionments for C.G.T. purposes (C.G.T.A. 1979, s.43).

It appears that a payment to secure planning permission by compensating the owner of land sterilised under an agreement under the Town and County Planning Act 1971, s.52, could be enhancement expenditure following *Taddale Properties Ltd.* v. *I.R.C.* [1988] S.T.C. 303.

Any assets owned at March 31, 1982, are deemed to have been sold at that date and immediately reacquired at open market value but not if this increases the gain or the loss, or converts a gain into a loss or vice versa. In order to avoid maintaining many old records, it is possible to elect that all assets disposed of shall be treated as having been sold and reacquired at March 31, 1982: F.A. 1988, s.96(5). Such an election is irrevocable, must be made before April 6, 1990, or within two years after the end of the tax year in which the first disposal after April 5, 1988, is made, and applies to all disposals, including those made after April 5, 1988, but before the election. In most cases such an election should be made; it will enable losses made since March 31, 1982, but not since April 6, 1965, to be relieved. It should not be made however if all in all there would be greater capital gains by reference to 1982 values than by reference to earlier cost on 1965 values.

Valuation for capital gains purposes

15–005 Market value is defined as the price which assets might reasonably be expected to fetch on a sale in the open market, with no reduction for flooding the market. The market value of quoted securities is in practice accepted as the lower of one-quarter of the difference between the lower and higher closing prices added to the lower price (one-quarter up) or midway between the highest and lowest prices at which the bargains were marked. There are exceptions where the quoted price is not a proper measure of the market value, but in view of *Crabtree* v. *Hinchcliffe* (1971) 47 T.C. 419 it would be necessary to find as a fact that special circumstances affecting the share price had been unjustifiably withheld from the stock exchange (C.G.T.A. 1979, s.150).

Although farming companies are unlikely to be quoted, many farmers hold shares in co-operatives or farming suppliers of goods such as foodstuffs and fertilizers as trade investments, which may be quoted.

Unquoted shares and securities

15–006 Unquoted shares or securities are valued for tax purposes on the basis of the price reasonably expected on a sale in the open market (*Holt* v. *I.R.C.* (1953) 32 A.T.C. 402) on the assumption that there is available to any prospective purchaser all the information which a prudent prospective purchaser might reasonably require for a purchase from a willing vendor by private treaty at arm's length (C.G.T.A. 1979, s.152). For C.T.T. these rules are repeated in I.H.T.A. 1984, s.168. This overrules the decision in *re Lynall deceased* (1971) 47 T.C. 375.

The information deemed to be available in each case will depend on the circumstances. For example, if a 20 per cent. shareholding is being valued it is probable that the information that could reasonably be required is far less than for the acquisition of a 60 per cent. controlling interest. If, however, the price to be paid for the 20 per cent. interest was substantial, additional information would reasonably be required compared with an investment of a much more modest amount. It is not to be envisaged that the prospective purchaser could reasonably ask the willing vendor to commit a breach of trust or disclose confidential information which the board, on request, would not be expected to authorise him to divulge.

When considering the valuation of shares in unquoted companies it is important to bear in mind the nature of a share. Following the case of *Borland's Trustee* v. *Steel Bros. & Co. Ltd.* [1901] 1 Ch. 279 it was held that a share is no more than a collection of rights between the shareholders *inter se* and between the shareholders and the company itself.

Many companies contain in their articles a restriction on the transfer of shares and this restriction has to be taken into account in the valuation. It is, however, necessary to assume that the hypothetical purchaser will in fact be registered as the new shareholder and it is not necessary to show that as a matter of fact the directors would not approve a transfer other than, for example, to a family member. The effect of a restriction on transfer is probably depreciatory in that it must be assumed that the shares acquired would be subject to the restrictions for any future transfer. On the other hand it could be argued that the existence of the restrictions makes it possible for the hypothetical purchaser to acquire further shares at less than their true value on any other party making a disposal, and, as such, the restrictions could tend to increase the value of the shares. The effect of a restriction on transfer has been considered in such cases as *Salvesen's Trustees* v. *I.R. Commissioners* [1930] S.L.T. 387 and *I.R.C.* v. *Crossman* [1926] 1 All E.R. 762.

Farm land and buildings

The valuation of farm land and buildings is complicated by the fact that there are different interests in land. Freehold farms with vacant possession tend to be based on the area of land and its quality at a price per acre on hectare for each quality of land, *Imrie* v. *I.R.* (1984) R. & V.R. 143, *Huckvale* v. *Gardner* (1985) R. & V.R. 148. This price normally includes the normal farm buildings for the area of land being disposed of with adjustment downwards if there are no buildings or if they are of poor quality and adjustments upwards for buildings of exceptional quality. In the case of a landed estate the main house and its grounds are often valued separately from the surrounding farm land and buildings. **15–007**

Recently the Revenue has tended to look more closely at the value of farm buildings and has suggested that values should be arrived at by notionally splitting up the estate if to do so would enhance the aggregate value, on the authority in *Duke of Buccleuch* v. *I.R.C.* [1967] A.C. 506, *Earl of Ellesmere* v. *I.R.C.* [1918] 2 K.B. 735. To estate agents this is known as optimum letting.

Obviously the value of the land will be increased, often dramatically, if there is planning permission or potential for development or if the land has possibilities for mineral extraction.

The value of farm land fluctuates considerably and in some areas has fallen to half what it was at the peak of a few years before.

In the case of tenanted land, the Agricultural Holdings Acts of 1948 and 1984, together with the Agriculture (Miscellaneous Provisions) Act 1976 and the Agriculture (Notices to Quit) Act 1977, provide the main legislation which in many cases gives the tenant substantial security of tenure. The value of the freehold reversion therefore depends to a large extent on the capitalised value of the net rent likely to be available from the land plus the possibility of obtaining vacant possession (*Willett* v. *I.R.C.* (1982) E.G. 636).

Historically, tenanted land has been worth considerably less than that with vacant possession. The tenant of an agricultural lease is not normally able to assign the tenancy to a third party and receive a premium on assignment, but he may agree to surrender his tenancy to the landlord for a substantial capital sum, which would reflect a substantial proportion of the difference between the tenanted value of the land and its value with vacant possession.

Where the freehold is owned by a shareholder who has a controlling interest in a company occupying the farm the Revenue argues that the value of the freehold reversion is the difference between the value of the freehold land with vacant possession and the value of the company's tenancy, on the grounds that the controlling interest could be used to secure the surrender of the tenancy. This argument would appear to be at variance with *Henderson* v. *Karmel's Executors* [1984] S.T.C. 572 under which it was held that a freehold interest in business premises occupied by a company controlled by the deceased could not be valued for C.G.T. purposes as if he had vacant possession. The Revenue also argues that an agricultural tenancy held by a company is directly assignable because the shares can be sold, and it is therefore a valuable asset of the company which should be reflected in the value of the shares.

In the case of a partnership, in practice, no value is usually attached to the value of the tenancy where the partners are unconnected with the landlord. Where the landlord is also a partner the Revenue may argue on the basis of *Newman* v. *Keedwell* (1977) 244 E.G. 469 that vacant possession could be obtained by serving notice to quit and, as a partner, refusing to sign the counter notice necessary to preserve security under the Agricultural Holdings Act 1984 but this might be an abuse of the Act and security of tenure preserved following *Sykes* v. *Land* (1984) 271 E.G. 1264, *Pyrah (Doddington) Ltd.* v. *Northamptonshire County Council* (1983) R. & V.R. 240, *Featherstone* v. *Staples* (1986) 273 E.G. 193, *Johnson* v. *Moreton* (1978) 3 All E.R. 37. However it is arguable that the tenancy has a value on the basis of *O'Brien* v. *Benson's Hosiery (Holdings) Ltd.* [1979] S.T.C. 735, see Chapter 18.

The whole question of valuation is a subject on which expert advice is essential.

Losses

15–008   A capital loss for the purposes of C.G.T. is calculated in exactly the same way as a chargeable gain (C.G.T.A. 1979, s.29(1)). For disposals after April 5, 1985, the indexation allowance applies to losses in the same way as it applies to gains (F.A. 1985, s.68(1)(*b*)) and can, therefore, be used to convert a profit into a loss or to augment a loss.

In certain cases, in particular assets held before April 6, 1965, where the gain or loss is less than calculated using the March 31, 1982, base value and no election under F.A. 1988, s.96(5), has been made, only part of the actual gain is a chargeable gain and similarly only part of the loss is an allowable loss (C.G.T.A. 1979, s.29(2)).

Similarly, gains accruing in a year in which an individual is neither resident nor ordinarily resident in the U.K. are not chargeable. Correspondingly, losses are not allowable losses. The exceptions are gains and losses arising through a U.K. branch or agency (C.G.T.A. 1979, s.29(3)).

A non-U.K. domiciled individual is not liable to C.G.T. in respect of gains attributable to the disposal of foreign assets unless they are remitted to the U.K. Furthermore, losses sustained on the disposal of non-U.K. assets are not allowable, whether or not the proceeds are remitted to the U.K. (C.G.T.A. 1979, s.29(4)).

An allowable loss can be set against any chargeable gain in the same or any subsequent tax year, or accounting period for a company, provided that relief has not been given in any other way. The only exception allowing a limited carry back of losses is in the year in which the taxpayer dies. Then any net losses in the tax year in which the death occurs so far as they are not set against chargeable gains in that period can be carried back against chargeable gains arising in the preceding three tax years using the latest gains first (C.G.T.A. 1979, s.49(2)).

Part disposal and disposal of a part

A part disposal, *i.e.* a disposal of less than the whole interest owned in an asset gives rise to an apportionment of the allowable expenditure. The cost is reduced by the fraction $\dfrac{A}{A + B}$ where A is the consideration for the disposal and B is the market value of the property retained (C.G.T.A. 1979, s.35). The formula is not to be applied to expenditure which can properly be identified with the part sold or retained (C.G.T.A. 1979, s.35(4)). For example, shares in unquoted companies can often be identified by reference to their numbers which gives scope for matching individual sales and purchases to reduce the capital gains arising. Similarly it may be possible to identify the actual cost of a particular field which is subsequently sold. If it is not possible, the Press Release of April 22, 1971 (see para. 15–010) may be of assistance. **15–009**

It frequently happens that a disposal of, *e.g* land, is for a fixed sum plus an amount related to future planning permission. This is a disposal with the right to that future sum being valued as part of the sale consideration. When in due course this sum is finally ascertained there is a further gain or loss of the excess over or shortfall from the value placed upon the right to the future sum at the date of the original disposal (*Marren* v. *Ingles* [1980] S.T.C. 500, *Marson* v. *Marriage* [1980] S.T.C. 177). The alternative of granting an option is considered later in this chapter.

The grant of a lease out of a freehold interest is a part disposal. However, the sale of a proportion of the total freehold land is a disposal of a part. Where there is a disposal of a part the taxpayer has the option of applying the part disposal formula $\dfrac{A}{A + B}$ already mentioned or of apportioning the cost to the part sold on some other reasonable basis. The Revenue published a Press Release on this point on April 22, 1971, now SP/D1, which provides as follows: **15–010**

"1. To save work for taxpayers and their advisers where part of an estate is disposed of (*e.g.* on the sale of a field) the Board of Inland Revenue will accept that the cost of the part can be calculated on the alternative basis set out in this note instead of under the general rule which requires the unsold part to be valued in order to apportion the total cost of the estate. Instructions about alternative basis are being issued to Inspectors of Taxes who will be glad to give information about its application to particular cases.

2. Under the alternative basis the part disposed of will be treated as a separate asset and any fair and reasonable method of apportioning part of the total cost to it will be accepted—*e.g.* a reasonable valuation of that part at the acquisition date. Where the market value at April 6, 1965, is to be taken as the cost, a reasonable valuation of the part at that date will similarly be accepted.

3. The cost of the part disposed of will be deducted from the total cost of the estate (or balance of total cost) to determine the cost of the remainder of the estate; thus the total of the separate amounts adopted for the parts will not exceed the total cost. The cost attributed to each part must also be realistic in itself.

4. The taxpayer can always require that the general rule should be applied (except in cases already settled on the alternative basis). If he chooses the general rule it will normally be necessary to apply this rule to all subsequent disposals out of the estate; but where the general rule has been applied for a part disposal before the introduction of the alternative basis and it produced a result broadly the same as under the alternative basis, the alternative basis may be used for subsequent part disposals out of the estate.

5. So long as disposals out of an estate acquired before April 6, 1965, are dealt with on the alternative basis, each part disposal will carry a separate right to elect for acquisition at market value on April 6, 1965. Similarly, where part is sold with development value the mandatory valuation at April 6, 1965 [C.G.T.A. 1979, Sched. 5, para. 9] will apply only to that part. Even where the part is to be treated as acquired as market value on April 6, 1965, however, it will still be necessary to agree how much of the actual cost should be attributed to the part disposed of; first, to ensure that any allowable loss does not exceed the actual loss, and second, to produce a balance of total cost for subsequent disposals.

6. . . . the Board reserve the right to apply the general rule if they are not satisfied that the apportionments claimed are fair and reasonable.

7. Taxpayers who wish to adopt the alternative basis will still be able to claim under existing statutory provisions that certain small [see below] disposals out of an estate should be deducted from cost instead of being assessed. The disposal proceeds will then be deducted from the total cost (or balance of total cost) available for subsequent disposals."

Some of the foregoing is no longer relevant following the enactment of F.A. 1988 bringing back the base date to March 31, 1982, in cases where the gain is calculated by reference to that date.

EXAMPLE

Robert Finney

	£	£
Sale proceeds of 15 acres of farm land		15,000
Value of remaining farm		200,000
Cost of farm: (indexation ignored)		
Farmhouse and buildings	50,000	
200 acres at £250 per acre	50,000	£100,000

	£
Part disposal:	
Proceeds	15,000
Less: cost × $\dfrac{A}{A + B}$	
£100,000 × $\dfrac{£15,000}{£15,000 + £200,000}$ =	6,977
Chargeable gain	£8,023

	£
Disposal of part:	
Proceeds	15,000
Less: cost – 15 acres at £250 per acre	3,750
(indexation has been ignored)	
Chargeable gain	£11,250

In this particular example the apportionment formula is more beneficial than a calculation based on the disposal of a part provision. If the part disposal apportionment provisions have previously been applied on the sale of part of an estate it is not normally possible to apply the alternative basis to subsequent sales. There is no restriction if the alternative basis has been applied on a previous sale and the apportionment formula is desired on a later sale. Whether or not the apportionment formula should be used will depend on the facts and values applicable to the specific case and alternative computations may be required before a decision is made.

Small part disposals

Where, with regard to a sale of land, the proceeds are less than £20,000 **15–011** for that and similar land disposals in the tax year and are small as compared with the market value of the total holding, the proceeds may be deducted from the cost of the asset (C.G.T.A. 1979, s.107). Less than 20 per cent. is regarded as small. This means that tax is not immediately payable on the chargeable gain. A relatively small disposal might be worth splitting into two separate sales at different times in order to come within these limits. A *de minimis* provision without monetary limit is also available for capital distributions on shares, such as on the sale of rights or fractions (C.G.T.A. 1979, ss.72–73).

EXAMPLE

Roy Saunders was a farmer who agreed to sell a small plot of land. In order to postpone the tax on the profit the disposal was made in two sales on March 28, 1989 and April 10, 1989. Indexation is ignored for the sake of simplicity, as is any Revenue argument that the sales can be merged by applying *Furniss* v. *Dawson* principles.

	£
Cost of 600 acres of land on 1.6.1982	660,000
Sale proceeds ½ acre 28.3.1989	16,000
Less: cost	550
Gain 1988–89	£ 15,450
Sale proceeds ½ acre 10.4.1989	16,000
Less: cost	550
Gain 1989–90	£ 15,450

Roy Saunders can claim to have no gain in either year, but to deduct the proceeds from cost in computing future gains, as follows:

Cost carried forward:	
As above re 1982 acquisition	660,000
Less: proceeds	32,000
Net cost carried forward	£628,000

In practice it is not necessary actually to calculate the gain where it is wished to deduct the proceeds from the cost.

Options and forfeited deposits

15–012   An option is treated as a separate asset for C.G.T. purposes, except where it is actually exercised, and the whole proceeds are chargeable as a gain (C.G.T.A. 1979, s.137), *Strange* v. *Openshaw* [1983] S.T.C. 416. A gain realised on the surrender of an option is chargeable to C.G.T.; see *Golding* v. *Kaufman* [1985] S.T.C. 152 and *Powlson* v. *Welbeck Securities Ltd.* [1986] S.T.C. 423.

Where it is exercised, the grantor adds the consideration for a put (sell) option to the sale proceeds and deducts the consideration for a call (buy) option from the acquisition cost. The cost to the grantee of an option which is exercised is added to the cost.

If an option in relation to quoted securities is exercised, the full cost of the option is added to the cost of the shares acquired under the normal pooling arrangements, but the option itself is a wasting asset and, if assigned, the cost has to be depreciated accordingly. If an option is not exercised, the grantee suffers a non-allowable loss, unless the option was a "quoted option" quoted on a recognised stock exchange to subscribe for shares in a company (not to buy shares already in existence) or a "traded option" quoted on a recognised stock exchange or futures exchange, or "financial option" or an option to acquire business assets, in which case a loss may be claimed under C.G.T.A. 1979, s.138.

So far as the grantor is concerned, if the option is abandoned, he is deemed to have made a disposal of the option and is chargeable on the proceeds, there being no disposal of the underlying asset. A forfeited deposit is treated as an option and the recipient is chargeable thereon, although the payer suffers a non-allowable loss unless the asset were to be used for trading purposes.

"Financial option" is one relating to currency, shares, securities or an interest rate and is granted by an authorised person or listed institution as defined by the Financial Services Act 1986, as well as an option on shares and securities granted by a member of a recognised stock exchange acting as agent in respect of quoted shares. Financial options are also excluded from being wasting assets within C.G.T.A. 1979, s.137, so that a loss arises on the expiry of an option without it being exercised.

EXAMPLE

A agreed to purchase land from B under a contract for sale and paid £1,000 as a deposit. The contract fell through and his deposit was forfeited. The transaction is treated as one of an option binding the grantor (the seller) to sell. A loses his £1,000 and cannot claim any loss relief, because there is no disposal by him. B is to be treated as disposing of an "asset" (the option) and the £1,000 is a capital receipt, so he is chargeable on the sum of £1,000, less allowable expenses, if any.

Assets held at April 6, 1965

Time apportionment

Although a new base date of March 31, 1982, was introduced by F.A. **15–013** 1988, s.96, it may still in some circumstances be beneficial to calculate the capital gain in accordance with the old rules as set out in the following paragraphs if they produce a smaller gain than by reference to the new 1982 base date; they would also apply if they produced a smaller loss. The chargeable gain in respect of an asset held at April 6, 1965, is subject to time apportionment (C.G.T.A. 1979, Sched. 5, para. 11) and only the proportion since April 6, 1965, is chargeable. Assets acquired before April 6, 1945, are deemed to have been acquired on that date (C.G.T.A. 1979, Sched. 5, para. 11(6)).

A new lease is a new asset and cannot be related back to a previous lease of the same premises (*Bayley* v. *Rogers* [1980] S.T.C. 544), nor can it be treated as if it were part of the goodwill when it has been sold as a separate asset (*Butler* v. *Evans* [1980] S.T.C. 613).

Time apportionment is applied separately to the original cost and to any allowable improvements (C.G.T.A. 1979, s.32(1)(*b*) and C.G.T.A. 1979, Sched. 5, para. 11) and the gain is apportioned to each item of allowable expenditure in accordance with the individual costs (C.G.T.A. 1979, Sched. 5, para. 11(4)(*a*)). Improvements consist of expenditure actually incurred on the asset (*Oram* v. *Johnson* [1980] S.T.C. 222).

Under the general law, compensation on the compulsory acquisition of premises would be deemed to be a disposal of the land. There are, however, provisions to enable the compensation to be apportioned and treated partly as a disposal of the land, goodwill, disturbance, etc. (C.G.T.A. 1979, s.110, *Stoke-on-Trent City Council* v. *Wood Mitchell & Co. Ltd.* (1979) S.T.C. 197, SP8/79 June 18, 1979). The Revenue is prepared to treat the disturbance element as relating to the sale of the land if so claimed.

Statutory compensation under the Agricultural Holdings Act 1948 on a disposal of farm land following a notice to quit should be tax free (*Davis* v. *Powell* [1977] S.T.C. 32), as should statutory compensation under the Landlord and Tenant Act 1954 on the termination of a tenancy (*Drummond* v. *Austin Brown* [1984] S.T.C. 321). It is important to ensure that a notice to quit is actually issued rather than the lessee merely surrendering his lease for a lump sum taking into account his right to statutory compensation, as this would be a normal chargeable disposal. The Revenue's view is that in most cases compensation is taxable on the recipient and in fixing the compensation no deduction should be made to take account of the tax which might have been charged on the lost profits giving rise to the compensation payment (Inland Revenue Press Release, December 13, 1972).

15–014 Where there is a part disposal of land to an authority exercising or having compulsory powers then if certain conditions are met and a claim is made it is possible for the proceeds to be deducted from the cost of the land rather than to calculate a capital gain (C.G.T.A. 1979, s.108). The conditions are:

(i) The consideration received is small as compared with the market value of the land immediately before the transfer.

(ii) The person making the disposal must not have taken steps by advertising or otherwise to dispose of the property.

(iii) The said person must not have made his willingness to dispose of the land known to the authority or others.

If the initial expenditure on the cost of the assets is disproportionately small having regard to the value of the asset immediately before the later expenditure, any part of the gain not attributable to the enhancement of the value of the asset due to the later expenditure is deemed to be attributable to the initial expenditure (C.G.T.A. 1979, Sched. 5, para. 11(5)).

Land reflecting development value

15–015 Time apportionment does not apply to an interest in U.K. land sold at a price reflecting development value, *i.e.* in excess of the current use value. The asset is deemed to have been sold and reacquired at the market value (including any hope value) on April 6, 1965. The cost is substituted if it would produce a smaller gain or loss (C.G.T.A. 1979, Sched. 5, para. 9 (*Mashiter* v. *Pearmain* [1985] S.T.C. 165)).

Allowance for betterment levy

Relief is available to exclude from a subsequent chargeable gain the **15–016** development value on which the betterment levy has already been paid on a notional disposal, provided that relief has not already been allowed as a trading expense under Schedule D, Case I: C.G.T.A. 1979, Sched. 5, para. 10.

Election for market value

It is possible to make an election within two years of the end of the year **15–017** of assessment in which the disposal took place, or such further time as the Commissioners of Inland Revenue may by notice in writing allow to have the gain computed on the basis of the excess of the disposal proceeds over the value of the asset at April 6, 1965 (C.G.T.A. 1979, Sched. 5, para. 12). An extension of time was refused in *Whitaker* v. *Cameron* [1982] S.T.C. 665. A lease and goodwill were regarded as separate assets in *Butler* v. *Evans* [1980] S.T.C. 613. If the land reflects development value the use of the April 6, 1965, value is mandatory and no election is required. The election for the April 6, 1965, value is irrevocable and should not be made without a careful appraisal of the value likely to be agreed with the Revenue.

In the case of land the value is dealt with by the District Valuer with appeal to the Lands Tribunal. The value of shares is dealt with by the Shares Valuation Division of the Inland Revenue with appeal to the Special Commissioners. Other assets are considered by the inspector with appeal to the Commissioners.

In practice, the Revenue will by concession sometimes allow an election **15–018** for the April 6, 1965, value to be withdrawn if there has been a fundamental misunderstanding of the basis of computation of the April 6, 1965, value. For example, a property might be sold with vacant possession which was tenanted at April 6, 1965. The April 1965 value would be based on the property as an investment with the tenant in possession and not the value of the property at that time with vacant possession (*Henderson* v. *Karmel's Exors.* [1984] S.T.C. 572).

The Board of Inland Revenue does not want to treat an irrevocable April 6, 1965, election as being irrevocable if the taxpayer makes this election under such a fundamental misconception. Form CG 21 was devised which explains the method of valuation and if the election is made on this form the taxpayer would be estopped from claiming fundamental misconception as to the basis of valuation. If, however, the election is made before the form CG 21 is issued, the fundamental misconception problem could still arise in which case the Board may be prepared to accept the withdrawal of the election.

Unquoted company reorganisations

With regard to unquoted shares in a farming company, if there has been **15–019** a reorganisation of the share capital between the date of acquisition and April 6, 1965, the use of April 6, 1965, value is compulsory (C.G.T.A.

1979, Sched. 5, para. 14(1)). This is not the case where there has been a bonus or rights issue of the same class of shares.

On such a reorganisation after April 6, 1965, the unquoted shares would be deemed to have been sold and reacquired at market value at the date of the reorganisation and the gain or loss so computed would be frozen until the new securities are sold. Time apportionment therefore ceases (C.G.T.A. 1979, Sched. 5, para. 14(2)).

On a take-over there is always deemed to be a reorganisation where shares are exchanged for shares in another company (C.G.T.A. 1979, Sched. 5, para. 14(3), *I.R.C.* v. *Beveridge* [1979] S.T.C. 592). The take-over is not regarded as a reorganisation along the lines of a bonus or rights issue of the same shares. This represents a change of practice by the Revenue (SP14/79, December 21, 1979).

By concession the Revenue will not charge a gain in excess of the actual profit realised (Extra-statutory Concession D10 (1988)).

15–020 Where there was a reorganisation before April 6, 1965, it is possible to obtain relief for a loss in excess of the actual loss suffered if the shares had a value at April 6, 1965, greater than cost and the shares were subsequently sold for less. The allowable loss in this case is calculated by reference to the value on April 6, 1965, and not the loss based on original cost as would normally have been the case if there had been no reorganisation and an election for April 6, 1965, value had been made (C.G.T.A. 1979, Sched. 5, para. 12(2)).

A greater C.G.T. loss than the commercial loss can also arise where there is a reorganisation after April 6, 1965, and the shares were originally acquired before April 6, 1965.

Investments

15–021 Most gilt-edged securities (C.G.T.A. 1979, ss.64 and 67) and qualifying corporate bonds (F.A. 1984, s.64 and Sched. 13) do not give rise to a chargeable gain.

Bed and breakfast

15–022 It is popular for investors holding securities, other than gilt-edged stocks, to sell them and reacquire the securities the next day in order to crystallise an allowable capital loss.

Indexation

15–023 In relation to disposals after April 5, 1985, the indexation allowance applies to expenditure on acquiring and enhancing the asset from the time that the expenditure is incurred, or March 1982 if later.

Calculation of indexation allowance

15–024 The allowance is calculated by increasing each of the separate items of base cost and improvement expenditure (C.G.T.A. 1979, s.32(1)(*a*) and (*b*)) by a decimal factor (to the nearest third decimal place) determined by a formula (F.A. 1982, s.87). For disposals after April 5, 1985, where the asset was acquired before April 1, 1982, the indexation allowance is calculated by reference to the market value at March 31, 1982, where that is greater than the original purchase price plus any subsequent expenditure incurred before that date on enhancing its value (F.A. 1985, s.68(4)).

The formula is $\dfrac{RD - RI}{RI}$

(a) Where RD is the retail prices index for the month in which the disposal occurs,

(b) RI is the retail prices index for the later of March 1982 or the date that the item of expenditure was incurred.

If RD is equal to or less than RI the factor is nil.

For the purpose of calculating the indexation allowance all costs relating to the acquisition of the asset are deemed to have been incurred at the date of acquisition. In addition enhancement expenditure is deemed to have been incurred when it becomes due and payable.

Part disposals

In the case of a part disposal where the base cost and enhancement **15–025** expenditure on the asset is apportioned between that part of the asset disposed of and that part retained, the indexation allowance applies to the expenditure after it has been so apportioned, and only in respect of the apportionment relating to the part disposed of (F.A. 1982, Sched. 13 para. 1).

EXAMPLE

As asset cost £10,000. Three years later a part disposal occurs. Part of the asset is sold for £30,000 the remainder being worth £20,000. The base cost attributable to the part disposal is:

$$£10,000 \times \frac{£30,000}{£30,000 + £20,000} = £6,000$$

The indexation allowance is calculated by reference to the £6,000. The balance of £4,000 is not indexed until some future disposal.

Disposals on a no-gain no-loss basis

The C.G.T. rules enable some transfers of chargeable assets to be **15–026** effected without crystallising the gain or loss that would otherwise accrue on the transfer. The gain or loss is rolled over or held over until a subsequent transfer by the recipient of the asset who takes over the original base cost of the asset (F.A. 1982, Sched. 13, para. 2). Such cases include gifts to charities (C.G.T.A. 1979, s.146), replacement of business assets (F.A. 1982, Sched. 13, para. 2), intra-group transfers (I.C.T.A. 1970, s.273), certain transfers to employee trusts (C.G.T.A. 1979, s.149) and so on.

In these cases, where the indexation allowance is available the transferee is not to take over the original base cost but is treated as acquiring the asset for a consideration equal to the base cost (including enhancement expenditure) plus the amount of the indexation allowance. The vendor is deemed thereby to make a gross gain equal to the amount of the indexation allowance. When this gain is then reduced by the allowance there is neither a gain nor a loss.

Subsequent disposals following no-gain/no-loss disposals

15–027 The transferee under a no-gain/no-loss disposal acquires the asset for a consideration equal to the base cost plus the indexation allowance.

In a case where the initial no-gain/no-loss transfer was one between spouses living together (C.G.T.A. 1979, s.44) the transferee is entitled to have the indexation allowance calculated from the time he acquires the asset.

Receipts, etc., which are not treated as disposals but affect relevant allowable expenditure

15–028 In certain cases a transaction which would otherwise constitute a part disposal for C.G.T. is treated instead as not being a C.G.T. disposal but as reducing the allowable expenditure on the asset, thus postponing the gain (or loss). This can apply where the consideration received is small. Examples are small part disposal of land (C.G.T.A. 1979, s.107), compensation and insurance money (C.G.T.A. 1979, s.21) and capital distributions in respect of shares (C.G.T.A., ss.72–73).

In such cases the indexation allowance will not be relevant on the initial part disposal, but it will apply on a subsequent actual disposal. On that later disposal, first the indexation allowance is calculated in the normal way ignoring any such reduction in allowable expenditure. Then a notional indexation rise for the amount of the reduction is calculated from the date that the reduction in question had effect up to the date of the actual disposal. The indexation allowance initially calculated is then reduced by the amount of the notional indexation rise. This then provides the amount of the indexation allowance available on the disposal (F.A. 1982, Sched. 13, para. 4).

Reorganisation, reconstructions, etc.

15–029 Where on a reorganisation, etc., shares are converted to a new holding and the new shares are treated as having been acquired when the old shares were acquired (C.G.T.A. 1979, s.78), then if new consideration is given or becomes due, that consideration commences to be indexed from the date it is incurred. For this purpose it is not treated as incurred at the same time as the original shares were acquired (F.A. 1982, Sched. 13, para. 5).

Calls on shares, etc.

15–030 Where some part of the consideration for the purchase of shares is paid more than 12 months after the acquisition the indexation allowance will be calculated from the date of payment (F.A. 1982, Sched. 13, para. 6). If the consideration is paid within 12 months of the acquisition then it is deemed to have been paid at the time of acquisition.

Options—indexation

In computing the indexation allowance, the option price and the sale **15–031** price are separately indexed from the date each was incurred despite the provisions which treat the acquisition and exercise of an option and underlying transactions as a single transaction (F.A. 1982, Sched. 13, para. 7 and C.G.T.A. 1979, s.137).

Appropriation to and from trading stock

If a capital asset is appropriated to trading stock it is deemed to have **15–032** been disposed of at market value which may well give rise to a chargeable gain or an allowable loss (C.G.T.A. 1979, s.122(1)). This will not necessarily be the position if the transfer is purely for the purpose of trying to obtain a taxation advantage (*Coates* v. *Arndale Properties Ltd.* [1984] S.T.C. 637, *Reed* v. *Nova Securities Ltd.* [1985] S.T.C. 124).

Conversely, if an asset is transferred from trading stock to become a capital investment it is deemed to have been acquired for the purpose of calculating any subsequent chargeable gain or loss at the figure brought into the accounts of the trade on the appropriation (C.G.T.A. 1979, s.122(2)). This would normally be market value (*Sharkey* v. *Wernher* (1955) 36 T.C. 275).

On the transfer of an asset to trading stock it is possible to elect for the transfer to be at market value reduced by the chargeable gain or increased by the allowable loss (C.G.T.A. 1979, s.122(3)). This means that no capital gain or loss would arise on the transfers, but would be merged with the trading profit or trading loss on the asset being brought into stock in the course of the trade.

It could be useful to convert the capital gain into an income profit if there are trading losses unrelieved, or there is a desire for a basic rate taxpayer to generate extra net relevant earnings for retirement annuity relief purposes. The capital gains exempt annual amount should not be wasted.

EXAMPLE

S. Sterling

Mr. Sterling appropriated land held as an investment to trading stock on July 1, 1983 and subsequently sold it, but not as part of a tax scheme.

	£	£	£
Sale proceeds – 1/11/88			200,000
Less: value at 1/7/83		180,000	
Add: notional loss:			
Cost 1/4/70	230,000		
Less: value at 1/7/83	180,000		
Notional loss		50,000	
			230,000
Trading loss			£30,000

If the election had not been made the position would have been as follows:

	£
Sale proceeds – 1/11/88	200,000
Less: value 1/7/83	180,000
Trading profit	£20,000
Value at 1/7/83	180,000
Less: cost 1/4/70	230,000
Capital loss	£50,000

Roll-over relief

15–033   An important capital gains relief is available in respect of the replacement of business assets (C.G.T.A. 1979, ss.115–121). If a taxpayer makes a chargeable gain on the disposal of a qualifying asset actually used (*Temperley* v. *Visibell* (1973) 49 T.C. 129) and, in the case of land and buildings, occupied (C.G.T.A. 1979, s.118 Head A, *Anderton* v. *Lamb* [1981] S.T.C. 43) only for the purpose of the trade throughout the period of ownership and a new qualifying asset is acquired, he may claim for the gain to be rolled over into the new asset. The old asset is treated as being sold for a price giving rise to neither a chargeable gain nor an allowable loss, and the acquisition value of the new asset is reduced by the amount of the chargeable gain.

The claim must be made within six years of the end of the tax year in which the disposal occurs.

EXAMPLE

A farm was sold and a new one purchased.

	£
Sale proceeds 1/7/89	600,000
Less: cost 1/6/82 (including indexation)	200,000
Gain set against reinvestment	£400,000
Cost of new farm	750,000
Less: gain on old farm	400,000
Deemed cost of new farm	£350,000

Note: On a future sale of the new farm the indexation allowance will be given only on £350,000 and not £750,000.

If the proceeds are not fully reinvested in qualifying assets the gain immediately chargeable is limited to the amount by which the funds reinvested fall short of the proceeds of disposal, *i.e.* there is no relief unless at least the base cost is reinvested.

The acquisition of the new asset must be contracted for during a period beginning one year before and ending three years after the disposal of the old asset, although an extension of this time limit may be given by the Revenue at its discretion (C.G.T.A. 1979, s.115(3), SP/D6). A conditional contract is acceptable if it becomes unconditional within the three-year period (C.G.T.A. 1979, s.27).

The Revenue wrote to the authors on August 23, 1983:

> "Each case has to be judged on its own merits and the new asset has to be a replacement for the old asset and the taxpayer has to show that he was prevented by some reasonable cause from complying with statutory time limits . . .
>
> I can advise you that the Board's policy is broadly to allow extensions of the normal time limits where the new asset was acquired not more than three years before or six years after the disposal of the old asset and there are acceptable reasons for such an extension. Where the acquisition preceded the disposal, these reasons might include difficulty in disposing of the old asset."

The assets must be qualifying assets, that is land and buildings occupied **15–034** for the purpose of the trade (except for a trade of dealing in or developing land, or providing services for the occupier of land). See Chapter 12 on woodlands. The other qualifying assets are fixed plant and machinery, ships, aircraft, goodwill hovercraft and spacecraft. Fixed plant includes such items as machinery fixed to the floor or structure such as milking equipment or silos as opposed to mobile or portable plant such as combine harvesters (*Williams* v. *Evans* [1982] S.T.C. 498). It is not necessary for the replacement asset to be in the same class as the asset sold, nor for it to be brought immediately into trading use (Extra-statutory Concession D24 (1988)).

Enhancement expenditure on improving an existing asset is allowed by concession, (Extra-statutory Concession D22 (1988)).

An apportionment is necessary where the old or new asset was used only partly for the purpose of the trade or has been so used for only part of the period of ownership, *Todd* v. *Mudd* [1987] S.T.C. 141, C.G.T.A. 1979, s.115(5). There are provisions allowing roll-over between two separate trades whether of a similar type or not (C.G.T.A. 1979, s.115(7)). Roll-over relief is also available to an individual who disposes of an asset used by his family company as if the asset were used in a trade carried on by him (C.G.T.A. 1979, s.120).

A farmhouse or farm cottage used by a sole trader or partner cannot qualify for roll-over relief as it is not occupied and only for the purposes of the trade, C.G.T.A. 1979, ss.115(1), 118 Class A(1), *Anderton* v. *Lamb* [1981] S.T.C. 43, although relief is available if the farmhouse is occupied by a manager, even if at a later date the proprietor occupies the property as his main residence.

The Revenue has confirmed that where two individuals exchange (not sell and purchase) qualifying assets, roll-over relief will be available on the disposal of the first asset and the acquisition of the second. An exchange of joint interests qualifies for a form of roll-over relief under Extra-statutory Concession D26 (1988) as does a partition of partnership land followed by a partnership dissolution under Extra-statutory Concession D23 (1988).

15–035   One difficult area is establishing whether or not roll-over relief can be claimed where a person with an interest in land acquires the freehold reversion. The view of the Revenue is that whilst each situation would depend upon a careful appraisal of its own facts, the following general points could be made in order to help establish the position:

(i) A tenant farmer, for example, has an interest in land which will be a qualifying asset if it is occupied, as well as used only for the purpose of his trade. The qualifying asset is "land" in which he has interest. If he buys the freehold reversion of his land, he has not strictly acquired a "new asset" but rather enlarged his existing interest in the same asset.

(ii) Roll-over relief is available on the basis that the proceeds of disposal of, or of an interest in, "old assets" have been reinvested in, or in an interest in, "new assets." That the assets, rather than the interest, must be distinct is clear from the words of the legislation "other assets."

The Board of Inland Revenue illustrates its view by citing the example of a farmer who buys 200 acres of which he was previously the tenant. If he sells 40 acres with vacant possession to help finance the acquisition, he cannot on a strict interpretation of the law claim roll-over in respect of the gain on the part disposal against the cost either of the freehold reversion of the 200 acres or of the 160 acres retained. Similar considerations apply if he gives up the tenancy of part of his land in exchange for the unencumbered freehold of the remainder.

Relief may, however, be available under Extra-statutory Concession D25 (1988) which relates to the acquisition of a further interest in an asset already in use for the purposes of the trade.

Milk and potato quotas

15–036   Roll-over relief is available in respect of milk and potato quotas with effect from October 30, 1987 (F.A. 1988, s.112). It also applies if the quota was disposed of before then but the replacement asset was acquired within the three-year time-limit after October 29, 1987.

If, exceptionally, payments had been made for dairy farming goodwill which became valueless on the disposal of the milk quota, a loss would arise which could be set against the gain on the quota. Quota arose by statute and there is no cost deductible unless it had actually been purchased, *Davenport* v. *Chilver* [1983] S.T.C. 426, C.G.T.A. 1979, s.29A(2) and therefore the entire proceeds will normally be taxed as a capital gain, unless roll-over relief is available. A sale of quota is not a part disposal of land and there is no deduction allowed for the diminution in value of land in respect of which the quota has been sold.

The actual mechanics of ensuring a valid transfer of quota on a change of occupation of the land as a result of, say, an 11-month grazing agreement, is not as simple as is sometimes thought and competent legal advice is essential if the transfer is to withstand a challenge from any interested party, particularly where the quota is sold in tranches. As milk quota came into existence under the Dairy Produce Quota Regulations 1984 (S.I. 1984 No. 1530) now superseded by the Dairy Produce Quota Regulations 1986 (S.I. 1986 No. 470) no indexation is available in respect of a March 1982 value, only on quota purchased.

On a disposal of the entire interest in land including the quota it is, in the Revenue's view, strictly necessary to apportion the consideration under C.G.T.A. 1979, s.43(4) between the land and the quota, although the apportionment is not in practice insisted upon if the tax charge is unaffected. It remains to be seen whether the courts would support an argument that no apportionment of any part of the proceeds to the quota is necessary as the existence of quota merely enhances the value of the land sold, in the same way as development hope value. The Australian case of *Baron-Hay* v. *Commissioner of Probate Duties* (1968) W.A.R. 81 suggests that the quota is still a separate asset from the land.

Apportionment is necessary where the land was acquired before April 6, 1965, or the quota was acquired other than by allocation or the introduction of the scheme. On a part disposal of land and quota, allocation is necessary and the small part disposal rules are applied by reference to the land values alone.

A sale of quota will be a disposal of a chargeable business asset for retirement relief under F.A. 1985, Sched. 20, para. 12 so retirement relief is available, if the remaining requirements are met.

A payment to an outgoing tenant for his share of the quota would be an acquisition of the payer and a disposal subject to C.G.T. of the tenant.

It is arguable that milk quota is a wasting asset, under C.G.T.A. 1979, s.37, as the quota scheme was originally scheduled to last five years, and therefore any quota purchased should be depreciated. However, this is not a point which the Revenue appears to take at present.

Gains deferred before March 31, 1982

The C.G.T. legislation contains several types of roll-over and hold-over **15–037** reliefs whereby the gains arising on disposal of assets are deferred until a subsequent disposal or event, or sometimes indefinitely. The rebasing of capital gains from April 6, 1965, to March 31, 1982 (F.A. 1988, s.96), would have caused inequitable results where the roll-over or hold-over had occurred before March 31, 1982, because the gains rolled over would have been subjected to tax on the sale of the replacement asset after April 5, 1988, thereby nullifying the exemption of gains arising before April 6, 1988.

Relief is therefore given where, after April 5, 1988, the taxpayer disposes of an asset which he had acquired after March 31, 1982, if, in calculating the capital gain, a deduction is made from the expenditure in respect of a rolled-over or held-over gain which accrued before April 6, 1988, and that deduction was directly or indirectly attributable to the disposal of an asset acquired before March 31, 1982. In such circumstances the relief is one-half of the rolled-over or held-over gain. The relief applies also where the deferred gain is attributable wholly or partly to a disposal by a person who did not hold the asset on March 31, 1982, but who acquired it by a no-gain/no-loss disposal from somebody who did hold it on March 31, 1982.

Compulsory acquisition

15–038   A capital gain can also be rolled over where land is sold to an authority exercising or having compulsory powers where:

(i) the landowner did not take steps, by advertising or otherwise, to dispose of the land in question;

(ii) he did not make his willingness to sell the land known to the authority or others.

(iii) the proceeds of sale are used by the landowner in acquiring other land—not being "excluded land" (C.G.T.A. 1979, s.111A).

For this purpose "excluded land" means a private residence which is exempt from C.G.T. either at the date of acquisition or within six years from that date (C.G.T.A. 1979, s.111B).

Basically, the normal roll-over rules apply for this relief. However, there is no need for the land to have been used for business purposes.

Hold-over relief

15–039   If the proceeds of sale of an asset qualifying for roll-over relief are reinvested in a wasting asset or one which will become a wasting asset within ten years of purchase (C.G.T.A. 1979, s.37) the gain is merely held over for up to ten years (C.G.T.A. 1979, s.117). If during the ten years the asset is sold or ceases to be used for the purpose of the business then the gain is assessable at that time. It is, however, possible to roll the gain through a wasting asset into a non-wasting asset if such is acquired at any time during the ten-year hold-over period so long as the gain has not already fallen to be assessed (C.G.T.A. 1979, s.117(3)). It is not necessary in such cases to dispose of the wasting asset into which the gain had been rolled.

EXAMPLE

John Anatra – farmer

	£	£
Sale of field 31/12/88		100,000
Cost 31/3/82 plus indexation		60,000
Gain		£40,000
Purchase of fixed computer 28/2/89	£90,000	
Chargeable gain limited to proceeds not reinvested (£100,000 less £90,000)		£10,000
Gain held over		£30,000

Purchase of goodwill of additional business		
1/5/95		125,000
Less gain held over		30,000
Deemed acquisition cost of goodwill for		
capital gains purposes		£95,000

Note the fact that although the purchase of the computer in 1989 qualifies for capital allowances it does not prevent the chargeable gain on the disposal of the premises being held over against the proceeds reinvested in the computer. For the purpose of the example it is assumed that the computer continues in use until after May 1, 1995.

Relief for gifts of business assets

15–040   A disposal of a business asset by an individual, by gift or otherwise than at arm's length, qualifies for C.G.T. hold-over relief (C.G.T.A. 1979, s.126). A business asset is one which is, or is an interest in, an asset used for the purpose of a trade, carried on by the transferor or by a company which is his family company. The relief will also apply to shares or securities of a trading company which is a family company (C.G.T.A. 1979, s.126(1A). Note that unlike for retirement relief there is no requirement for the asset to have been used for 10 years ending with the disposal. Hold-over relief is not given if the gain is already exempt owing to retirement relief (C.G.T.A. 1979, s.126(2)).

The hold-over relief is the amount of the chargeable gain which would have accrued to the transferor on the disposal and this held-over gain is deducted from the deemed cost of acquisition of the transferee.

The held-over gain is also reduced in certain cases such as on a disposal of shares in a company, the assets of which include investments, and the resultant gain is known as the unrelieved gain on disposal (C.G.T.A. 1979, s.126(5) and Sched. 4, Part II). Where the actual consideration, although less than the market value, exceeds the cost and any improvement expenditure, the held-over gain is the excess of the unrelieved gain over the excess of the actual consideration compared with the cost, *i.e.* the excess of the unrelieved gain over the actual gain.

Family company is defined as for retirement relief (C.G.T.A. 1979, s.126(7)(*a*), F.A. 1985, Sched. 20, para. 1), *Davenport* v. *Hasslacher* (1977) 51 T.C. 497, *Hepworth* v. *William Smith Group* [1981] S.T.C. 354) which is a company where at least 25 per cent. of the voting rights are held by the transferor or not less than 5 per cent. of the voting rights are held by the transferor, where more than 50 per cent. of the voting rights are held by the transferor or his spouse and brother, sister, ancestor or lineal descendant of the transferor or his spouse.

15–041   A similar relief is available for gifts of business assets by U.K. resident trustees (C.G.T.A. 1979, Sched. 4, para. 2).

On a joint election of the transferor and the transferee the transferor's chargeable gain is reduced to nil and the transferee's base value is reduced by the amount of the transferor's gain. If there is a disposal at under value the smaller of the gain element or the chargeable gain may be held over, subject to the prior utilisation of retirement relief if appropriate.

The cost, as reduced, is deducted from the proceeds in order to calculate the chargeable gain.

EXAMPLE

Mr. Farrow sold at an undervalue two parcels of agricultural land farmed by his family company.

	Plot A £	Plot B £
Market value on disposal, July 1989	80,000	80,000
Less cost, 1983	21,000	12,000
Unrelieved gain	£59,000	£68,000
Cash received	50,000	50,000
Less: cost	21,000	12,000
Chargeable gain assessed	£29,000	£38,000
Held-over gain	£30,000	£30,000
Transferee's allowable cost:		
Market value	80,000	80,000
Less held-over gain	30,000	30,000
	£50,000	£50,000

15–042   For non-business assets the availability of relief has been severely restricted for disposals on or after March 14, 1989. Even for business assets, relief is not available for gifts to non-residents or to foreign-controlled companies, and a deemed disposal arises if trustees who control a company become non-resident while the company holds rolled-over assets: C.G.T.A. 1979, ss.126A, 126B, 126C (F.A. 1989, s.124, Sched. 14).

Agricultural property

15–043   C.G.T. hold-over relief for business assets applies to a gift of agricultural property even though it was not used for a trade carried on by the transferor, so long as inheritance tax relief would have been available on the disposal or would have been available had the disposal been a chargeable transfer (C.G.T.A. 1979, Sched. 4, para. 1). This would cover, for example, where the land is occupied by the transferor's family (C.G.T.A. 1979, Sched. 4, para. 4).

Settled property

Where trustees are deemed to have disposed of and immediately **15–044** reacquired settled property on a beneficiary becoming absolutely entitled to the trust property or on the termination of an interest in possession in the trust property, the trustees may claim C.G.T. hold-over relief or business assets provided that the asset was used for the purpose of the trade, profession or vocation carried on by the trustees or by a relevant beneficiary, or the deemed disposal was of shares or securities in a trading company where the trustees at that date controlled at least 25 per cent. of the voting rights (C.G.T.A. 1979, Sched. 4, paras. 2–3).

The trustees are deemed to be the transferor and transferee. The provisions dealing with an excess of the unrelieved gain on the disposal over the actual gain which apply where there is actual consideration does not apply as they are inappropriate to a deemed disposal, even though there would be an actual transfer of assets to the beneficiary on his becoming absolutely entitled as against the trustees.

A relevant beneficiary is the one becoming absolutely entitled as against the trustees, or a beneficiary the termination of whose life interest gave rise to the charge.

Reductions peculiar to disposal of assets

Where an asset is disposed of and relief is claimed as a gift of business **15–045** assets which was used for only part of a period of ownership for the purposes of a trade, the held-over gain is proportionately reduced on a time basis calculated according to the number of days of non-business use compared with the number of days of ownership (C.G.T.A. 1979, Sched. 4, para. 5).

If a substantial part of a building or structure was not used for the purposes of a trade, a just and reasonable proportion of the unrelieved gain is arrived at which relates to the non-business usage and this proportion of the total is applied to reduce the held-over gain. It is possible to have a reduction both for non-business usage during part of the period of ownership and for partial business usage during the period when it would qualify as a business asset (C.G.T.A. 1979, Sched. 4, para. 6).

Reductions peculiar to disposal of shares

Where shares are disposed of and the company owns non-business **15–046** chargeable assets the held-over gain is reduced by multiplying it by the value of the chargeable business assets at the date of disposal divided by the total value of the chargeable assets at that date (C.G.T.A. 1979, Sched. 4, para. 7). This is a similar fraction to that which applies for retirement relief.

Reduction where gain partly relieved by retirement relief

If retirement relief is available on the disposal of a chargeable business **15–047** asset and the held-over gain exceeds the chargeable gain after retirement relief, the held-over gain is reduced by the excess so that it cannot exceed the amount of the gain which would be chargeable after retirement relief. Similar provisions apply where retirement relief is given in respect of a disposal of shares in a family company (C.G.T.A. 1979, Sched. 4, para. 8).

Wasting assets

15–048   Wasting assets are those with a predictable life not exceeding 50 years (C.G.T.A. 1979, s.37) other than freehold land and buildings. Plant and machinery is deemed always to have a predictable life of less than 50 years (C.G.T.A. 1979, s.37(1)(*d*)).

Those assets which are tangible movable property are generally exempt from the capital gains provisions unless capital allowances were due on the assets in question (C.G.T.A. 1979, s.127(2)) in which case a chargeable gain could arise on the excess of the sale price over the indexed cost, although this is unlikely to arise often in practice.

The cost, less residual scrap value, of a wasting asset other than a lease is written off equally over its anticipated life as is any enhancement expenditure (C.G.T.A. 1979, s.38). The cost is reduced in accordance with the formula:

$$C \times \frac{T}{L} \text{ where:}$$

T = The period from the date of acquisition to disposal.
L = The predicted life of the asset.
C = Cost.

15–049   However, wasting assets which qualify for capital allowances are not depreciated which makes it somewhat difficult to find an asset to which the time-apportionment depreciation formula would apply (C.G.T.A. 1979, s.39). If an asset qualifying for capital allowances is sold for more than cost, then the cost would be deemed to be the disposal proceeds for capital allowances purposes which would be deducted from the capital allowances pool, possibly giving rise to a balancing charge. The excess over the indexed cost would be the chargeable gain. The exemption for tangible movable wasting assets does not apply where capital allowances were available (C.G.T.A. 1979, s.127(2)).

EXAMPLE

Gwen Gobles

Gwen Gobles bought a combine harvester in 1988 and sold it at a profit in 1989.

		£
Sale proceeds		26,000
Less: cost	23,200	
	800	
	———	24,000
Chargeable gain		£2,000

226

The credit on sale to the capital allowances pool would be limited to £23,200, *i.e.* the historical cost that was taken into the capital allowances computation.

Chattels

Chattels sold for £6,000 or less (£3,000 before April 6, 1989) are exempt **15–050** from tax on any chargeable gain (C.G.T.A. 1979, s.128). There is marginal relief which fixes the maximum chargeable gain as being five-thirds of the excess of the proceeds over £6,000 (C.G.T.A. 1979, s.128(2)).

EXAMPLE

Mr. Powers sold the following tangible movable assets on June 30, 1989. Indexation ignored.

		A £	B £	C £
Sale proceeds		8,000	10,000	12,900
Less: Cost		6,000	4,000	20
Gain	(A)	£2,000	£6,000	£12,880
Excess of proceeds over £6,000		2,000	4,000	6,900
5/3rds excess (maximum gain)	(B)	£3,333	£6,667	£11,500
Chargeable gain – lower of (A) or (B)		£2,000	£6,000	£11,500

Capital allowances

Expenditure is not disallowed merely because it attracts capital allow- **15–051** ances. If plant were sold at a profit, C.G.T. could be due, although in most cases the sale would be for less than cost and the proceeds would be brought into the capital allowances pool and would therefore be excluded for C.G.T. as being otherwise brought within the tax net.

There are provisions which prevent a loss being claimed where capital allowances have been granted (C.G.T.A. 1979, s.34). In such circumstances, there is usually merely a balancing allowance or adjustment to the capital allowances pool. Any expenditure is reduced by the amount of capital allowances granted. If the written-down value has been taken over from a connected person, allowances granted to him are also included. If it were not for these provisions, it would be possible to claim a capital loss in respect of depreciation which has already been allowed by means of capital allowances. Renewals are included and capital allowances are reduced by any balancing charges.

Leases

15–052   The most usual wasting asset encountered in practice is a lease of less than 50 years. This is subject to curved-line and not straight-line depreciation under which the rate of depreciation increases as the period of the lease reduces (C.G.T.A. 1979, Sched. 3, para. 1). If a premium is paid on the assignment of a lease or capital expenditure is incurred on improving the leasehold property the cost is reduced in accordance with the following formula:

$$\frac{P(1) - P(3)}{P(1)} \text{ where:}$$

P(1) is the depreciation percentage taken from the leasehold depreciation table, at the date of acquisition;
P(3) is the percentage at the time of disposal.

If the premium for a lease is paid to the landlord he would be chargeable on a proportion under Schedule A, I.C.T.A. 1988, s.34, and the purchaser would claim relief for the chargeable proportion spread over the term of the lease either against rents received, I.C.T.A. 1988, s.37, or against trading profits, I.C.T.A. 1988, s.87. In such circumstances the landlord's disposal for capital gains purposes is restricted to the part of the proceeds not assessable under Schedule A (C.G.T.A. 1979, Sched. 3, para. 5).

Where a lease is granted out of a freehold or long leasehold interest, *i.e.* a lease in excess of 50 years, the normal part-disposal rules apply using the $\frac{A}{A + B}$ formula (C.G.T.A. 1979, Sched. 3, para. 2 and C.G.T.A. 1979, s.35).

The granting of a lease out of a short lease requires the curved line depreciation table to be applied to the cost to arrive at the allowable proportion (C.G.T.A. 1979, Sched. 3, para. 4).

EXAMPLE
O. Thetford

Mr. Thetford granted a sub lease for 10 years out of a 40-year lease.

	£	£
Premium on grant of sub-lease 1/4/85 (30 years left to run)		20,000
Less: assessed under Schedule A		16,400
		3,600
Capital portion (10–1) × 2% × £20,000		
Less: cost (premium on assignment) 1/4/75	10,000	
Less: depreciation		
$\frac{95.457 - 87.330}{95.457} \times £10,000 =$	851	
		£9,149

228

Value of reversion	£30,000	
$£9,149 \times \dfrac{A}{A + B}$		
$£9,149 \times \dfrac{£20,000}{£20,000 + £30,000}$		3,660
Allowable loss		£ 60

Envelope system

Where a person is acquiring a wasting asset it is often sensible to apply **15–053** the envelope system whereby the wasting asset is acquired by a company specially formed for the purpose. It will be appreciated that if at any time it is desired to dispose of the wasting asset the same effect can be achieved by selling the shares in the company. Any resultant loss would be a loss on the shares of the company and not a loss on the wasting asset itself. The loss on the sale of the shares is then allowed in full instead of being restricted under the wasting asset provisions already outlined. If the lease in fact appreciates in value to the date of sale the appreciation is not increased by the allowable cost being deemed to waste away. The company used should have some purposes for its existence other than the seeking of a tax advantage.

EXAMPLE

Mr. Stuka

A premium of £50,000 was paid on assignment of a 10-year lease which was placed in a wholly owned company Stuka Ltd. At the end of 10 years the lease was not renewed and Stuka Ltd. was liquidated.

	£	£
Liquidation proceeds		Nil
Less: cost of shares in Stuka Ltd.		50,000
Allowable loss on shares		£50,000
Proceeds on lease termination		Nil
Less: cost	50,000	
Deduct depreciation:		
$£50,000 \times \dfrac{46.695 - 0}{46.695} =$	50,000	
		Nil
Allowable loss to company		£Nil

229

Mr. Stuka will need to re-charge the outgoings on the property. Also, if Stuka Ltd. is not a trading company it must charge a commercial rent (I.C.T.A. 1988, s.770).

Venture capital relief

15–054 In certain circumstances, where an individual subscribes for shares in an unquoted "trading company" and incurs a loss in relation to that investment he can claim to have that loss allowed against his income for income tax purposes rather than under the normal C.G.T. provisions: I.C.T.A. 1988, ss.574–576. This is advantageous if the individual has insufficient gains (above the annual exemption) against which to set the loss.

Conditions—generally

15–055 The individual must have subscribed for "ordinary shares," I.C.T.A. 1988, s.574(1), for a consideration of money or money's worth in a "qualifying trading company." Ordinary shares are all the issued shares (by whatever name called) of the company, other than shares the holders whereof have a right to a dividend at a fixed rate but have no other right to share in the profits of the company: I.C.T.A. 1988 ss.576(5) and 832(1) and I.C.T.A. 1970, s.526(5). This means that participating preference shares can potentially qualify for relief. For the purpose of the relief an individual is treated as having subscribed for the shares if his spouse, who is living with him, did so and then transferred the shares to him *inter vivos*: I.C.T.A. 1988, ss.574(3) and 576(5).

The venture capital loss is computed as for C.G.T. purposes but with certain modifications. If the loss is to be allowable then it is provided that the loss must be sustained on a disposal: I.C.T.A. 1988, s.575:

    (i) by way of a bargain made at arm's length for full consideration;
    (ii) by way of a distribution in a liquidation;
    (iii) as a result of making a claim and establishing that the shares are of nil or negligible value (C.G.T.A. 1979, s.22(2)).

Anti-avoidance

15–056 If the transferor and transferee on the disposal of a capital asset are connected the transaction is deemed to take place at market value (C.G.T.A. 1979, ss.29A and 62(2)). There is also a deemed consideration equal to the market value of the asset where it is acquired or disposed of otherwise than by way of a bargain made at arm's length (C.G.T.A. 1979, s.29A), in particular by way of gift on a transfer into settlement (*Berry* v. *Warnett* [1982] S.T.C. 396), or by way of distribution from a company in respect of shares in the company. The market value is also substituted where the disposal or acquisition is for a consideration that cannot be valued. However, market value shall not be taken as the acquisition cost where there is no corresponding disposal and the consideration given is less than market value. This prevents the use of the market value rules as a tax avoidance mechanism (overruling *Harrison* v. *Nairn Williamson* (1978) 51 T.C. 135, used in reverse).

Any loss arising to the transferor on a disposal to a connected person can only be deductible from gains on future transactions between the same connected parties (C.G.T.A. 1979, s.62(3)). This is the position even where the transfer took place at full market value. This is a most important point where farm land values have fallen since acquisition and the owner desires to sell but wishes to keep the land in the family.

Certain rights or restrictions over asset are to be ignored for calculating the market value for transfers between connected persons (C.G.T.A. 1979, s.62(5) and (6)).

The persons who together control a company are connected with each other in connection with the company. Trustees are connected with the settlor and also with persons connected with the settlor and a body corporate connected with the settlor (C.G.T.A. 1979, s.63).

Artificial schemes

The features of an artificial tax avoidance scheme as identified by the House of Lords (*W.T. Ramsay Ltd.* v. *I.R.C.*, *Eilbeck* v. *Rawling* [1981] S.T.C. 174 at p. 179) include, but cannot be limited to, the situation where, the gain or profit has already been made by the taxpayer who then engages consultants who produce a preconceived and ready-made plan designed to produce an equivalent allowable loss. There are usually a number of steps in the scheme supported by documents and often involving payments of real money in accordance with a rapid timetable. The transactions are carried out in close proximity to one another, the effect of which results in avoidance, reduction or deferral of tax. The proximity does not necessarily need to be a few days; months or longer may be sufficient. There is a clear intention to complete all steps in the scheme, whether this is contractual or not and the real money could well be provided by loans repaid at the end of the scheme.   **15–057**

The sole or main motive is to avoid or defer tax. Normally, but certainly not necessarily (*Furniss* v. *Dawson* [1984] S.T.C. 153), the transactions are of a circular nature and at the end of the scheme leave the taxpayer in the same position as at the commencement, minus costs representing fees. Whether the transactions are of a circular or straight line nature they can be equally caught by the principle set out by the House of Lords.

Where the transactions are of a straight-line nature, rather than circular, it is necessary to look at the first and last transactions, and if the acts of persons in between are in any way under control or influence of the person avoiding or deferring tax then it is that person who is deemed to have carried out the last transaction.

In the case of *I.R.C.* v. *Burmah Oil Co. Ltd.* [1982] S.T.C. 30, which was a bespoke tax scheme rather than a ready-made plan, Lord Diplock confirmed that it would be: **15–058**

> "disingenuous to suggest, and dangerous on the part of those who advise on elaborate tax avoidance schemes to assume that Ramsay's case did not mark a significant change in the approach adopted by the House of Lords, in its judicial role to a pre-ordained series of transactions (whether or not they included the achievement of a legitimate commercial end) into which are inserted steps that have no commercial purpose apart from the avoidance of a liability to tax which in the absence of these particular steps would have been payable."

It is important to remember that the Courts have not yet necessarily set out an extensive list of the badges of an artificial tax scheme and therefore variations may well be caught. Also it is not clear how many characteristics such a scheme needs to have before it is ruled ineffective. The result of such an artificial scheme which has no real commercial justification apart from tax mitigation is a fiscal nullity.

> "In each case two assets appear, like particles in a gas chamber with opposite charges, one of which is used to create the loss, the other of which gives rise to an equivalent gain which prevents the taxpayer from supporting any real loss, and which gain is intended not to be taxable. Like particles, these assets have a very short life. Having served their purpose they cancel each other out and disappear. At the end of the series of operations the taxpayer's financial position is precisely as it was in the beginning, except that he has paid a fee, and certain expenses, to the promoter of the scheme." (Lord Wilberforce) (*W.T. Ramsay Ltd.* v. *I.R.C.*, *Eilbeck* v. *Rawling* [1981] S.T.C. 174 at p. 179).

Principles of legitimate avoidance

15–059 The taxpayer is only to be taxed on clear words, not the apparent intention of the legislation (*W.T. Ramsay Ltd.* v. *I.R.C.*, *Eilbeck* v. *Rawling* [1981] S.T.C. 174 at p. 179, *I.R.C.* v. *Wesleyan and General Assurance Soc.* (1946) 30 T.C. 11, *Mangin* v. *I.R.C.* [1911] 1 All E.R. 179). The subject is entitled to arrange his affairs to reduce tax. A sham is something different from what it professes to be, which is a question of fact. Given that a document is genuine, the court cannot go behind it to some supposed underlying substance (*I.R.C.* v. *Duke of Westminster* (1936) 19 T.C. 490).

It is the task of the court to ascertain the legal nature of any transaction to which it is sought to attach a tax consequence and if that emerges from a series or combination of transactions intended to operate as such, it is that series or combination which may be regarded (*Chinn* v. *Collins* [1981] S.T.C. 1, *I.R.C.* v. *Plummer* [1979] S.T.C. 793, *Black Nominees Ltd.* v. *Nicol* (1975) 50 T.C. 229, *Floor* v. *Davis* (1979) 52 T.C. 609).

There may be a commercial reality in such transactions, *e.g.* "a covenant for a capital sum to make annual payments coupled with security arrangements for the payments" (*I.R.C.* v. *Plummer* [1979] S.T.C. 793). On the other hand there may be no commercial and therefore no fiscal effect (*Knetsch* v. *United States* (1960) 364 U.S. 361, *Gilbert* v. *C.I.R.* (1957) 248 2d 399, *W.T. Ramsay Ltd.* v. *I.R.C.* and *Eilbeck* v. *Rawling* [1981] S.T.C. 174, *Rubin* v. *United States* (1962) 304 F2d 766, *MacRae* v. *C.I.R.* (1961) 34 T.C. 20).

The House of Lords has confirmed that a strategic rearrangement of a taxpayer's affairs for a tax advantage at some future time is permissible, *Baylis* v. *Gregory*, *Bowater Property Developments Ltd.* v. *I.R.C.*, *Craven* v. *White* [1988] S.T.C. 476.

Sale and lease back

15–060 In the case of a sale and lease back transaction the rent must be limited to a commercial rent: I.C.T.A. 1988, s.799. If the sale is in respect of a short lease having less than 50 years to run and the lease back is for a term not exceeding 15 years, part of the capital proceeds are deemed to be income: I.C.T.A. 1988, s.780.

The part treated as income is:

$\dfrac{16 \text{ years} - \text{n}}{15 \text{ years}}$ where n is the term of the new lease back.

If the sale and lease back is of trading premises the assessable amount is a trading profit; otherwise it falls to be assessed under Schedule D, Case VI.

There are anti-avoidance provisions to prevent a sale by one party and a lease back by an associated party: I.C.T.A. 1988, s.779(11). In addition there are provisions preventing the term of the lease being treated as artificially longer than is really the case.

Bad debts and loans

Bad debts in respect of trading transactions are allowable as a trading expense for tax purposes: I.C.T.A. 1988, s.74(*j*). If the item is not a trading debt, *e.g.* a loan to a company which has become worthless, the loss is a capital loss. However, unless it is a debt on a security (C.G.T.A. 1979, s.82(3)(*b*), *W.T. Ramsay Ltd.* v. *I.R.C.* [1981] S.T.C. 174) it is a non-allowable capital loss (C.G.T.A. 1979, s.134) unless specifically relieved by statute as a loan to a trader (C.G.T.A. 1979, s.136). Conversely, a profit on a non-trading debt which is not a debt on a security does not give rise to a chargeable gain. **15–061**

Relief in respect of loans to traders

If money has been lent after April 11, 1978, to a U.K. resident borrower for the purpose of his trade, other than a debt on a security (C.G.T.A. 1979, s.82(3)(*b*), *W.T. Ramsay Ltd.* v. *I.R.C.* [1981] S.T.C. 174) and the loan becomes irrecoverable, the lender may claim an allowable loss for capital gains purposes (C.G.T.A. 1979, s.136). The relief is not available if the lender and the borrower are spouses or if the lender has assigned his right to recover the loan. **15–062**

If a guarantor of a loan has to make a payment under a guarantee given after April 11, 1978, he is entitled to an allowable loss as if he were the original lender.

Where the loan is recovered in whole or part after loss relief has been given it is treated as giving rise to a chargeable gain equivalent to the amount recovered.

A loan by a money lender or person able to charge the loss as a trading expense does not also give rise to a capital loss. Should the loan become irrecoverable under the terms of the loan or similar arrangements or as a result of any act or omission by the lender or guarantor no loss relief is available.

Foreign exchange losses

Foreign currency (C.G.T.A. 1979, s.19(1)(*b*) and foreign bank accounts (C.G.T.A. 1979, s.135(1)) are chargeable assets for the purpose of C.G.T. and any profit or loss on disposal would give rise to a chargeable gain or **15–063** allowable capital loss in the normal way. However, foreign currency and foreign bank accounts are not chargeable assets where the sum concerned relates to money for the personal expenditure outside the U.K. of the taxpayer or his family or dependants (including expenditure on the provision or maintenance of any residence outside the U.K.) (C.G.T.A. 1979, ss.133 and 135(2)).

Apart from these, and the specific provisions relating to the 1967 devaluation (C.G.T.A. 1979, Sched. 6, para. 17), an exchange loss on repayment of a foreign currency loan arising from the depreciation in value of sterling is not a loss allowable as a capital loss if the loan is for capital purposes such as the purchase of a farm overseas. The corresponding profit would similarly not be subject to tax if the loan were for capital purposes. This has to be taken into account when considering perhaps borrowing from overseas in foreign currency at what appear to be attractive rates of interest compared with borrowing costs in the U.K.

Where the purpose of the loan has been to finance stock, the profit on exchange has been held to be a trading receipt in most cases (*Landes Bros.* v. *Simpson* (1934) 19 T.C. 62 and *Imperial Tobacco Co. (of Great Britain and Ireland) Limited* v. *Kelly* (1943) 25 T.C. 292). However, a temporary deposit pending the purchase of stock was held to give rise to a capital profit (*McKinlay* v. *Jenkins (H.T.) & Son Limited* (1926) 10 T.C. 372). Profits on non-trading loans have been held to be capital (C.G.T.A. 1979, s.77(2)(a); *I.R.C.* v. *Burmah Oil Co. Ltd.* [1982] S.T.C. 30) and it is permissible to apportion a loan between capital and revenue (*Firestone Tyre & Rubber Co. Ltd.* v. *Evans* (1977) 51 T.C. 615). The trading proportion of the loss relating to such a loan is allowable as a trading expense (*Radio Pictures Ltd.* v. *I.R.C.* (1938) 22 T.C. 106).

The Revenue has unsuccessfully argued that it was possible to borrow long-term and have an exchange loss on the loan disallowed as capital, and to use the funds for trading purposes giving rise to a taxable exchange profit. The commercial realities of matched loans and borrowings were ultimately upheld with the result that the Revenue lost the argument (*Pattison* v. *Marine Midland Bank Ltd.* [1984] S.T.C. 10): see SP 1/87.

Double tax relief

15–064   Relief from a double charge to C.G.T. on the disposal of an overseas asset by a U.K. resident may be given under the appropriate bilateral treaty with the foreign country. If no such relief is available, credit for the overseas tax against the U.K. C.G.T. liability would be given unilaterally (C.G.T.A. 1979, s.10). If there is no U.K. tax liability the foreign tax may be treated as an expense (C.G.T.A. 1979, s.11).

Private residence

15–065   The disposal of an interest in an individual's main residence, which includes a caravan (C.G.T.A. 1979, Sched. 5, para. 14(2); *Makins* v. *Elson* (1977) 51 T.C. 437) and including grounds of up to one acre, or such larger area as the Commissioners may determine as being required for the reasonable enjoyment of the residence, is exempt from C.G.T. (C.G.T.A. 1979, Sched. 5, para. 14(1); C.G.T.A. 1979, s.101(1) to (4)). This is provided that it has been occupied as such throughout the period of ownership except for:

(i) the final two years (*Green* v. *I.R.C.* [1982] S.T.C. 485; C.G.T.A. 1979, s.102(1); or

(ii) while living in job-related accommodation either as self-employed or as an employee provided that the individual intends to live in the property (C.G.T.A. 1979, s.101(8) as his main residence in due course.

A lodger who takes meals with the family, etc., does not cause the relief to be lost SP/D 15. The house must not be sold before the land (*Varty* v. *Lynes* [1976] S.T.C. 508) unless there is no development value (C.C.A.B. Press Release June 1976). A caretaker's lodge can be exempt, in addition to the main house (*Batey* v. *Wakefield* [1981] S.T.C. 521), but was not held to be so in *Markey* v. *Sanders* [1987] S.T.C. 256 and *Williams* v. *Merrylees* [1987] S.T.C. 445.

Vinelott J. reconciled these cases in *Williams* v. *Merrylees* [1987] S.T.C. 445 at 454. This case emphasises the importance of the Commissioners' hearing and the reluctance of the courts to disturb a finding of fact.

The area of land in excess of one acre which is required for the reasonable enjoyment of the house as a residence, regard being had to the size and character of the dwelling house, C.G.T.A. 1979, s.101, will depend on the circumstances in each case. If part of the garden is sold for development it is difficult to argue that it was required for the reasonable enjoyment of the house.

If an individual has two residences he can, within two years of acqui- **15–066** sition of the second residence, nominate either as his main residence. Late elections will be accepted in certain cases: Extra-statutory Concession D21 (1988). If he does not so elect then the inspector may, subject to appeal, determine the matter (C.G.T.A. 1979, s.101(5); C.C.A.B. Press Release March 31, 1980).

A husband and wife living together may have only one main residence, but any transfer of an interest in the property from one to the other is backdated to the acquisition of the other spouse (C.G.T.A. 1979, s.101(6) and (7)). An interest in a matrimonial home transferred to a spouse as a result of a breakdown of a marriage may still be regarded as the main residence for the purposes of this section, provided it continues to be occupied by the transferor's spouse, although this concession is lost if, in the meantime, the transferor acquires a further house occupied as his main residence (Extra-statutory Concession D6 (1988)).

Periods of ownership before April 6, 1965, are ignored (C.G.T.A. 1979, s.102(4)) for disposals before April 6, 1988, and, unless the calculation is based on the alternative "old" rules, for disposals after April 5, 1988, periods of ownership before March 31, 1982, are ignored.

If there is a non-exempt chargeable gain on the disposal of a residence which has been partly let, the gain is chargeable only to the extent that it exceeds the lower of the exemption on the unlet proportion or £20,000 (F.A. 1980, s.80).

It is possible to exempt a proportion of the gain, (C.G.T.A. 1979, **15–067** s.101(9)) being the period of ownership and use as the main residence, together with the last two years (*Green* v. *I.R.C.* [1982] S.T.C. 485; C.G.T.A. 1979, s.102(1)) as a fraction of the total period of ownership (C.G.T.A. 1979, Sched. 5, para. 14(2), *Makins* v. *Elson* (1977) 51 T.C. 437).

For this purpose periods of absence not exceeding three years, any period where the owner was employed wholly outside the U.K., and periods not exceeding four years during which his living in the residence was prevented by the owner's employment, will be treated as periods of use as the main residence (C.G.T.A. 1979, s.102(2)). This applies even if let, provided that it is the main residence before and after—but subject to having to move elsewhere for employment (Extra-statutory Concessions D3 and D4 (1988)). Only any excess period would be chargeable (C.C.A.B. Press Release March 10, 1983, para. 13). This extension would not apply where another house was being claimed as the main residence.

Farmhouse

15–068   Where part of the residence has been used *exclusively* for a trade or business such as a farmhouse, a proportionate part of the exemption is lost (C.G.T.A. 1979, s.103(1)). For example, part of a farmhouse used exclusively for business purposes would not qualify for private residence relief, although it would qualify for roll-over relief as an asset used for the purposes of the trade of farming. The mere use of the room for business purposes for which an income tax claim for expenses has successfully been made does not necessarily mean that the exemption is lost if there is occasional use for other purposes. The Commissioners are given discretion to arrive at any just and reasonable adjustments where there have been changes to the property occupied as the main residence as the result of, for example, business occupation of a part (C.G.T.A. 1979, s.103(2)).

15–069   The C.G.T. exemption for owner-occupied houses does not apply where the house was acquired, or expenditure was incurred on it, with a view to realising a gain on its disposal, (C.G.T.A. 1979, s.103(3)) but it cannot be taxed as an artificial transaction in land: I.C.T.A. 1988, s.776(9). See Chapter 17.

Beneficiaries and dependants

15–070   The C.G.T. exemption is extended to a main residence owned by a trust occupied by a beneficiary (C.G.T.A. 1979, s.104, *Sansom and Another* v. *Peay* [1976] S.T.C. 494) or legatee (Extra-statutory Concession D5 (1988)).

C.G.T. exemption formerly applied to the sole residence of a dependent relative provided rent free (C.G.T.A. 1979, s.105) Extra-statutory Concession D20 (1985) but this exemption was abolished from April 6, 1988: F.A. 1988, s.111. However, as a transitional measure, the relief remains for residences occupied by a dependent relative at April 5, 1988, as long as that occupation continues. A husband and wife were only allowed one dependent relative's residence between them. Dependent relative is defined as any relative of the taxpayer or spouse incapacitated by old age or infirmity from maintaining himself or herself, or a widowed, separated or divorced mother or mother-in-law whether incapacitated or not. It should be noted that the income of the dependent relative is irrelevant.

Death

The personal representatives and legatees are deemed to take over the **15–071** assets of the deceased at the probate value at the date of death before any I.H.T. deduction, *e.g.* for agricultural property (C.G.T.A. 1979, s.49(1), *Bentley* v. *Pike* [1981] S.T.C. 360). However, the notional disposal of these assets at the date of death does not give rise to a C.G.T. charge. Sales by the personal representatives will give rise to chargeable gains or allowable losses based on the probate value and the sale proceeds (C.G.T.A. 1979, s.49(3)). Any latent gains in the assets of the deceased at the date of death therefore escape C.G.T. Losses sustained by the deceased in the year of death may be carried back to the preceding three tax years (C.G.T.A. 1979, s.49(2)). Unused personal representatives' losses cannot be transferred to the beneficiaries, unlike under a settlement (C.G.T.A. 1979, s.54(2)).

The personal representatives are treated as a single body of persons with the domicile, residence and ordinary residence of the deceased. A legatee is deemed to take over an asset from the personal representatives at the probate value at the date of death irrespective of when actually transferred (C.G.T.A. 1979, s.49(4); C.C.A.B. Press Release June 1967).

A deed of family arrangement or other variation or disclaimer made within two years of death is retroactive to the date of death, but otherwise takes effect on the date of the deed (C.G.T.A. 1979, s.49(6), *Stephenson* v. *Barclays Bank Trust Co. Ltd.* (1975) 50 T.C. 374). Written notice of election must be given to the Revenue within six months of the date of the instrument or such longer time as the Board may allow (C.G.T.A. 1979, s.49(7)). Deeds of variation or disclaimer made for consideration in money or money's worth are ignored (C.G.T.A. 1979, s.49(8)).

A *donatio mortis causa*, a gift in contemplation of death, is not a C.G.T. disposal (C.G.T.A. 1979, s.49(5)). The donee of a *donatio mortis causa* takes over the value at the date of death as if he were a legatee. Under an entailed estate, in Scotland, on the date of the liferenter, the property is deemed to pass at market value to the next heir (C.G.T.A. 1979, s.50).

Husband and wife

A disposal between a husband and wife does not give rise to a chargeable **15–072** gain or loss as the acquisition value of the transferor is deemed to be that of the transferee (C.G.T.A. 1979, s.44(1)). This does not apply to appropriations to or from trading stock (C.G.T.A. 1979, s.44(2)) or by way of *donatio mortis causa*, which is already exempt (C.G.T.A. 1979, s.49(5)). The transferor's acquisition date is that of the transferee (C.G.T.A. 1979, Sched. 5, para. 17).

Transfers in the tax year of separation are included, but not thereafter (*Aspden* v. *Hildesley* [1982] S.T.C. 206). The freedom from the C.G.T. charge extends to a transfer by a resident to a non-resident spouse

(C.G.T.A. 1979, s.155(2), I.C.T.A. 1970, s.42, *Gubay* v. *Kington* [1984] S.T.C. 99).

Tax on married woman's gains

15–073   For 1990–91 husbands and wives are treated as independent taxpayers, so each will be eligible for the annual exemption and each will be responsible for declaring the gains and paying the tax.

Up to 1989–90 chargeable gains of a married woman living with her husband are, except in the year of marriage, assessed on the husband (C.G.T.A. 1979, s.45(1)). An election may be made by either the husband or the wife, before July 6 following the end of the year of assessment, to have the gains of the husband and wife separately assessed (C.G.T.A. 1979, s.45(2)). Such an election does not affect the total tax payable, but means the tax of the wife's gains will be collected from her instead of from her husband.

The Revenue may collect unpaid C.G.T. in respect of the wife's gains from her and the husband may disclaim his deceased's wife's C.G.T., thus making it a charge on her estate for inheritance tax (C.G.T.A. 1979, s.45(4)).

An election can be made to prevent the set-off of one spouse's losses against the other's gains, under C.G.T.A. 1979, s.4(2): this can make better use of the losses.

Exemption for first £5,000 of gains

15–074   For 1988–89 onwards gains in excess of £5,000 are charged to tax at income tax rates. (C.G.T.A. 1979, s.5). Losses brought forward from earlier periods need only be used up sufficiently to reduce the gain to £5,000 on which the liability is nil (C.G.T.A. 1979, s.5(4)).

Unless Parliament specifically determines otherwise, for each tax year the previous December's retail prices index is compared with that of the December before and the percentage increase in the index is applied to the 'exempt amount for the year', with any necessary rounding-up to the nearest £100. The amount is announced in advance each year by statutory instrument (C.G.T.A. 1979, s.5(1C) and 5(5)).

Similarly, the amount relevant for trusts is index-linked. The exempt amount for a single trust for a mentally disabled person is for 1988–89 onwards £5,000, and in any other case £2,500 for a single trust, *i.e.* one-half of the "exempt amount for the year" (C.G.T.A. 1979, Sched. 1, paras. 5 and 6).

Husband and wife

15–075   Until 1989–90 the C.G.T. small gain exemption limit "the exempt amount for the year" is divided between husband and wife living together in proportion to their respective chargeable gains, as calculated ignoring any small gains relief. If the gains in aggregate do not exceed the exempt amount and allowable losses are brought forward from a previous year, the relief can be apportioned between husband and wife as they desire. If the husband and wife are only living together at the beginning of the year, the wife's gains for that part are aggregated with the husband for the small gains relief but for the remainder of the year she will, in her own right, be entitled to small gains relief as if it were a separate year of assessment (C.G.T.A. 1979, Sched. 1, para. 2).

In most cases the husband will return both his own and his wife's chargeable gains and they will be treated as if they were his gains.

From 1990–91 the foregoing rules are not relevant, following independent taxation of husbands and wives.

Personal representatives

Small gains relief is extended to personal representatives in respect of **15–076** chargeable gains made by them in respect of the year in which an individual dies and during the next two years of assessment (C.G.T.A. 1979, Sched. 1, para. 4).

Trustees

Trustees of a trust for the benefit of a mentally or physically disabled **15–077** person qualify for an annual exemption provided that at least half the property is applied for the benefit of the disabled person who is also entitled to at least half the income or, otherwise provided that no such income may be applied for the benefit of any other person. There are anti-avoidance provisions to prevent the fragmentation of settlements (C.G.T.A. 1979, Sched. 1, para. 5).

In respect of trusts originally formed before June 7, 1978, the annual exemption for chargeable gains is an amount equal to one-half of the C.G.T. "exempt amount for the year." For qualifying settlements formed after June 6, 1978, the exemption is divided by the number of related settlements with a minimum figure of one-tenth of the exempt amount. A qualifying settlement is any settlement, other than a non-resident or charitable settlement or superannuation fund. Related settlements are those created by the same settlor and in the case of will trusts this includes the testator. There are provisions apportioning the relief where there are multiple settlors (C.G.T.A. 1979, Sched. 1, para. 6).

Commodity and financial futures and traded options

Non-trading transactions in commodity or financial futures dealt in on a **15–078** recognised futures exchange and in traded options quoted on a recognised stock exchange are taxed as capital gains or losses under F.A. 1985, s.72. Commodity dealings carried on as part of a trade assessed under Schedule D, Case I. Traded options are not wasting assets: C.G.T.A. 1979, s.138.

Farmers dealing on the commodity markets, *e.g.* in potato futures or cereal futures, will normally have the profits or losses regarded as part of the farming trade under Schedule D, Case I.

Miscellaneous provisions

A transfer of title by way of mortgage is not a disposal for C.G.T. pur- **15–079** poses (C.G.T.A. 1979, s.23, *Aspden* v. *Hildesley* (1982) 5 T.C. 206).

Where consideration is payable by instalments tax may be paid over a period of up to eight years provided that hardship can be established (C.G.T.A. 1979, s.40(1)).

Trusts

15–080   Settled property is defined by C.G.T.A. 1979, s.51, as property held by trustees other than as nominees. See *Kidson* v. *Macdonald and another* (1974) S.T.C. 54, *Crowe* v. *Appleby* [1976] S.T.C. 301.

Trustees

15–081   The trustees of settled property are treated as a single and continuing body of persons resident and ordinarily resident in the U.K., unless the general administration of the trust is carried on outside the U.K. and a majority of the trustees are non-resident or are professional trustees (C.G.T.A. 1979, s.52). A settlement on trustees is a disposal of the entire assets transferred.

If the tax due by the trustees is not paid within six months the Revenue may, within two years, recover the tax from any beneficiary up to the value of assets transferred to him. A life interest is defined to include the right to the use of settled property, but to exclude a contingent or discretionary right or an annuity. Single settlements may have more than one set of trustees, for example, if there is settled land: a life interest may be over part only of the settled property. See also *Chinn* v. *Collins* [1981] S.T.C. 1, *Berry* v. *Warnett* [1980] S.T.C. 631 and *Roome and another* v. *Edwards* [1981] S.T.C. 96.

Gifts in settlement

15–082   A gift in, and any transfer (including a sale: F.A. 1981, s.86) into, settlement is a disposal of the entire property settled (C.G.T.A. 1979, s.63, *Berry* v. *Warnett* [1982] S.T.C. 396).

Person becoming absolutely entitled to settled property

15–083   On a beneficiary becoming absolutely entitled as against the trustees, the appropriate trust assets are deemed to have been disposed of at market value by the trustees and reacquired by them as nominees of the beneficiary (C.G.T.A. 1979, s.54, *Crowe* v. *Appleby* [1975] S.T.C. 502, *Pexton* v. *Bell & another* [1976] S.T.C. 301; *Stephenson* v. *Barclays Bank Trust Co. Ltd.* [1975] S.T.C. 151; *Hoare Trustees* v. *Gardner* [1978] S.T.C. 89, *Chinn* v. *Collins* [1981] S.T.C. 1, *Roome and another* v. *Edwards* [1981] S.T.C. 96; as to resettlements, see *Hart* v. *Briscoe* and others [1978] S.T.C. 89 and *Bond* v. *Pickford* [1982] S.T.C. 403.)

On a beneficiary becoming absolutely entitled as against the trustees any unused losses of the trustees may be carried forward and are available to the beneficiary. A partial release by a life tenant is treated as a termination of the interest in that part (SP/D10). A remainderman becoming absolutely entitled on the death of a life tenant during the period of administration takes the assets on at the date of death as if he were a legatee (C.C.A.B. Press Release, June 1967).

Termination of life interest, etc.

No chargeable gain accrues on the termination of a life interest (where the **15–084** settled property remains intact in the trust) whether the termination is on the death of a life tenant or not (C.G.T.A. 1979, s.55).

On a death, the settled property is treated as having been disposed of and immediately reacquired by the trustees at market value but no chargeable gain is deemed to accrue.

There is a deemed disposal of the settled property on the death of the life tenant even if his interest is *pur autre vie* and does not therefore terminate.

A life interest includes an annuity where it is charged on part of the settled funds.

Death of life tenant: exclusion of chargeable gain

On the death of a life tenant there is no chargeable gain on the deemed **15–085** disposal, but if the property then reverts to the settlor he takes over the trustees' acquisition values, and if the trustees had acquired assets before April 6, 1965, the settlor also takes over the acquisition dates (C.G.T.A. 1979, s.56).

After April 5, 1982, if the deceased only had an interest in part of the trust assets in question the exemption only applies to that proportion.

For this purpose, as for C.G.T.A. 1979, s.55 above, a life interest in part of the income of settled property is a life interest in a corresponding part of that property. If no right of recourse to other property in the settlement exists, the life interest is deemed to be a settlement separate from other property.

Death of annuitant

There is also a C.G.T. free disposal and reacquisition on the death of an **15–086** annuitant as if it were the death of a life tenant (C.G.T.A. 1979, s.57).

Disposal of interest in settled property

No chargeable gain arises to an original beneficiary on disposing of an **15–087** interest under a settlement unless that interest was acquired for money or money's worth (C.G.T.A. 1979, s.58). Such an acquirer would be deemed to have disposed of his interest for the market value of any settled property on his becoming absolutely entitled as against the trustees. The exemption to an original beneficiary under this paragraph is extremely important (*Chinn* v. *Collins* [1981] S.T.C. 1, *Berry* v. *Warnett* [1982] S.T.C. 396, *Eilbeck* v. *Rawling* [1981] S.T.C. 174).

Cases where there has not been a sale of an interest in settled property include *Harthan* v. *Mason* [1980] S.T.C. 94 and *Kidson* v. *MacDonald* [1974] S.T.C. 54.

The exemption does not apply on the disposal of an interest in settled property to non-residents (F.A. 1981, s.88).

There are number of anti-avoidance provisions related to C.G.T. including migrant settlements and dual resident trusts which are outside the scope of this book.

CHAPTER 16

Partnerships and Companies—Capital Gains

Partnerships

16–001 The C.G.T. legislation dealing with partnership is, to say the least, somewhat deficient. The legislation in C.G.T.A. 1979, s.60 merely provides that:

> "Where two or more persons carry on a trade or business in partnership
> (a) tax in respect of chargeable gains accruing to them on the disposal of any partnership assets shall, in Scotland as well as elsewhere in the United Kingdom, be assessed and charged on them separately, and
> (b) any partnership dealings shall be treated as dealings by the partners and not by the firm as such, and
> (c) section 112(1), (2) of the Taxes Act 1988 (residence of partnerships) shall apply in relation to tax chargeable in pursuance of this Act as it applies in relation to income tax."

It is also provided by C.G.T.A. 1979, s.63(4) that:

> "Except in relation to acquisitions or disposals of partnership assets pursuant to bona fide commercial arrangements, a person is connected with any person with whom he is in partnership, and with the husband or wife or a relative of any individual with whom he is in partnership."

A relative for this purpose includes brother, sister, ancestor, or lineal descendant, but unlike for inheritance tax purposes, does not include uncles and aunts, nephews and nieces.

The only asset that a partner actually possesses in connection with a partnership is a chose in action but the C.G.T. legislation reconstitutes the assets in the hands of the partners by providing that gains attributable to partners are allocated to them in the proportion in which they share surplus assets. This is usually known as the asset sharing ratio and in many cases, although not invariably, will be the same as the profit sharing ratio for dividing income profits among the partners.

It will be noted that the legislation does not provide that the partners will cease to have their interest in the partnership as a separate chose in action, which is important when considering aspects of partnership goodwill.

16–002 As a result of the inadequate legislation to deal with C.G.T. on partnerships the Revenue published a Statement of Practice on January 17, 1975, (now SP/D12) and although this has no statutory force it nonetheless deals with most partnership C.G.T. situations. The Statement of Practice followed discussions with the Law Society and the Allied Accountancy Bodies and provides as follows (as updated for C.G.T.A. 1979 references and with examples supplied by the authors):

> "1. *Nature of the asset liable to tax* 16–003
> CGTA 1979, s.60 treats any partnership dealings in chargeable assets for capital gains tax purposes as dealings by the individual partners rather than by the firm as such. Each partner has, therefore, to be regarded as owning a fractional share of each of the partnership assets and not for this purpose an interest in the partnership.
>
> Where it is necessary to ascertain the market value of a partner's share in a partnership assets for capital gains tax purposes, it will be taken as a fraction of the value of the total partnership interest in the asset without any discount for the size of his share. If, for example, a partnership owned all the issued shares in a company, the value of the interest in that holding of a partner with a one-tenth share would be one-tenth of the value of the partnership's 100 per cent. holding."

The following examples of partnership disposals ignore indexation relief unless otherwise stated.

EXAMPLE

		£
1982 Simon bought a small farm for		80,000
1985 Simon took Nick into equal partnership when the farm was worth		140,000
1989 Sale of entire farm		120,000

Simon	£	
1985 Sale of 50% of £140,000	70,000	
Less: 50% of £80,000	40,000	
Gain		30,000
1989 Sale 50% of £120,000	60,000	
Less: 50% of £80,000	40,000	
Gain		20,000
Nick		
1989 Sale 50% of £120,000	60,000	
Cost 50% of £140,000	70,000	
Loss		(10,000)

If the farm had not been revalued in 1985 there would have been no chargeable gain at that stage, unless the parties were related, in which case there would normally have been a disposal at market value.

16–004

"2. *Disposals of assets by a partnership*

Where an asset is disposed of by a partnership to an outside party each of the partners will be treated as disposing of his fractional share of the asset. Similarly if a partnership makes a part disposal of an asset each partner will be treated as making a part disposal of his fractional share. In computing gains or losses the proceeds of disposal will be allocated between the partners in the ratio of their shares in asset surpluses at the time of the disposal. Where there is not specifically laid down the allocation will follow the actual destination of the surplus as shown in the partnership accounts; regard will, of course, have to be paid to any agreement outside the accounts. If the surplus is not allocated among the partners but, for example, put to a common reserve, regard will be had to the ordinary profit-sharing ratios in the absence of a specified asset-surplus-sharing ratio. Expenditure on the acquisition of assets by a partnership will be allocated between the partners in the same way at the time of theacquisition. This allocation may require adjustment, however, if there is a subsequent change in the partnership sharing ratio. (see para. 4)."

EXAMPLE

Mikoyan acquired his interest in a farming partnership from a former partner Polikarpov for £70,000. Value in books not adjusted.

	Firm 100% £	Sukhoi 50% £	Tupolev 30% £	Mikoyan 20% £
Cost of freehold property (1982)	200,000	100,000	60,000	
Acquisition from former partner				70,000
1989 Part disposal: Proceeds	160,000	80,000	48,000	32,000
Value retained	180,000	90,000	54,000	36,000
Computation Cost		100,000	60,000	70,000
Attributable portion: $\dfrac{(A)}{(A+B)}$ $\dfrac{\text{proceeds}}{(\text{proceeds + value retained})}$		$\dfrac{8}{17}$	$\dfrac{48}{102}$	$\dfrac{32}{68}$
Attributable cost		47,059	28,235	32,941
Proceeds		80,000	48,000	32,000
Chargeable gain		£32,941	£19,765	£ 941

16–005

"3. *Partnership assets divided in kind among the partners*

Where a partnership distributes an asset in kind to one or more of the partners, for example, on dissolution, a partner who receives the asset will not be regarded as disposing of his fractional share in it. A computation will first be necessary of the gains which would be chargeable on the individual partners if the asset had been disposed of at its current market value. Where this results in a gain being attributed to a partner not receiving the asset the gain will be charged at the time of the distribution of the asset. Where, however, the gain is allocated to a partner receiving the asset concerned there will be no charge on distribution. Instead, his capital gains tax cost to be carried forward will be the market value of the asset at the date of distribution as reduced by the amount of his gain. The same principles will be applied where the computation results in a loss."

EXAMPLE

Sikorsky, Gurevich and Lavochkin shared assets and surplus profits 40:35:25. On July 1, 1989, Sikorsky retired and took over the ownership of a farm cottage and paddock.

	Book value (and cost) £	Market value £
Farm cottage and paddock	80,000	120,000
Remaining farm property	500,000	400,000

Chargeable gains	Gurevich (35%) £	Lavochkin (25%) £
Proceeds (% of value of cottage etc.)	42,000	30,000
Cost (%)	28,000	20,000
Gain	£14,000	£10,000

Allowable loss – Sikorsky	£
disposal of 40% remaining farm property	160,000
Cost (40% of £500,000)	200,000
Allowable loss[i]	£(40,000)

Acquisition cost of cottage etc.–Sikorsky	
Market value	120,000
Less: notional gain (40% × (£120,000 − £80,000))	16,000
Cost carried forward	£104,000

(or 40% at cost £80,000 (£32,000) plus 60% at market value £120,000 (£72,000) = £104,000)

(i) Losses with connected persons are restricted and can only be set against gains with the same persons (C.G.T.A. 1979, s. 63(3)). Partners are connected persons except in relation to disposals of partnership assets pursuant to bona fide commercial considerations. (C.G.T.A. 1979, s. 63(4)). It is considered that Sikorsky's disposal is such as to escape restriction under these provisions.

16–006

"4. Changes in partnership sharing ratios

An occasion of charge also arises when there is a change in partnership sharing ratios, including changes arising from a partner joining or leaving the partnership. In these circumstances a partner who reduces or gives up his share in asset surpluses will be treated as disposing of part or the whole of his share in each of the partnership assets and a partner who increases his share will be treated as making a similar acquisition. Subject to the qualifications mentioned at 6 and 7 below the disposal consideration will be a fraction (equal to the fractional share changing hands) of the current balance sheet value of each chargeable asset provided that there is no direct payment or consideration outside the partnership. Where no adjustment is made through the partnership accounts (for example, by revaluation of the assets coupled with a corresponding increase or decrease in the partner's current or capital account at some date between the partner's acquisition and the reduction in his share) the disposal is treated as made for a consideration equal to his capital gains tax cost and thus there will be neither a chargeable gain nor an allowable loss at that point. A partner whose share reduces will carry forward a smaller proportion of cost to set against a subsequent disposal of the asset and a partner whose share increases will carry forward a larger proportion of cost.

The general rule in C.G.T.A. 1979, s.35, for apportioning the total acquisition cost on a part disposal of an asset will not be applied in the case of a partner reducing his asset-surplus share. Instead, the cost of the part disposed of will be calculated on a fractional basis."

EXAMPLE

BALANCE SHEET

	Capital A/C £	Current A/C £	Surplus asset share Old £	Surplus asset share New £
Antanov	250,000	20,000	50	40
Beriev	150,000	20,000	30	30
Chetverikov	100,000	15,000	20	30
	£500,000	£55,000	100%	100%

	Assets £
Freehold property at cost	650,000
Milk quota at cost	Nil
Non-chargeable assets	40,000
	£690,000

246

Disposal by Antanov

	£
Proceeds (50% − 40% × £650,000)	65,000
Cost (50% − 40% × £650,000)	65,000
Gain	Nil

Acquisition by Chetver-ikov

Cost (30% − 20% × £650,000)	£65,000

"5. *Adjustments through the accounts*   **16–007**
Where a partnership asset is revalued a partner will be credited in his current or capital account with a sum equal to his fractional share of the increase in value. An upward revaluation of chargeable assets is not itself an occasion of charge. If, however, there were to be a subsequent reduction in the partner's asset-surplus share, the effect would be to reduce his potential liability to capital gains tax on the eventual disposal of the assets without an equivalent reduction of the credit he has received in the accounts. Consequently at the time of the reduction in sharing ratio he will be regarded as disposing of the fractional share of the partnership asset represented by the difference between his old and his new share for a consideration equal to that fraction of the increased value at the revaluation. The partner whose share correspondingly increases will have his acquisition cost to be carried forward for the asset increased by the same amount. The same principles will be applied in the case of a downward revaluation.
A revaluation can be avoided by providing by agreement that the profit on an asset such as a freehold property should be divided in the old profit sharing ratio up to the value at the date of change and thereafter any future increase in value would be divided in the new profit sharing ratio."

EXAMPLE

Shavrov and Moskalev share assets and profits 60:40

	£
Freehold property: cost 1982	40,000
revalued 1987	120,000

Profit and asset sharing ratios changed to 50:50 on 1/10/89 when the property was worth £140,000 but was not again revalued.

	£
Shavrov	
1987 – no disposal	
1989 – Proceeds (60% – 50% of £120,000)	12,000
Cost (60% – 50% of £40,000)	4,000
Gain	£8,000
Cost carried forward 50% of £40,000	£20,000

	£
Moskalov	
1982 – Cost 40% of 40,000	16,000
1989 – Cost 10% of 120,000	12,000
Cost carried forward	£28,000

16–008
"6. *Payments outside the accounts*
Where on a change of partnership sharing ratios payments are made directly between two or more partners outside the framework of the partnership accounts, the payments represent consideration for the disposal of the whole or part of a partner's share in partnership assets in addition to any consideration calculated on the bases described in 4 and 5 above. Often such payments will be for goodwill not included in the balance sheet. In such cases the partner receiving the payment will have no capital gains tax cost to set against it unless he made a similar payment for his share in the asset (for example, on entering the partnership) or elects to have the market value at April 6, 1965 treated as his acquisition cost. The partner making the payment will only be allowed to deduct the amount in computing gains or losses on a subsequent disposal of his share in the asset. He will be able to claim a loss when he finally leaves the partnership or when his share is reduced provided that he then receives either no consideration or a lesser consideration for his share of the asset. Where the payment clearly constitutes payment for a share in assets included in the partnership accounts, the partner receiving it will be able to deduct the amount of the partnership acquisition cost represented by the fraction he is disposing of. Special treatment, as outlined in 7 below, may be necessary for transfers between persons not at arm's length."

(The Finance Act 1988 reorganised the C.G.T. legislation in that for sales after April 5, 1988, only the gains arising since March 31, 1982, are taxed.)

It is unusual for a farm to have any goodwill as this is usually reflected in the value of the land.

EXAMPLE

Nieman and Kalinin were in a farming partnership sharing profits 50:50. On January 1, 1989, Putilov was admitted into partnership with a 20 per cent. share for which he paid £12,000 to Nieman.

The profit shares become:

	Old%	New%
Nieman	50	30
Kalinin	50	50
Putilov	–	20
	100	100

Nieman	£
Sale proceeds – share of farm	12,000
Cost nil; March 31, 1982, value	4,000
Gain	£8,000

Putilov	
Cost of share of farm	£12,000

"7. Transfers between persons not at arm's length   **16–009**
Where no payment is made either through or outside the accounts in connection with a change in partnership sharing ratio, a capital gains tax charge will only arise if the transaction is otherwise than by way of a bargain made at arm's length and falls therefore within C.G.T.A. 1979, s.29A(1) as extended by C.G.T.A. 1979, s.62(2) for transactions between connected persons. Under C.G.T.A. 1979, s.63(4) transfers of partnership assets between partners are not regarded as transactions between connected persons if they are pursuant to bona fide commercial arrangements. This treatment will also be given to transactions between an incoming partner and the existing partners.
Where the partners (including incoming partners) are connected other than by partnership (for example, father and son) or are otherwise not at arm's length (for example, uncle and nephew) the transfer of a share in the partnership assets may fall to be treated as having been made at market value. Market value will not be substituted, however, if nothing would have been paid had the parties been at arm's length. Similarly if consideration of less than market value passes between partners connected other than by partnership or otherwise not at arm's length, the transfer will only be regarded as having been made for full market value if the consideration actually paid was not less than that which would have been paid by parties at arm's length. Where a transfer has to be treated as if it has been taken for market value, the deemed disposal proceeds will fall to be treated in the same way as payments outside the accounts."

EXAMPLE

Petlyokov, a farmer, introduced his son Nikitin into partnership, without charge, as a 25 per cent. partner. The transfer could not be justified on commercial grounds.

	Book value (and cost) £	Market value £
Freehold property	200,000	600,000
Milk quota	25,000	40,000
Non-chargeable assets	30,000	30,000
Capital a/c	£255,000	£670,000

Petlyokov

	£	£
Freehold property: proceeds (25% of £600,000)	150,000	
cost (25% of £200,000)	50,000	
		100,000
Milk quota: proceeds (25% of £40,000)	10,000	
cost (25% of £25,000)	6,250	
		3,750
Chargeable gain		£103,750

16–010

"8. *Annuities provided by partnerships*

A lump sum which is paid to a partner on leaving the partnership or on a reduction of his share in the partnership represents consideration for the disposal by the partner concerned of the whole or part of his share in the partnership assets and will be subject to the rules in 6 above. The same treatment will apply when a partnership buys a purchased life annuity for a partner, the measure of the consideration being the actual cost of the annuity.

Where a partnership makes annual payments to a retired partner (whether under covenant or not) the capitalised value of the annuity will only be treated as consideration for the disposal of his share in the partnership assets under C.G.T.A. 1979, s.31(3) if it is more than can be regarded as a reasonable recognition of the past contribution of work and effort by the partner to the partnership. Provided that the former partner had been in the partnership for at least ten years an annuity will be regarded as reasonable for this purpose if it is not more than two-thirds of his average share of the profits in the best three of the last seven years in which he was required to devote substantially the whole of his time to acting as a partner. In arriving at a partner's share of the profits regard will be had to the partnership profits assessed before deduction of any capital allowances or charges. The ten year period will include any period during which the partner was a member of another firm whose business has been merged with that of the present firm. For lesser periods the following fractions will be used instead of the two-thirds:

Complete years in partnership	Fraction
1–5	1/60th for each year
6	8/60ths
7	16/60ths
8	24/60ths
9	32/60ths

Where the capitalised value of an annuity is treated as consideration received by the retired partner, it will also be regarded as allowable expenditure by the remaining partners on the acquisition of their fractional shares in partnership assets from him."

This provision was extended by SP1/79 published on January 12, 1979.

"Paragraph 8 of the Statement of Practice issued by the Board of Inland Revenue on January 17, 1975 explains the circumstances in which the capitalised value of an annuity paid by a partnership to a retired partner will not be treated as consideration for the disposal of his share in the partnership assets. The Board have now agreed that this practice will be extended to certain cases in which a lump sum is paid in addition to an annuity. Where the aggregate of the annuity and one-ninth of the lump sum does not exceed the appropriate fraction (as indicated in the Statement) of the retired partner's average share of the profits, the capitalised value of the annuity will not be treated as consideration in the hands of the retired partner. The lump sum, however, will continue to be so treated.

This extension of the practice will be applied to all cases in which the liability has not been finally determined at the date of this Notice."

"9. *Mergers* **16–011**
When the members of two or more existing partnerships come together to form a new one, the capital gains tax treatment will follow the same lines as that for changes in partnership sharing ratios. If gains arise for reasons similar to those covered in 5 and 6 above, it may be possible for roll-over relief under C.G.T.A. 1979, ss.115–121, to be claimed by any partner continuing in the partnership in so far as he disposes of part of his share in the assets of the old firm and acquires a share in other assets put into the 'merged' firm. Where, however, in such cases the consideration given for the shares in chargeable assets acquired is less than the consideration for those disposed of, relief will be restricted under C.G.T.A. 1979, s.116(1)."

EXAMPLE

Ilyushin and Kamov (50:50 partners) merged with Yakovlev and Mil (60:40 partners) on March 1, 1988. Profits are shared equally after the merger.

	I & K £	Y & M £
Milk quota		
Cost	Nil	£10,000
Market value	£25,000	£30,000

Cost in balance sheet of merged practice £55,000

Computation	Ilyushin £	Kamov £	Yakovlev £	Mil £
Disposal				
25% × £25,000 (each)	6,250	6,250		
35% × £30,000			10,500	
15% × £30,000				4,500
Cost				
25% × Nil (each)	–	–		
35% × £10,000			3,500	
15% × £10,000				1,500
Gain	£6,250	£6,250	£7,000	£3,000
Acquisition				
25% × £30,000 (each)	7,500	7,500		
25% × £25,000 (each)			6,250	6,250
Roll-over relief (C.G.T.A. 1979, ss. 115–121)	6,250	6,250	2,750[i]	3,000
	1,250	1,250	3,500	3,250
Original milk quota retained				
25% × Nil (each)	–	–		
25% × £10,000 (each)			2,500	2,500
Revised cost for milk quota	£1,250	£1,250	£6,000	£5,750
Chargeable gain	Nil	Nil	£4,250	Nil

(i) Under C.G.T.A. 1979, s. 116, where the reinvestment (£6,250) is less than the consideration for the earlier disposal (£10,500), and is also less than the gain (£7,000), that deficit (£4,250) is treated as the gain, and the difference (£2,750) between that and the actual gain is deducted from the consideration for the reinvestment.

"10. *Shares acquired in stages*

Where a share in a partnership is acquired in stages wholly after April 5, 1965, the acquisition costs of the various chargeable assets will be calculated by pooling the expenditure relating to each asset. Where a share built up in stages was acquired wholly or partly before April 6, 1965 the rules in C.G.T.A. 1979, Sched. 5, para. 13 will normally be followed to identify the acquisition cost of the share in each asset which is disposed of on the occasion of a reduction in the partnership's shares; *i.e.* the disposal will normally be identified with shares acquired on a 'first in, first out' basis. Special consideration will be given, however, to any case in which this rule appears to produce an unreasonable result when applied to temporary changes in the shares in a partnership, for example, those occurring when a partner's departure and new partner's arrival are out of step by a few months."

EXAMPLE

Orion acquired an interest in a farming partnership in stages, as follows:

Acquisition 6 April	Total value of farm £	%	Orion's share Cost £
1982	100,000	10	10,000
1983	150,000	5	7,500
1984	200,000	4	8,000
1985	250,000	8	20,000
		27%	£45,500

Disposal			Proceeds
1986	300,000	7	21,000
1989	200,000	5	10,000
Share remaining		15%	

The computation ignores the part disposal rules and works on a first in, first out basis.

	1986		1989
	£ (7%)	£ (3%)	£ (2%)
Proceeds	21,000	6,000	4,000
Cost (7/10ths × £10,000)	7,000		
(3/10ths × £10,000)		3,000	
(2/5ths × £7,500)			3,000
Chargeable gain subject to indexation	£14,000	£3,000	£1,000

16–013
"11. *Elections under C.G.T.A. 1979, Sched. 5, para. 4*
Where the assets disposed of are quoted securities eligible for a pooling election under C.G.T.A. 1979, Sched. 5, para. 4, partners will be allowed to make separate elections in respect of shares or fixed interest securities held by the partnership as distinct from shares or securities which they hold on a personal basis. Each partner will have a separate right of election for his proportion of the partnership securities and the time limit for the purpose of Sched. 5 will run from the earlier of:
(a) the first relevant disposal of shares or securities by the partnership; and
(b) the first reduction of the particular partner's share in the partnership assets after March 19, 1968."

Note: F.A. 1985, Sched. 19, para. 4 gives a further opportunity to make a pooling election with the substitution of April 5, 1985, for March 19, 1968.

16–014
"12. *Transitional arrangements*
The practices set out in this statement will be applied to all capital gains tax assessments after January 17, 1975. Where tax liabilities have already been settled on other reasonable bases the assessment will not be upset. Often gains and losses on previous disposals will have been computed on a different basis and the acquisition costs allocated as a result to partners to set against future disposals will be different from those which would have arisen had the new practices been followed. Such costs will only be recalculated in exceptional circumstances."

Companies

Value shifting

16–015
If a shareholder exercises his control over a company to allow value to pass out of his shares or those held by a connected person he is treated as making a disposal of the shares at the market value before the values passed out (C.G.T.A. 1979, s.25(2)).

Any resultant loss when the shares are ultimately sold is not allowable to the extent that value has passed to other assets (C.G.T.A. 1979, s.25(3)). An agreement not to enforce a restriction, such as a restrictive covenant, is treated as a disposal at market value. The exercise of control does not require positive action. Inaction, for example, in failing to subscribe for a favourable rights issue, may well come within the provisions (*Nichols* v. *I.R.C.* [1973] S.T.C. 497 overruled by *Floor* v. *Davis* (1979) 52 T.C. 609).

Complicated schemes which result in an indirect value shift are also caught (C.G.T.A. 1979, s.26).

EXAMPLE

James Liberator
Mr. Liberator owned 65 per cent. of the shares in Harry (Farms) Ltd. The remaining shares were held by his daughter Helen. The share rights were changed to give Helen's shares twice the voting and income rights of Mr. Liberator's shares. Indexation has been ignored.

	£
Deemed disposal by James Liberator at market value:	
Before share rights changed	25,000
Less: cost	15,000
Chargeable gain	£10,000
Deemed reacquisition value	25,000
Less: ultimate disposal proceeds	20,000
Loss disallowed (CGTA 1979, s.25(3))	£ 5,000

If a close company transfers an asset other than at arm's length the **16–016** amount by which the proceeds fall short of market value is apportioned amongst the shareholders and is treated as a deduction from the acquisition cost of their shares (C.G.T.A. 1979, s.75). Should the amount exceed the cost of the shares then the cost is reduced to nil; it is not possible to have a negative figure for cost. There could also be distribution or benefit-in-kind problems.

EXAMPLE

Grapple Limited
A property worth £25,000 was transferred to an associated but non-group company, Vulcan Ltd, for £15,000. All the shares in Grapple Ltd. were owned by Linda who subsequently sold them. Indexation has been ignored.

	£	£
Sale proceeds		60,000
Cost	30,000	
Less: amount of under value transferred	10,000	
		20,000
Chargeable gain		£40,000

Such a transfer might also give rise to an inheritance tax liability unless Linda also owns all the shares of Vulcan Ltd. (I.H.T.A. 1984, s.94).

Transfer of business to a company

The transfer of the whole of a business carried on by one or more pro- **16–017** prietors to a company (not necessarily a U.K. company) in exchange for an issue of shares in that company ("the transferee company") is an attractive method of turning a business into a limited company or amalgamating the business which is being transferred with the business which is being carried on by the transferee company. The C.G.T on the disposal of the business assets at the date of the transfer is postponed (possibly indefinitely). Deferment is given in respect of chargeable gains arising on the disposal of the business as a going concern to the company wholly or partly in exchange for shares issued to the transferor (*Radio Pictures Ltd.* v. *I.R.C.* (1938) 22 T.C. 106). The shares issued are described as the new assets and the business assets transferred are known as the old assets.

The postponement of C.G.T. will be either temporary or permanent depending upon a number of factors. These include whether or not the shares are sold during the lifetime of the transferor or are exempt from C.G.T. as a result of death, whether retirement relief is due on a sale of the shares and the residence status of the individual at the time of disposal of the shares.

For the relief to operate it is necessary for all the old assets to be transferred to the company, other than cash which may be excluded if desired. Normally it would be recommended that the cash should not be transferred to the company but lent as required. Stamp duty saving schemes, whereby the debts are left with the transferor for collection should be avoided as this would prevent the relief being available. Liabilities may be transferred to the company without being regarded as non cash consideration (Inland Revenue Press Release September 10, 1971). However, there would be scope for the business to borrow funds and for the partners/sole trader to extract those funds by reducing the current and capital accounts—however great care is needed as there can be certain tax problems of following this course of action.

Initially it is necessary to calculate the aggregate chargeable gains, less losses, which would accrue on the disposal of the business at open market value. That part of the total consideration for the disposal received from the company which is attributable to the shares issued is deferred by way of a deduction from the allowable base cost of those shares. Therefore it is necessary to apportion the chargeable gains between the shares and any other consideration such as cash or a loan.

The apportionment formula to be applied is:

Chargeable gains $\times \dfrac{M}{T}$ where:

M = The market value of the shares issued on the transfer of the business;
T = The total consideration received by the transferor for the business.

16–018 This formula is applied to the chargeable gains less losses on the whole of the business assets being transferred (C.G.T.A. 1979, s.123(2)). The amount of the deferral so calculated is deducted from the cost of the shares in the company issued for the business. Where the shares are eventually sold the gain is computed by reference to the reduced cost with the consequent reduction on the future indexation allowances.

EXAMPLE

Dacata Farms

Dacata Farms Ltd. was formed to take over the business of Mrs. Dacata's on January 1, 1989. Mrs. Dacata's balance sheet of December 31, 1988, was as follows:

	£		£	£
Capital account	124,000	Freehold property (at cost 1.4.82)		100,000
		Current assets:		
Current liabilities:				
Creditors	15,000	Stock	12,000	
		Debtors	18,000	
		Cash	9,000	39,000
	£139,000			£139,000

The freehold property was worth £225,000 and milk quota £10,000. Mrs. Dacata received 10,000 shares of £1 each, retained the £9,000 cash and left £20,000 on loan account with the company.

The balance sheet of the company at January 1, 1989, was as follows:

	£		£	£
Share capital	10,000	Freehold property		225,000
Share premium	220,000	Milk quota		10,000
		Current assets:		
Current liabilities:				
Creditors	15,000	Stock	12,000	
Loan	20,000	Debtors	18,000	30,000
	£265,000			£265,000

The chargeable gains on the transfer, ignoring indexation, were:

	£
Freehold property	225,000
Less: cost	100,000
Gain	£125,000

Milk quota	10,000
Less: cost	Nil
Gain	£10,000
Total chargeable gains	£135,000

$$\text{Chargeable gain} \times \frac{\text{Market value of shares in transferee company}}{\text{Total consideration received}}$$

$$£135,000 \times \frac{£10,000 + £220,000}{\substack{£10,000 + £220,000 + £20,000 \\ \text{share capital} + \text{premium} + \text{loan}}}$$

$$£135,000 \times \frac{£230,000}{£250,000} = £124,200$$

Chargeable gains on disposal	135,000
Less: deferred	124,200
Assessable capital gain	£10,800
Cost of shares for C.G.T. purposes	
Market value	230,000
Less: deferred gain	124,200
Net cost	£105,800

16–019   The example shows that increasing the value of the assets at the date of transfer increases the value of the shares, *i.e.* the deemed cost of the shares for future use, without causing a corresponding increase in the capital gains, whether deferred or not. If shares only had been received for the business it would have been desirable to endeavour to secure the inspector's agreement to the maximum value for the assets.

If the company were to sell the farm land then the cost to be deducted, subject to indexation, would be the market value at the date of transfer, *i.e.* £225,000 and not the original cost of £100,000. Therefore, where a sale of business assets by the proprietors is a future possibility, it may be useful to incorporate the business, because the capital gain is deferred against the shares received and does not crystallise on the sale of the assets by the company.

If the consideration for the transfer of the business is wholly shares, C.G.T. will be deferred completely until the shares are eventually disposed of; or until the transferor dies and the deferment becomes an exemption (C.G.T.A. 1979, s.49).

If it is desired not to transfer any asset, for example the farm land, to avoid the potential double charge on any future gain the company and personally on the increase in value of the shares, it must be taken out of the business before the transfer to the company. This could, perhaps, be done by a gift to a spouse followed by a lease back on commercial terms with the lease being assigned to the company on the transfer.

Paper for paper roll-over relief

Where a company's share capital is reorganised and the shareholders **16–020** receive new shares or debentures for their old securities the new assets are deemed to have been acquired at the same date, and for the same cost, as the original shares. If further consideration is payable the amount paid is added to the cost of acquisition, but is notionally backdated to the date of acquisition of the original shares (C.G.T.A. 1979, ss.77–80).

Replacement of securities

A paper for paper transaction involving conversion of securities, such as **16–021** unsecured loan stock converted into share capital, or the replacement of maturing loan stock by further stock, would not give rise to a disposal for C.G.T. purposes. As a result, no capital gain or loss will arise. The cost and date of acquisition of the original securities are carried over to the new securities (C.G.T.A. 1979, s.82).

Take-overs

No chargeable capital gains arise on a paper take-over of one company **16–022** by another provided that the acquiring company obtains more than a 25 per cent. interest as a result of the transaction (C.G.T.A. 1979, s.85). The roll-over relief is also available where shares or debentures are issued in exchange for shares as a result of a general offer made to members which is conditional, in the first instance, on the acquiring company obtaining control as a result of the exchange. Relief remains available if the offer is declared unconditional, even if less than 25 per cent. of the shares actually pass to the acquiring company (C.G.T.A. 1979, s.85(1)).

Before April 20, 1977, it was necessary, except in the case of a general offer, etc., for the acquiring company to obtain control and not merely a 25 per cent. interest. However, after that date it is also necessary to show that the amalgamation is effected for bona fide commercial reasons and does not form part of a scheme or arrangement a main purpose of which was the avoidance of tax (C.G.T.A. 1979, s.87(1)). It is possible to obtain advance clearance in respect of a proposed reconstruction or take-over on application in writing to the Inland Revenue, Technical Division, Capital Gains, Somerset House, London WC2R 1LB (Inland Revenue Press Release April 19, 1977, C.G.T.A. 1979, s.88). It has now become standard practice to apply for such clearances in the same way as clearances are requested in respect of transaction in securities: I.C.T.A. 1988, s.707.

Reconstruction

If shares or debentures are received in the new company as a result of a **16–023** scheme of reconstruction or amalgamation under which the assets are passed to the new company, but the original shares are retained or cancelled, no chargeable gain arises. This is provided that the assets received are shares or debentures and the reconstruction is effected for bona fide commercial reasons and does not form part of a scheme, a main purpose of which is to avoid tax (C.G.T.A. 1979, ss.85 and 86).

Demergers

16–024   Demerger relief, I.C.T.A. 1988, ss.213–218, facilitates the division of the trading activities of a company or group so that these may be carried on by two or more independent companies or groups.

On a demerger exempt distributions are excluded from being treated as distributions for corporation tax purposes. An exempt distribution consists of a distribution *in specie* by a company to some or all of its members of the shares in its 75 per cent. subsidiaries. This is a simple demerger.

A 75 per cent. subsidiary is defined as one in which at least 75 per cent. of the ordinary share capital is owned. Ordinary share capital means any share capital other than fixed rate preference shares. An indirect ownership of shares is disregarded for demerger purposes.

An exempt distribution is also one in which a trade or trades are transferred or shares in a 75 per cent. subsidiary are distributed *in specie* to another company in exchange for the issue of shares by the transferee company to some or all the members of the transferor company ("the distributing company"). This would apply in the case of a reconstruction with the consent of the court which does not require liquidation of the transferor company (Companies Act 1985, ss.425 and 427). Such a distribution is known as a three-cornered distribution. A liquidation distribution (C.G.T.A. 1979, s.38) would already be an exempt distribution I.C.T.A. 1988, s.209(1).

Conditions for a demerger

16–025   All companies in the demerger must be U.K. resident. The distributing company which is demerging must be either a trading company or a member of a trading group and the subsidiary being distributed must be either a trading company or a holding company of a trading group. It is therefore not possible to separate say the land from the farming activities through a demerger, because the land owning company would not be trading. It would however be possible to separate say arable farming from dairy farming, or farming from cheesemaking.

Where shares are being distributed in a simple demerger the shares must be irredeemable and must comprise substantially the whole of the distributing company's holding and voting rights in the subsidiary. After the distribution the distributing company must remain a trading company or holding company of a trading group unless all the shares in subsidiaries are distributed leaving no net assets in the distributing company, which is thereafter dissolved.

In a three-cornered distribution, if a trade is being transferred the transferor must retain no interest or only a minor interest in the trade. If shares are transferred they must constitute substantially the whole of the distributing company's holding and votes. The main activity of the transferee company must be carrying on the trade or the holding of shares transferred. The shares issued by the transferee must be irredeemable and constitute substantially the whole of the equity shares and voting rights in issue. The distributing company must remain a trading company or holding company of a trading group after the distribution unless the whole of the trade or shares are transferred to two or more transferee companies and the transferor is dissolved without assets available for distribution.

The distribution must be for the benefit of the trading activities and not as a means of reconstruction for the avoidance of tax or which results in the making of a chargeable payment, nor must it constitute part of the sale of the company or result in a cessation or sale of a trade.

Where the distributing company is itself a 75 per cent. subsidiary of another company, the group must be a trading group and there is no requirement that the distributing company remains a trading or holding company, but the distribution must be followed by another distribution of shares in one or more 75 per cent. subsidiary to members of the holding company so that there is, in effect, a demerger and not merely the disposal of part of a trade.

Capital gains

An exempt distribution is not a capital distribution taxed as a partial dis- **16–026** posal or deducted from the cost of the shares (F.A. 1980, Sched. 18, paras. 9 and 10). However, it will be treated as a reconstruction on a paper for paper basis to that the C.G.T. base value is apportioned to the shares in the transferee company and in the distributing company in proportion to their values at the time of the distribution. Provided that there is no chargeable payment within five years there is no C.G.T. charge on the company ceasing to be a member of a group as a result of an exempt distribution. This means that intra group transfers of assets or shares in a subsidiary disposed of without giving rise to corporation tax on the chargeable gain (because of being an intra group disposal) will not crystallise a charge which would otherwise arise (I.C.T.A. 1970, s.278 or 279).

Stamp duty

Documentation executed solely for the purposes of effecting an exempt **16–027** distribution used to be relieved from stamp duty but this relief was abolished by F.A. 1986, s.73(5). The Controller of Stamps has confirmed that documents which are executed in preparation for, or to facilitate, or with a view to the later effecting of an exempt distribution by a further transaction, are not themselves exempt from stamp duty.

Anti-avoidance

If a company makes a payment other than for bona fide commercial pur- **16–028** poses or as part of a tax avoidance scheme it is a chargeable payment if made by any company involved in a demerger in connection with the shares of such a company, other than to a group company, unless it is an exempt distribution or a taxable distribution. The definition is widened in the case of an unquoted company or one under the control of five or fewer persons and not under the control of a company which is not under the control of five or fewer persons.

If there is a chargeable payment within five years of the making of an exempt distribution that payment will be treated as income chargeable to a tax under Schedule D, Case VI, and unless it is a payment for moneys worth, tax at the basic rate will be paid by the company as an annual payment. The chargeable payment will be regarded as a distribution for corporation tax purposes and not a repayment of capital.

The C.G.T. exemption is revoked if there is a chargeable payment within five years of an exempt distribution and the time limit for raising an assessment is extended to three years after the chargeable payment.

Clearance procedure

16–029   Application for clearance may be made to the Board of Inland Revenue by the distributing company before the distribution is made. Clearance may also be applied for by a person considering making what could be a chargeable payment or by a connected company: I.C.T.A. 1988, s.215. The application must be in writing containing particulars of the proposed transactions and the Board has 30 days within which to request further information. The Board must give a decision within 30 days of the application or receipt of further information. The applicant may appeal to the Special Commissioners. If full particulars are not given the clearance notice is void. There are provisions as to returns to be made to the Revenue following various transactions carried out relating to demergers and power to enable the Revenue to obtain information.

Unquoted companies—reconstructions

16–030   Where there has been a reconstruction of an unquoted company's share capital, other than merely a bonus or rights issue of similar shares, the securities are deemed to have been sold and immediately reacquired at market value (C.G.T.A. 1979, Sched. 5, para. 14(2); *Makins* v. *Elson* (1977) 51 T.C. 437).

Groups of companies

16–031   It is permissible to transfer capital assets within a group of companies without giving rise to any chargeable gains. Transfers such as a sale of a building from a farming company to a cheesemaking subsidiary are treated for corporation tax purposes as taking place at such a price as gives rise to neither a gain nor a loss to the disposing company with a similar value being placed on the asset for any future disposals by the acquiring group member (I.C.T.A. 1970, s.273). The treatment is automatic and not optional. No claim is required. These provisions do not apply to a disposal of a debt on repayment, a disposal of redeemable preference share on redemption, or a capital distribution on liquidation (I.C.T.A. 1970, s.273(2)). Although for these provisions the commencement of liquidation does not break the group relationship it is necessary to take great care in transferring assets post liquidation. Basically, a distribution *in specie* normally gives rise to a capital gains disposal.

A group for the purpose of intra group transfers dealt with on a no gain/

no loss basis, consists of a company and its 75 per cent. subsidiaries. The 75 per cent. test is based on ordinary share capital, I.C.T.A. 1988, ss.838(1)(b) and s.832(1), which is all capital other than fixed interest preference share capital (*Tilcon Ltd.* v. *Holland* [1981] S.T.C. 365). The shares may be held directly or indirectly.

It is not necessary to control 75 per cent. of the votes in order to constitute a group for transferring assets in accordance with these provisions.

Either both companies or neither must be close investment-holding companies. Either both companies must be resident in the U.K. (I.C.T.A. 1970, s.272(1)(a)) or both companies must be resident outside the U.K. (C.G.T.A. 1979, s.16). Non-resident companies would not normally be liable to tax on chargeable gains unless in respect of U.K. branch assets. However, a chargeable gain can nonetheless be apportioned through to the shareholders in certain circumstances (C.G.T.A. 1979, s.15) and therefore the relief is extended to non U.K. resident groups. The intra group transfer provisions do not normally apply when one company is resident in the U.K. and one outside.

Disposals through designated companies

The absence of specific group relief relating to chargeable gains means **16–032** that a group of companies should designate one company through which capital disposals outside the group are made. Naturally only those assets which might give rise to a material chargeable gain or allowable loss need be passed through the designated company. In this way capital losses can be set against capital profits in the same company so that the group suffers corporation tax only on the net chargeable gains. It is usually convenient to ensure that such assets as leasehold and freehold property are acquired by the designated company in the first instance in order to avoid the necessity of an intra group transfer prior to a sale outside the group.

It has been suggested that the Revenue may try to frustrate such moves by referring to certain *obiter* comments in *Floor* v. *Davis* [1979] S.T.C. 379 and *W.T. Ramsay* v. *I.R.C.* [1981] S.T.C. 174, but losses which have arisen during a period of group ownership should be all right (C.C.A.B. Press Release September 25, 1985, T.R. 588 paras. 4–5).

The price at which the assets are to be transferred within a group depends on which company it is desired should make the actual profit. This could be either the designated company acquiring the asset or the company which has previously owned the asset. Intra group transfers of this kind are specifically excluded from the value shifting anti-avoidance provisions (C.G.T.A. 1979, s.26(6)).

So far as stamp duty is concerned there is no duty payable within a 90 per cent. group, measured on the basis of the nominal amount of the issued share capital of whatever class, if the appropriate claim is made (F.A. 1930, s.42 and F.A. 1967, s.27(2)).

There could be duty payable if there is a less than 90 per cent. group as **16–033** measured for stamp duty but still a 75 per cent. group as defined for companies' chargeable gains. This is usually avoided by the benefit of the sale contract between the group member owning the asset and the designated company being assigned to the third party purchaser. In this way only one lot of stamp duty is payable. As stamp duty is in most cases 1 per cent. the avoidance of a double charge to such duty can be significant.

It has been argued that where interests in land are concerned, the transfer to the designated company should take place at the same price as will be received from the unconnected purchaser. This is to ensure that no book profit will arise and therefore frustrate any Revenue arguments that the land had been acquired by the designated company with the sole or main object of realising a gain from disposing of the land: I.C.T.A. 1988, s.776(2)(a). It is understood however that the Revenue does not normally take this point in practice.

Losses on transfers by the designated company must be reduced by any capital allowances granted to any other company in the group (I.C.T.A. 1970, s.275) not recovered by a balancing charge.

EXAMPLE

Wellington Farms Limited

Wellington Farms Ltd. acquired an agricultural building from its fellow subsidiary Thunderbolt Farms Ltd. and sold it to an unconnected third party, Mustang Farms Ltd.

	£
Sale proceeds	10,000
Cost (to Thunderbolt Farms Ltd.)	50,000
	40,000
Less: agricultural buildings allowances (to Thunderbolt Farms Ltd.) 5 years at 10% p.a.	25,000
Allowable loss to Wellington Farms Ltd.	£15,000

Roll-over relief

16–034   Roll-over relief is granted on a group basis by providing that all trades carried on by members of a group shall be treated as a single trade for the purposes of roll-over relief on the replacement of business assets (I.C.T.A. 1970, s.276(1), C.G.T.A. 1979, ss.115–121). This applies equally to hold-over relief in respect of wasting assets with an anticipated life of less than 60 years when acquired (C.G.T.A. 1979, s.117(2)).

There is a problem in that, for maximum roll-over relief to apply, it is necessary for the asset to have been used for the purpose of the trade throughout the period of ownership (C.G.T.A. 1979, s.115(1)). If the company has been acquired by the group after the date of purchase of the asset now sold, the asset has not been used for the notional group trade throughout the period of ownership. It is understood that the Revenue will in practice normally allow the relief in such circumstances.

Different trades of a single company may be treated as a single trade for roll-over relief purposes (C.G.T.A. 1979, s.120). The Revenue accepts that where an asset is transferred intra group, then it is the use of the asset for the period from date of acquisition to the date of sale that is relevant for computing maximum roll-over relief. This could be used to advantage where the asset has been an investment.

EXAMPLE

Lincoln Farms Limited

Agricultural land has been let to an unconnected third party. The property is now vacant and a sale is possible.

	£
Estimated sale proceeds	500,000
Cost (to Lincoln Farms Ltd.)	100,000
Capital gain	£400,000

No roll-over relief would be due as it has been held as an investment since purchase.

A 100 per cent. subsidiary Shackleton Farms Ltd. is formed and the property is transferred to it for £500,000. No gain and no loss arises as it is intra-group. The subsidiary uses the land for a bona fide trade of farming for (say) 12 months. At the end of that period it is sold to a third party for (say) £500,000. If all the proceeds are reinvested in qualifying assets within the necessary period of one year before and three years after the date of sale then Shackleton Farms Ltd. should be entitled to full roll-over relief thereby saving the company a potential tax liability of £140,000 (£400,000 at 35 per cent.). As with most situations involving tax mitigation there are various important matters to watch in order to ensure that Shackleton Farms Ltd. is itself using and occupying the land for its trade.

Recovery of tax due

If the corporation tax on a company's chargeable gain is not paid within **16–035** six months of the due date the Revenue may collect the tax from the principal company in the group or from any company which was a member of the group within two years of the time when the tax became payable if that company had previously owned the assets in question: I.C.T.A. 1988, s.347(1). Where the principal company pays the tax it has a right of recovery against the liable company, and if another company previously owning the asset pays the tax it has a right of recovery against both the principal company and the liable company: I.C.T.A. 1988, s.347(3).

Company leaving a group

16–036 Assets owned by a company leaving a group which were acquired within the previous six years are treated as having been sold and reacquired by that company at market value when they were originally acquired intra group, so that the chargeable gain falls on the company leaving the group (I.C.T.A. 1970, s.278(3)). There is an exemption for two or more companies leaving the group at the same time in respect of assets transferred between those companies (I.C.T.A. 1970, s.278(2)). If the corporation tax is not paid within six months of the due date it may be recovered from what was the principal company of the group, or from the company which owned the asset when the tax became due. There is a right of recovery against the chargeable company (I.C.T.A. 1970, s.278(5)). It is not clear, where there is only a parent and a single subsidiary, whether if the subsidiary is sold both companies leave the group. It is considered that this interpretation, although possible from the legislation, would be resisted by the Revenue.

There is an exemption from the provisions relating to a company leaving the group where this happens on a bona fide commercial merger and where the avoidance of tax was not a main purpose (I.C.T.A. 1970, s.278A).

16–037 Similar provisions prevent the avoidance of corporation tax on chargeable gains where shares in a subsidiary company change hands within a group prior to that company leaving the group within a six year period (I.C.T.A. 1970, s.279(1)). The shares in the subsidiary company are deemed to have been sold and reacquired at market value immediately before the earliest intra group share transfer (I.C.T.A. 1970, s.279(2)). If the chargeable company, *i.e.* the previous owner of the shares, not the company leaving the group, has ceased to exist at the time the shares are sold outside the group, the principal company in the group is made liable (I.C.T.A. 1970, s.279(3)). There are again provisions for recovery of the tax if unpaid (I.C.T.A. 1970, s.279(4)).

Depreciatory transactions

16–038 If shares or securities in a company are sold after there has been a depreciatory transaction, such as a disposal at an undervalue to a connected company, any loss arising on the disposal is adjusted as is fair and reasonable to restrict the loss (I.C.T.A. 1970, s.280). The depreciatory transaction provisions merely disallow a loss and do not charge the profit that would otherwise have arisen, nor do they increase a profit which has been reduced by the transaction. The non-payment for group relief or surrendered advance corporation tax could be depreciatory transactions although Inland Revenue Technical Division has stated that this point is not normally taken by the Revenue.

EXAMPLE

Mosquito Farms Limited
Mosquito Farms Ltd's balance sheet at December 31, 1988, was as follows:

	£		£
Share capital	120,000	Freehold property at market value	200,000
Reserves	120,000	Net current assets	40,000
	£240,000		£240,000

The freehold property is sold to a fellow subsidiary Hornet Farms Ltd. for £20,000. This is an intra group sale (I.C.T.A. 1970, s.273) and a depreciatory transaction (I.C.T.A. 1970, s.280(1)(*a*)). The balance sheet is now:

	£		£
Share capital	120,000	Net current assets	60,000
Accumulated deficit	(60,000)		
	£60,000		£60,000

Mosquito Farms Ltd. is now sold for its asset value of £60,000 and the parent company which paid for the shares has incurred a capital loss of £60,000, which is disallowed (I.C.T.A. 1970, s.280(4)). The gain that could have been realised if the property had been transferred at market value is not assessed either under the depreciatory transaction (I.C.T.A. 1970, s.280) or value shifting provisions (C.G.T.A. 1979, s.26(6)).

Non-resident companies

Transfers between companies in a group cannot take advantage of the **16–039** exemption from liability to corporation tax on chargeable gains if one of the companies is non-resident. Where a non-resident company, which would be a close company if resident in this country, realises a chargeable gain it may be apportioned through to any U.K. resident shareholder, (C.G.T.A. 1979, s.15(2)) or non-resident trustee shareholder, (F.A. 1981, s.85) and therefore through to the U.K. resident beneficiary of a non-resident trust (F.A. 1981, s.80). This does not apply if the chargeable gain is distributed to the shareholders within two years of realisation, nor if it relates to assets used in the U.K. and therefore subject to tax already, or overseas business assets, subject to certain limitations (C.G.T.A. 1979, s.15(5)).

However, if a non-resident company incurs a capital loss there is no relief for that loss and it cannot be apportioned through to the U.K. resident shareholders. Should the non-resident company pay the tax owing on the gain apportioned to the resident shareholder, this is not regarded as a distribution. Where the tax on the capital gains is paid by a shareholder it is allowed as a deduction in computing any chargeable gain arising on the ultimate disposal of the shares (C.G.T.A. 1979, s.15(7)).

Dealing losses

16–040   If a capital loss is going to arise for which relief is unlikely to be available it might be possible within a group of companies to transfer the asset to a dealing company which will then appropriate the asset to stock, an election is made to bring the asset into account at the market value as increased by the allowable loss (C.G.T.A. 1979, s.122(3)). The asset is then sold or brought into stock at market value and a trading loss results. It is important that the dealing company realises a book profit on the transactions, as otherwise it might be contended that relief is not due for the loss which would not be in the course of a trade carried on with a view to profit. The Revenue will endeavour to show that the group company acquiring the asset did not appropriate it to trading stock and therefore great care is needed when taking these steps, (*Coates* v. *Arndale Properties Ltd.* (1984) S.T.C. 637, *Reed* v. *Nova Securities Ltd.* [1984] S.T.C. 124).

Share valuation

16–041   It is important to consider the factors involved in valuing shares in farming companies. The importance of particular factors in any given set of circumstances will depend very much on the facts of the case and the percentage shareholding being considered. Clearly the purchaser of the whole of the share capital is much less interested in the ability of the directors, who are in any event likely to retire, than the purchaser of 10 per cent. of the shares who would himself be unable to influence the board to any marked degree and is therefore very dependent upon the managerial ability within the company.

Other factors which should be considered as having a greater or lesser bearing on the valuation which would be the state of the farming industry at the time of valuation and the company's significance within that industry. The general economic and political climate must also be considered.

The company's profit record and trends are important points to consider but a minority shareholder is likely to be even more interested in the dividend record and cover, which in turn determines the yield of his investment. It is important to consider the general commercial standing of the company such as its dependence on a small number of suppliers or customers; in farming there should be no difficulty in finding alternative suppliers, but alternative customers could present a problem.

It is also necessary to consider whether the company is properly capitalised and whether the gearing between loan capital and equity capital is reasonable. The short- and long-term cash requirements of the business should be considered in the light of the resources available.

The asset backing of the shares is important for both minority and majority shareholders, albeit for different reasons. The majority shareholder, on the one hand, looks to the asset backing to give security to his investment and therefore reduce the yield required compared with a more speculative investment without such backing.

The asset backing could take the form of the company's land and buildings, the value of which would be affected by whether they were freehold or leasehold and the physical age and condition of any buildings, development plans and potential, and in the case of leasehold premises the length of lease and security of tenure under the Agricultural Holdings Acts. Agricultural tenancies are not normally assignable but when owned by a company can be a valuable asset which can be realised by a disposal of the shares. The value of an agricultural tenancy in this way is normally regarded as half the difference in value between the land with vacant possession and its value as tenanted. The extent to which a shareholder will be able to realise this value will depend on the size of his shareholding and the interests of any other shareholders.

With regard to other fixed assets such as plant and machinery the age, **16–042** condition and replacement policy are all of considerable relevance to a potential purchaser of shares and the underlying value of any investments or non-trading assets held by the company could also have considerable significance.

Any abnormal income or expenditure should be reviewed and it may be necessary to make a deduction for contingent liabilities. If, for example, a majority shareholding was being valued on the basis of the company's realisable assets on a proposed liquidation it would be necessary to provide for the corporation tax liability on any chargeable gain on property or balancing charges which would result from the liquidation, following the decision in *Winter (Executor of Sutherland)* v. *I.R.C.* [1961] 3 All E.R. 855. If, however, the assets are being considered only on the basis of supporting the possible sale of the business as a going concern or as backing for an earnings or yield basis valuation, it would merely be necessary to include a discounted figure for contingent liabilities including corporation tax on unrealised gains the degree of discount reflecting the probability of those liabilities materialising within the foreseeable future.

Other special factors can be extremely important in the question of share valuation such as the relationship between the remaining shareholders, because a non-family purchaser of say, 10 per cent. of the shares would be unlikely to have much influence in the company where the remaining 90 per cent. of the shares were all held within a closely knit family.

It has been stressed that the value of shares depends on the rights attaching to those shares. In the simplest of cases there is but one class of shares with equal rights to dividends, votes and any surplus on winding up. It is, however, possible to have many classes of shares and it is common to have, for example, preference shares where the right to dividend is limited and the voting rights and entitlement to a surplus on winding up are severely limited. It is by no means uncommon to find companies where there are various classes of ordinary, preference and deferred shares with different rights as to voting, entitlement to dividend and surplus on winding up. In each case it is necessary to look at the rights attaching to the shares in order to arrive at a valuation. A fundamental entitlement would be to the voting powers at a company meeting and the number of voting shares held would normally determine the relative value attached to them.

269

90 per cent. or more

16–043   The holder of 90 per cent. or more of the shares could not only sell his shares but could give the purchaser compulsory purchase powers under the Companies Act 1985, s.428 to enable him to buy the remaining shares outstanding, whether the other shareholders wished to sell or not.

75 per cent. or more

The holder of 75 per cent. or more of the shares can pass a special resolution which would enable him to place the company into liquidation, or to sell the business as a going concern. Because of such ability the lowest value which would apply to such a shareholding would be on the basis of the realisable value on a liquidation, including all contingent liabilities such as additional taxation and redundancy payments and subject to a further discount to enable the person actually doing the liquidation to make a reasonable profit. Such a discount might be in the order of, say 25 to 33 per cent. (*re Courthope* (1928) 7 A.T.C. 538).

50 per cent. or more

A holder of more than 50 per cent. of the voting share capital has day-to-day control over the company in that he can appoint himself or his nominees as the directors in charge of his business. He also has control over the remuneration policy and can probably pay a large proportion of the profits to himself as a director. He also has control over the dividend policy and may decide to pay small or large dividends or even decide to pay no dividend at all, but to roll up the profits within the company. The holder of more than 50 per cent. of the shares therefore will value his shares on the basis of the potential earnings of the company. It is normal to measure the potential earnings on the basis of, say, the last three years available accounts, but any profit trend may be taken into consideration as can any special factors such as non-recurring items or excessive directors remuneration, which in reality may be a profit distribution.

50 per cent. exactly

Once the shareholding drops to 50 per cent. there is no longer any control but at exactly 50 per cent. no other party has control either. In the circumstances it depends very much on how the other shares are held. Clearly if the remaining 50 per cent. are held by a single individual the possibilities of deadlock are reasonably high, whereas if the remaining shares are held by 30 different individuals the possibility of securing an ally or at least an abstainer and thereby achieve effective control might be very considerable. If effective control is available the shares would still be valued on an earnings basis, even if the holding were marginally less than 50 per cent.

Under 50 per cent.

If effective control is not available the value of the shares is going to be based on what is known as the yield basis which would depend on the dividends paid on the shares. On such a basis the assets and earnings are only of importance in so far as they indicate the probability of the dividend being maintained or increased and therefore the likely future yield on the shares. It very often happens, however, that an unquoted company does not pay a dividend and in such a case the valuation has to proceed on the basis of a notional dividend. It may be appropriate to decide the dividend which a reasonable board of directors would in the circumstances pay on the shares in view of the company's trading results; this might well be a dividend of, say, one-third of the available profit after corporation tax and the valuation would be based on such a notional dividend. It would, however, be further discounted for the fact that it is merely a notional dividend and the company has no dividend history, perhaps by 50 per cent.

25 per cent. or more

A shareholding in excess of 25 per cent. is sufficient to block a special resolution and as such is marginally more valuable than a shareholding of 25 per cent. or less. Depending on the manner in which the shares are held it may be that a shareholding in excess of say 33 per cent. may begin to give powers equivalent to control. On the other hand a shareholding as high as 49 per cent. may be totally without influence, if the remaining shares are held by a single party.

Under 25 per cent.

A shareholding of 25 per cent. or below is also valued on a yield basis, but a slightly higher yield would be required in view of the vulnerability of not being able to block a special resolution and therefore to prevent a liquidation or sale of the business. A holding of 10 per cent. or less is subject to compulsory purchase under the Companies Act 1985, ss.429–430 and the required yield would therefore be marginally higher than that required for a larger shareholding.

Price earnings ratio

The valuation of the earnings basis is often computed on the basis of what is known as the price earnings ratio. The price earnings ratio is basically the number of years purchase of the after tax profits of the company. Expressed in a different way this becomes: **16–044**

$$\text{share capital} \times \frac{\text{price per share}}{\text{maintainable profit after tax}}$$

Yield

The dividend yield is usually expressed as a percentage and the gross equivalent yield (*i.e.* on the dividend declared plus tax credit) that is required for a minority interest in a private company would usually be within the region of 25 to 30 per cent. a year for a trading company and 10 to 15 per cent. a year for a property investment company. These percentages are merely broad guide-lines. It will be appreciated that the earnings **16–045** divided by the dividend gives the number of times the dividend is covered by the profits and valuations on the basis of yield and earnings are to that extent related to each other.

In ascertaining the yield required by a prospective purchaser it is often suggested that reference should be made to comparable quoted companies but there are major objections to such a procedure. In the first instance the prices at which quoted shares change hands are subject to very much wider fluctuations than those applicable to unquoted shares, because shares in a quoted company are in a volatile market of easily marketable assets, whereas a purchaser of shares in a private company tends to hold on to them for a considerable period.

The other problem with any comparison with a quoted company is the improbability of finding a comparable quoted company, particularly for a normal farming company. If the unquoted company being valued is generally comparable in size, structure, history and expectations to a quoted public company such comparisons may well be meaningful. It is not however realistic to value the shares in Joe Bloggs (Farms) Ltd. with three employees, by reference to the share price of a public company merely because both companies are engaged in food production. The basis of valuation of shares by reference to an addition for non-marketability of, say, 35 per cent. to the value of a quoted company is unlikely to give a realistic value, at least in the case of ordinary shares (*McNamee* v. *Revenue Commissioners* [1954] I.R. 214).

If preference shares and debentures are being valued however, a comparison with quoted public companies is likely to be very much more meaningful, as it would probably be reasonable to assume that both companies would meet their dividend liabilities and a purchaser of preference shares would be influenced by the yield available on similar shares in respect of any investment on the stock market. It has been held in the case of *Attorney-General of Ceylon* v. *Mackie* [1952] 2 All E.R. 775 that if a transferor or deceased person holds preference and ordinary shares or various classes of shares, such that the combined value is greater than the value of each class in isolation, it is necessary to value the holdings together.

There is also a tendency with companies such as farming companies with a high land value to value the shares by applying some suitable method of discount to the underlying value of the assets. If, however, a minority interest is being considered it is most unlikely that this basis will produce a reasonable valuation, *Re Smith, Lloyds Bank plc.* v. *Duker* [1987] 3 All E.R. 193 and even for a holding of between say, 50 and 75 per cent. the basis is probably unrealistic when considering the rights of the shareholders. It might, however, be argued, if the break-up value is considerably in excess of the earnings basis valuation, that a shareholder with a controlling interest could probably be able to obtain sufficient support to place the company into liquidation, which might make a discounted assets basis appropriate in the circumstances. However, each case must be considered on its merits to establish whether such an argument is realistic and in many family farming companies it is not, and should be resisted.

16–046   In practice, the valuation of shares in unquoted companies for tax purposes has to be agreed by the Shares Valuation Division of the Inland Revenue. If such a valuation is required it is highly desirable to ascertain all the relevant facts relating to the shares to be valued and on that basis arrive at what is considered to be a fair and reasonable valuation such as could be recommended to both a hypothetical purchaser and a hypothetical vendor. If the valuation could be recommended to both parties with conviction it is likely that this value would ultimately be accepted by the Shares Valuation Division.

It is desirable to formulate a detailed and supportable valuation which may be submitted to the Shares Valuation Division in the first instance, rather than arrive at a figure which is incapable of rational support and then haggle with the Shares Valuation Division to try and reach an acceptable valuation. Appeals on the question of share valuation for unquoted companies may be made to the Special Commissioners, whose decision on a matter of fact is final (*McBrearty* v. *I.R.C.* [1975] S.T.C. 614).

For I.H.T. purposes, the value transferred is normally the reduction in **16–047** the transferor's estate as a result of the transfer. For C.G.T. and stamp duty the relevant value is the value of the shares actually transferred, subject to the series of transactions provisions, see below.

The market value for I.H.T. is also used for C.G.T. The acquisition value of agricultural property following a death is the full value not the value on which I.H.T. is paid following agricultural relief under I.H.T.A. 1984, ss.115–124B. I.H.T. reliefs for lifetime transfers are ignored for C.G.T. purposes.

I.H.T. business relief on death (Chapter 18) under I.H.T.A. 1984, ss.103–114 is ignored in computing the C.G.T. value at death. However, I.H.T. reliefs that recalculate the value at death, such as securities sold within 12 months of death under I.H.T.A. 1984, ss.178–189, interests in land sold within three years of death under I.H.T.A. 1984, ss.190–198 and related property sold within three years of death under I.H.T.A. 1984, s.176 will be taken into account for C.G.T. purposes in fixing the appropriate value at death.

Assets disposed of in a series of transactions

Transfers by a transferor, irrespective of the identity of the transferee, **16–048** may be linked with any earlier transaction within a six-year period. Once a series of linked transactions has been identified, it is necessary to compare the original market value of the assets disposed of and the aggregate market value in accordance with F.A. 1985, Sched. 21. If the original market value is less than a reasonable proportion of the aggregate market value, the proportion of the aggregate market value is applied and the C.G.T. recomputed for all the linked transactions in the series. If there are several transfers a disposal may have to be recomputed on a number of occasions.

Inter-spouse transfers are not themselves adjusted as they are treated on a no-gain/no-loss basis under C.G.T.A. 1979, s.44 but may be included in a series of linked transactions in determining the aggregate market value to be applied proportionately to the other transfers.

Transactions between companies in the same 75 per cent. C.G.T. group within I.C.T.A. 1970, s.273 are not included in a linked series of transactions. There are anti-avoidance provisions under which assets transferred in a series of intra group transfers may be linked if the ultimate disposal is to a person who is connected with both companies under C.G.T.A. 1979, s.63 and I.C.T.A. 1988, s.345(3). It would otherwise have been possible to fragment assets within a group and make transfers by several companies of the group, all at a proportionately low value. As a result of the linking, the disposal is valued on the basis of the aggregate market value, although the tax remains payable by the company actually making the transfer.

CHAPTER 17

Dealing and Development Profits

Alternative assessments

When land is sold it may currently be taxed either as a capital transaction **17–001** subject to C.G.T. as explained in Chapter 15, as a capital profit subject to income tax under Schedule D, Case VI under what are known as the artificial transactions in land provisions of I.C.T.A. 1988, s.776 dealt with later in this chapter, or as a trading transaction assessable under Schedule D, Case I as a trade of land dealing and/or developing, which would be a separate trade from that of farming or any other trade carried on by the taxpayer.

Trade of dealing or development

In order to determine whether or not a trade is being carried on it is **17–002** worthwhile considering the badges of trade. As a trade for the purposes of taxation is not defined anywhere in I.C.T.A. it is necessary to look at the case law on the subject.

The normally recognised badges of trade are as follows:

(1) **Profit seeking motive** (*Californian Copper Syndicate Ltd.* v. *Harris* (1904) 5 T.C. 159)
Has the taxpayer gone out of his way to try and produce a profit?
(2) **Method of acquisition** (*Hudson Bay Co. Ltd.* v. *Stevens* (1909) 5 T.C. 424)
Did the taxpayer intend to dispose of the land at a profit when he originally acquired it?
(3) **Nature of the asset** (*I.R.C.* v. *Fraser* (1942) 24 T.C. 498)
Is the asset one which is commonly bought and sold by way of trade?
(4) **Modification pending sale** (*I.R.C.* v. *Livingston* (1926) 11 T.C. 538)
Did the taxpayer do anything with the property to make it more saleable?
(5) **Interval between purchase and sale** (*Rutledge* v. *I.R.C.* (1929) 14 T.C. 490)
Did the taxpayer sell the property soon after acquiring it?
(6) **Method of securing sale** (*Martin* v. *Lowry* (1926) 11 T.C. 297)
Did the taxpayer go out of his way to attract potential purchasers for the property?

(7) **Number of transactions** (*Pickford* v. *Quirke* (1927) 13 T.C. 251)
Does the taxpayer consistently buy and sell property by way of trade?

(8) **Trading interest in the same field** (*Gloucester Railway Carriage & Wagon Co. Ltd.* v. *I.R.C.* (1925) 12 T.C. 720)
Has the taxpayer any other interest or expertise in dealing in land, *e.g.* as a builder or surveyor, or (possibly) farmer?

(9) **Method of financing** (*Harvey* v. *Caulcott* (1952) 33 T.C. 159)
Was the acquisition made from the taxpayer's own resources or did he borrow in order to finance the purchase?

If the answer to the questions raised above is "yes" on some or all of the points there is fair likelihood that the taxpayer's activities would amount to trading. It is worth noting that the destination of the proceeds of sale as such is irrelevant in determining whether or not the activities constituted trading, except that it might explain what could otherwise be interpreted as a trading motive. For example, if there was a pressing need for funds for some other purpose which necessitated the sale of land which had recently been acquired, it might be possible to show that were it not for this need the land would have been retained as a long-term investment.

It is necessary to look at these badges of trade in rather more detail.

Motive and method of acquisition

17–003   In *Californian Copper Syndicate Ltd.* v. *Harris* (1904) 5 T.C. 159 property was acquired for £24,000 in 1901 and sold for £300,000 in 1902 and 1903 which was held to be a trading transaction. Exploiting concessions in South West Africa was held to be trading in *Hugh* v. *South West Co. Ltd.* (1924) 9 T.C. 141 as was the acquisition of land in a single tranche on behalf of creditors and its subsequent sale in small tranches in *Alabama Coal, Iron, Land & Colonisation Co. Ltd.* v. *Mylam* (1926) 11 T.C. 232.

Trust lands which were transferred to a company for development resulted in trading in *Balgownie Land Trust Ltd.* v. *I.R.C.* (1929) 14 T.C. 684, *St. Aubyn Estates Ltd.* v. *Strick* (1932) 17 T.C. 412 and *Temperly Estates Ltd.* v. *Walmsley* (1975) 51 T.C. 305. Land acquired by a company from a controlling shareholder at cost, by assignment of the contract, gave rise to a trading profit in *Jacgilden (Weston Hall) Ltd.* v. *Castle* (1969) 45 T.C. 685.

A devisee who acquired a half interest in the land under a will acquired the remaining interest in the land, which was then developed and sold giving rise to a trading profit in *Pilkington* v. *Randall* (1966) 42 T.C. 662. The acquisition of a farm followed by its subsequent sale of auction gave rise to a trading profit in *Clark* v. *Follett* [1973] S.T.C. 240.

Somewhat surprisingly, the sale of a woodland by a housing society which had held the land for many years gave rise to a trading profit in *Hayards Heath Housing Society Ltd.* v. *Hewison* (1966) 43 T.C. 321. Equally surprising is the case of *I.R.C.* v. *Reinhold* (1953) 34 T.C. 389 where four houses were acquired in 1945 with an admitted intention of reselling them. They were subsequently sold in 1947 but it was held that the profit was a capital one as the intention to re-sell at a profit was not necessarily trading. The fact that property has been acquired originally as trading stock does not necessarily prevent it becoming a fixed investment as is shown in the case where oil concessions acquired in 1935 which were sold as investments in 1964 giving rise to a capital profit (*British Borneo Petroleum Syndicate Ltd.* v. *Cropper* (1968) 45 T.C. 201).

Where the original motive was clearly to hold the property as an invest- **17–004** ment rather than for exploitation at a profit the disposal gives rise to a capital profit as in the case of *Hudson's Bay Co. Ltd.* v. *Stevens* (1909) 5 T.C. 424 where concessions were surrendered to the Crown for a share in the sale proceeds when these were ultimately resold by the Crown at a profit. Trust land transferred to a company for sale gave rise to a capital profit in *Rand* v. *Alberni Land Co. Ltd.* (1920) 7 T.C. 629 as did rubber estates which were bought and planted and then immediately sold in *Tebrau (Johore) Rubber Syndicate Ltd.* v. *Farmer* (1910) 5 T.C. 658.

The liquidation of an investment company did not convert properties originally acquired with a view to holding them as investments into trading stock (*Simmons (as liquidator of Lionel Simmons Properties Ltd.)* v. *I.R.C.* [1980] S.T.C. 350). The joint ownership of property does not necessarily result in the subsequent gain being trading (*Dodd & Tanfield* v. *Haddock* (1964) 42 T.C. 229) and it is possible to hold property in a partnership for investment purposes (*I.R.C.* v. *Dean Property Co.* (1939) 22 T.C. 706).

In *Taylor* v. *Good* [1974] S.T.C. 148 the taxpayer attended an auction of a property in which his parents had worked and on impulse he bid successfully for the property. He considered owner-occupation but the property was far too large for practical occupation by his family. He then obtained planning consent and sold the property to a developer. The court held that where the asset had originally been acquired, as in this case, without a trading motive, subsequent steps taken to enhance its value prior to sale does not constitute trading. If, however, there is an established trade of property development it is probable that the property would have been treated as having been transferred to trading stock at its market value at the time it became clear that it was not suitable for the original non-trading purpose.

Nature of the asset

If the asset is one which is commonly bought and sold by way of trade a **17–005** single purchase and sale can constitute trading as in *I.R.C.* v. *Fraser* (1942) 24 T.C. 498 which in fact referred to whisky purchased in bond but could have equally related to land. Indirect sales such as the sale of shares in a land owning company by a property dealing company gave rise to a trading profit in *Associated London Properties Ltd.* v. *Henriksen* (1944) 26 T.C. 46. Conversely, in the case of *Fundfarms Developments Ltd.* v. *Parson* (1969) 45 T.C. 707 a loss on the sale of shares in similar circumstances did not give rise to a trading loss.

Interval between purchase and sale

Although it can be said that the longer the interval between a purchase **17–006** and sale the less the likelihood of it being treated as a trading transaction it is nonetheless impossible to draw any conclusion from the case law as to how long a property would have to be held to avoid the inference of trading. In *I.R.C.* v. *Hyndland Investment Co. Ltd.* (1929) 14 T.C. 694 flats bought in 1900 were sold between 1920 and 1926, having been let in the meantime, and this produced a capital profit. However, in *Reldim Ltd.* v. *Vise* (1951) 32 T.C. 254 properties bought in 1938 and sold between 1945 and 1947 resulted in a trading profit as did a purchase in 1942 followed by sales between 1946 and 1949 in *I.R.C.* v. *Toll Property Co. Ltd.* (1952) 34 T.C. 13. Properties bought in 1915 and developed between 1918 and 1937 and sold between 1935 and 1937 gave rise to a trading profit in *Cayzer, Irvine & Co. Ltd.* v. *I.R.C.* (1942) 24 T.C. 491.

In *Orchard Parks Ltd.* v. *Pogson* (1964) 42 T.C. 442, property was bought in 1934 and partly developed between 1934 and 1936 and then abandoned as uneconomic to develop further. It was ultimately sold in 1961 under threat of compulsory acquisition and the resultant profit was held to be trading.

Land sold during the Second World War, with pre-emption rights if the land ceased to be required for Crown purposes, was re-acquired in July 1962 and sold in August 1962 to give rise to a trading profit in *Bath & West Counties Property Trust Ltd.* v. *Thomas* [1978] S.T.C. 30. A purchase and resale of a farm in 1932, and again in 1934, resulted in trading profits in *Reynold's Executors* v. *Bennett* (1943) 25 T.C. 401. In *MacMahon & MacMahon* v. *I.R.C.* (1951) 32 T.C. 311 properties were bought and sold jointly between 1943 and 1948 which resulted in trading profits. The joint ownership of an estate bought and sold within six months confirmed a trading venture in *Burrell & Others* v. *Davis* (1948) 38 T.C. 307.

Where land is acquired which is surplus to the requirements the on-sale of the surplus land is likely to give rise to a trading profit (*Iswera* v. *Ceylon Commissioner of Inland Revenue* [1965] 1 W.L.R. 663; *Snell* v. *Rosser, Thomas & Co. Ltd.* (1967) 44 T.C. 343). It is very common for farmers to acquire additional land at auction and to resell whatever is surplus to their requirements. If the resale were at a profit it is likely to be taxed as a trading profit. It is also common for a consortium of farmers to buy land at auction and divide it among themselves but this would not normally be regarded as an on-sale at a profit, merely a joint acquisition with each farmer paying a proportion of the total price for his parcel of land.

A purchase followed by a sale within a few months, having in the meantime obtained additional planning permission, resulted in a trading profit in *Turner* v. *Last* (1965) 42 T.C. 517. Multiple ownership gave rise to a trading profit in *Reeves* v. *Evans, Boyce & Northcott Syndicate* (1971) 48 T.C. 495 where land was purchased by a syndicate in December 1960 and sold in January 1961. On the other hand a purchase by wives of property developers followed by an on-sale to their husbands' development companies gave rise to a capital profit in *Williams* v. *Davies* (1945) 26 T.C. 371. A capital profit also arose in *Wrigley* v. *Ward* (1967) 44 T.C. 491 where property was bought in 1956 and sold in 1959.

17–007   Where there is an established business of building it is extremely difficult to transfer from stock to a capital investment. In *J & C Oliver* v. *Farnsworth* (1956) 37 T.C. 51 a house built in 1921 and let until sale in 1953, by a firm of builders, resulted in a trading profit. In *James Hobson & Sons Ltd.* v. *Newall* (1957) 37 T.C. 609 houses built before the war were sold postwar as and when they became vacant, being let in the meantime, and gave rise to trading profits. In *W.M. Robb Ltd.* v. *Page* (1971) 47 T.C. 465 a factory which was built in 1951 by a builder and let until sale in 1963 gave rise to a trading profit. In *Bowie* v. *Reg Dunn (Builders) Ltd.* [1974] S.T.C. 234 land was acquired by a builder in 1955 and the property built thereon subsequently let. The business was transferred to a company in 1957 and the sale of the property in 1966 gave rise to a trading profit.

On the other hand if there was no original trading motive a very short period of time between purchase and acquisition may result in a capital profit as in *Marson* v. *Morton* [1986] S.T.C. 463 where land was acquired as an investment in June 1977 and sold in September 1977.

In *Forest Side Properties (Chingford) Ltd.* v. *Pearce* (1961) 39 T.C. 655 properties had been built by builders which were then transferred to an investment company owned by them. This gave rise to a trading profit on the disposal by the shareholders. The properties were let and subsequently sold by the investment company but it was held that there was a trading profit, although a contrary decision was reached in the similar case of *Glasgow Heritable Trust Ltd.* v. *I.R.C.* (1954) 35 T.C. 196. In *Lucy & Sunderland Ltd.* v. *Hunt* (1961) 40 T.C. 132, properties were built by a subsidiary company and then transferred to a parent company which let and held the properties as investments. The profit on eventual sale was held to be capital.

Method of financing

In *Cooke* v. *Haddock* (1960) 39 T.C. 64 land was purchased with planning permission with the benefit of a mortgage. It was sold in building plots and the profit was held to be trading. Similarly where there was a bare sale of the property, before purchase, due to lack of funds by the buyer, the resultant profit was trading, *Johnston* v. *Heath* (1970) 46 T.C. 463. In *Smart* v. *Lowndes* (1978) S.T.C. 607 a builder bought potential development land in his wife's name and used the property to secure his business overdraft. **17–008**

When the land was subsequently sold for development it was held to result in a trading profit. In each of these cases the fact that the land did not produce any income and was effectively purchased on borrowed money inevitably led to the conclusion of trading. The mere fact that the land does not produce income is not of itself necessarily decisive in determining that the land is held as trading stock (*Marson* v. *Morton* [1986] S.T.C. 463).

Modification pending sale

In *Parkstone Estates Ltd.* v. *Blair* (1966) 43 T.C. 246 the taxpayer installed services and sub-let plots in a trading estate. This was not unnaturally a trading activity. However, it also had a factory built on the site which was let for 10 years before sale, nevertheless the disposal gave rise to a trading not a capital profit. **17–009**

Method of securing sale

In *Eames* v. *Stepnall Properties Ltd.* (1966) 43 T.C. 678 the taxpayer sold property at agricultural value to his company at a time when negotiations for the development of the site were already under way. Not unnaturally **17–010** the profit was held to be trading, which emphasises the importance of structural planning well in advance of any disposal. In *Marshall's Executors and Others* v. *Joly* (1936) 20 T.C. 256 however, two of the partners in a land development partnership died and their executors forced the winding up of the business. The disposals caused by the executors gave rise to capital profits as it was held that there was no evidence that the executors had themselves been engaged in trading.

Number of transactions

17–011   In *Hudson* v. *Wrightson* (1934) 26 T.C. 55 three properties were bought and sold which were held to give rise to trading profits even though other property was accepted as being held as investment. It is, however, normally undesirable to have both trading and investment properties in the same ownership in view of the danger that the trading activities will taint all the land transactions. In *Page* v. *Pogson* (1954) 35 T.C. 545 the taxpayer built a property, lived in it for a short time and then sold it. He then did the same with another property and it was held that the profit on the second sale was taxable as trading. This case is often used as evidence of the taxpayer's right to have at least one property deal treated as a capital venture before becoming regarded as a trader in land. This is a very dangerous interpretation as it is well held that a single deal can give rise to a trading profit; see *I.R.C.* v. *Fraser* (1942) 24 T.C. 498 above.

In *Leach* v. *Pogson* (1962) 40 T.C. 585 a taxpayer who bought and immediately on-sold a number of driving schools was held to be trading.

Where properties are acquired with controlled tenancies the practice of holding these until they can be sold with vacant possession on the tenancies terminating can give rise to a trading profit as in *Parkin* v. *Cattell* (1971) 48 T.C. 462. A similar decision was reached in *Mitchell Bros.* v. *Tomlinson* (1957) 37 T.C. 224 where 299 houses were bought by builders between 1930 and 1948 and subsequently let. 58 houses were sold during the period 1946 to 1952 on becoming vacant and the profits were taxed as trading.

Trading interests in same field

17–012   There are a number of cases which indicate that once a taxpayer has been recognised as a builder or developer it is very difficult to buy and sell land without having any profit taxed as trading. A builder and grocer who bought a farm and land in 1930 and 1932 and sold it in 1935 and 1936 was held to be trading in *Gray & Gillitt* v. *Tiley* (1944) 26 T.C. 80. Even where land was acquired for the purpose of hobby farming by a builder which was subsequently sold on his deterioration in health the profit was held to be trading in *Sharpless* v. *Rees* (1940) 23 T.C. 361. The sale properties which were originally trading stock, by builders, after cessation of active building, gave rise to trading profits in *Gladstone Development Co. Ltd.* v. *Strick* (1948) 30 T.C. 131, *Speck* v. *Morton* (1972) 48 T.C. 476, *Granville Building Co. Ltd.* v. *Oxby* (1954) 35 T.C. 245 and *West* v. *Phillips* (1958) 38 T.C. 203.

A former builder who sold houses he had either bought or built was held to be trading in *Foulds* v. *Clayton* (1953) 34 T.C. 382, and a former builder who dealt in land through syndicates was trading in *Broadbridge* v. *Beattie* (1944) 26 T.C. 63. Where the executor of a deceased syndicate member actively continued dealing in land he was held to be trading in *Newbarns Syndicate* v. *Hay* (1939) 22 T.C. 461. An alteration to the method of trading still amounted to the same trade in *Spiers & Sons Ltd.* v. *Ogden* (1932) 17 T.C. 117 and the sale by a builder of an abandoned site produced a trading profit in *Shadford* v. *H. Fairweather & Co. Ltd.* (1966) 43 T.C. 291.

This is not to say that it is totally impossible for a developer to produce a capital profit, for example, in *Cooksey & Bibby* v. *Rednall* (1949) 30 T.C. 514 a farm was bought in 1924 and let until it was sold in 1938 and this gave rise to a capital profit, even though the taxpayers were property developers. Where a building business is transferred to a company and properties are retained for investment it is possible to produce a capital gain, as in *Seaward & Others* v. *Varty* (1962) 40 T.C. 523 and *Bradshaw* v. *Blunden* (No. 1) (1956) 36 T.C. 397, but see *Bowie* v. *Reg Dunn (Builders) Ltd.* and *Forest Side Properties (Chingford) Ltd.* v. *Pearce* above.

It is not impossible for a builder to show that properties are held as investment, for example, in *Harvey* v. *Caulcott* (1952) 33 T.C. 159 a builder built and let some shops in 1927 which were subsequently sold in 1946 and 1948 to give rise to a capital profit. Similarly the sale of a house built in 1939 for his foreman to live in gave rise to a capital profit as a disposal of investment.

In *Andrew* v. *Taylor* (1965) 42 T.C. 557 the court held that sales in the few years after cessation of active building were disposals of trading stock giving rise to trading profits and agreed a subsequent final cessation of trade at which time the remaining stock was regarded as investment properties. In *West* v. *Phillips* (1958) 38 T.C. 203, above, certain properties that were clearly identified as investment properties were accepted as giving rise to capital profits on disposal after cessation of actual building activities even though there were contemporary disposals of other properties which continued to give rise to trading profits as a disposal of stock.

Destination of proceeds irrelevant

In *Laver & Laver* v. *Wilkinson* (1944) 26 T.C. 105 the taxpayer argued that the sale of land and ground rents in order to produce finance for a film venture meant that there was no trading motive in the disposal and therefore the profits should be capital. This was rejected by the court which argued that there was still a trading profit as the purpose for which the taxpayer required the proceeds of sale was not relevant in determining whether the disposal was by way of trade.   **17–013**

Other trading considerations

It does appear that a property venture cannot be taxed under Schedule D, Case VI as an adventure in the nature of trade. It is either a trading transaction taxable under Schedule D, Case I or it is a capital profit taxable as a capital gain, or under Schedule D, Case VI if within the special anti-avoidance provisions of I.C.T.A. 1988, s.776, (see below), (*Pearn* v. *Miller* (1927) 11 T.C. 610, *Leeming* v. *Jones* (1930) 15 T.C. 333).   **17–014**

In the case of a dealer in land there is no need to regard the date of the contract as being binding for tax purposes under Schedule D, Case I. For C.G.T. the date of disposal and acquisition is the contract date under C.G.T.A. 1979, s.27(1) or if it is a conditional contract the date the condition is satisified (C.G.T.A. 1979, s.27(2)). However, for Schedule D, Case I purposes it is quite common to regard the contract date as being the appropriate date for the acquisition of property but the completion date as being the date of disposal giving rise to the trading profit. The consistent adoption of this basis could serve to delay slightly the realisation of profits but is in accordance with prudent accounting principles (*Symons* v. *Weeks & Others* [1983] S.T.C. 195, *Heather* v. *P-E Consulting Group Ltd.* (1972) 48 T.C. 293). From the date of exchange of contract the acquirer is liable to pay for the property and should recognise the liability by taking the acquisition into his account. On the other hand although a signed contract for disposal may give rise to a claim against the purchaser for specific performance and or damages it may well turn out that the purchaser has insufficient funds to complete or to be worth suing and it is therefore prudent to defer bringing in the credit for the sale until completion takes place and the proceeds are received.

Where a developer grants long leases at a premium it is necessary to bring into credit in the calculation of the trading profit the cost of the reversion which is arrived at by apportioning the expenditure of the property and carrying forward as stock the expenditure multiplied by the market value of the reversion (A) as a proportion of the premium (B) plus the market value of the reversion (A) which is a similar calculation to the $\frac{A}{A+B}$ calculation for a C.G.T. part disposal (*B.G. Utting & Co. Ltd.* v. *Hughes*) (1940) 23 T.C. 174, *Heather* v. *Redfern & Sons* (1944) 26 T.C. 119). On the other hand a sale subject to the creation of ground annuals or feu duties is an outright sale and it is necessary to include as additional consideration to market value of the right to receive the ground annuals or feu duties (*I.R.C.* v. *John Emery & Sons* (1936) 20 T.C. 213, *MacMillan* v. *I.R.C.* (1942) 24 T.C. 417).

The allowability of interest for tax purposes is considered in Chapter 21.

Expenses wholly and exclusively laid out for the purpose of the trade are deductible under I.C.T.A. 1988, s.74 in the same way as for any other trade as explained in Chapter 3. Capital allowances are not available if the expenditure can be deducted as a trading expense (C.A.A. 1968, s.82(1)(a), F.A. 1971, s.50(3)). Where a lessee incurs capital expenditure on fittings which would qualify as plant, capital allowances are now available under F.A. 1985, s.59 even though they technically belong to the landlord.

It is possible for the lessor and lessee of an industrial building to treat the grant of a long lease as a sale for industrial buildings allowances purpose under F.A. 1978, s.37.

There is an anti-avoidance provision in I.C.T.A. 1988, s.774 which applies to transactions between a dealing company and an associated non-dealing company which can result in a Schedule D, Case VI charge or trading receipt in the non-dealing company if the dealing company claims a deduction or loss as a result of the transaction.

Close investment-holding companies

After March 31, 1989, a company is a close company if five or fewer par- **17–015**
ticipators or participators who are directors together possess or are entitled
to acquire such rights as would entitle them to receive the greater part of
the assets of the company available for distribution among the participators
on a winding up, or would do so if rights of loan creditors were disregarded. These rights can be carried through a chain: I.C.T.A. 1988, s.414.

For accounting periods beginning after March 31, 1989, a close investment-holding company is a close company other than one which exists
wholly or mainly to trade on a commercial basis or to make investments in
land to be let to unconnected persons or to act as a group holding or co-
ordinating company: I.C.T.A. 1988, s.13A. A close investment-holding
company is chargeable to corporation tax at the standard 35 per cent. rate
of corporation tax and does not qualify for the small companies' rate:
I.C.T.A. 1988, s.13.

Previously, in the case of a company holding land as an investment, the net
income from property was classified as estate income and 50 per cent. of it
was distributable together with 100 per cent. of any distributable investment
income from securities or deposits: I.C.T.A. 1988, Sched. 19, paras. 2(1)
and 4(2). There was a reduction for the smaller of £3,000 or 10 per cent. of the
estate or trading income: I.C.T.A. 1988, Sched. 19, para. 2(2) subject to an
overriding *de minimis* limit of £1,000: I.C.T.A. 1988, s.424(1)(*b*). It was also
possible to argue against an apportionment of income on the grounds that the
funds were required for the maintenance and development of the company's
business, I.C.T.A. 1988, Sched. 19, para. 1(1).

Sheltering trading profits

If it is accepted that the activities in relation to land amount to trading it **17–016**
is worth considering ways in which these profits might be sheltered.

It may be possible to move profits by means of a management charge
provided that this can be justified (*Stephenson* v. *Payne Stone Fraser & Co.
Ltd.* (1967) 44 T.C. 507).

If the trading is financed through borrowed money a deduction may be
claimed for interest as explained in Chapter 21. If the trading is carried out
through a company it might be possible to shelter some of the profits
through a small self-administered pension scheme as explained in Chapter
23. It might even be possible to arrange for the land dealing to be carried
out from overseas. If, for example, there is a company in Jersey or some
other suitable area with an appropriate double taxation agreement, such as
Guernsey or the Isle of Man, it might be possible to arrange that the property dealing is carried on by a company resident in one of the Islands without there being a permanent establishment, branch or agent in the U.K.
The profits would then be subject to tax in say, Jersey, but should be
exempt from tax in the U.K. under the Double Taxation Agreement, S.I.
1952 No. 1216, para. 3(2). The advantage of Jersey is that it is possible, by
agreement with the Controller, to agree in advance a level of management
charges between the Jersey resident company and a non-resident Jersey
company, whereby up to 90 per cent. of the profits may be paid up by way
of management charge. The non-resident company bears tax in Jersey at a

flat rate of £500. The resident company bears tax in Jersey at 20 per cent. on 10 per cent. of the income giving an effective rate of tax at 2 per cent. plus £500.

It is important to avoid a branch or agency in the U.K. as there would be a charge to U.K. tax in the name of the branch or agent under T.M.A. 1970, s.78. There is an exception for a broker or general commission agent under T.M.A. 1970, s.82 and it may be possible to arrange for the Jersey company to appoint an estate agent in Jersey who appoints a sub-agent in the U.K. who is a general commission agent protected by T.M.A. 1970, s.82. It is also arguable in this structure that the U.K. agent cannot be regarded as the agent of the Jersey company under the maxim *delegatus non potest delegare, i.e.* that an agent of an agent is not an agent of the principal.

If the Jersey company were owned directly by the taxpayer the anti-avoidance provisions of I.C.T.A. 1988, s.739–742 would enable the profits of the Jersey company to be apportioned through and taxed as the income of the U.K. resident taxpayer unless he could escape under I.C.T.A. 1988, s.741 as being a transaction for bona fide commercial purposes without the avoidance of tax being a main motive. This is likely to prove difficult and it would be sensible to ensure that the Jersey company were owned, for example, by a discretionary trust for the taxpayer's family in which case there would only be a U.K. tax charge under I.C.T.A. 1988, ss.740–745 on income distributed to the beneficiaries.

17–017   Another method of sheltering a profit suitable for a property development is a low-high-high-low partnership arrangement as explained in Chapter 2. Property development by its nature often results in a low or negligible profit in year one while the property is held and the planning permission for development finalised. In year two the building will hopefully have commenced and it should be possible, if the timing is carefully controlled, to take in approximately half the eventual profit in the work-in-progress valuation in accordance with the Statement of Standard Accounting Practice No. 9. The balance of the development profit would be taken into account in the next year on completion of the development and the partnership would be continued for the final year during the maintenance period. A carefully timed cessation should then result in approximately half of the development profit falling out of assessment under the opening and closing year rules of I.C.T.A. 1988, ss.60–63. If this procedure is to be used more than once it would be necessary to have a different partnership for each property and to ensure that there is not a single partnership with changes in the constituent partners in view of the provisions of I.C.T.A. 1988, s.61(4).

It is not possible to use a Jersey partnership and argue that under the overseas partnership provisions of I.C.T.A. 1988, s.112 the profit share of the U.K. partner is not subject to tax on the basis of the decision in *Padmore v. I.R.C.* [1986] S.T.C. 744 as this is countered by I.C.T.A. 1988, s.112(4) and (5).

It may be possible to set off losses against property dealing profits as explained in Chapter 4 but it is hardly sensible planning to seek losses in order to obtain tax relief.

If it is desired to retain a property originally acquired as trading stock as an investment it would be sensible to transfer this to an investment company at the full market value (*Petrotim Securities Ltd.* v. *Ayres* (1963) 41 T.C. 389, *Sharkey* v. *Wernher* (1955) 36 T.C. 275). This is likely to be effective in producing a capital profit on eventual sale in the investment company, although as explained previously this cannot be guaranteed (see *Forest Side Properties (Chingford) Ltd.* v. *Pearce* (1961) 39 T.C. 655 above).

Artificial transactions in land

Even if it is accepted that a profit relating to a land transaction is a capi- **17–018**
tal profit it may nonetheless still be taxed as income under the provisions of I.C.T.A. 1988, s.776. The provisions apply not only to land itself but to buildings and any estate or interest in land or buildings: I.C.T.A. 1988, s.776(13)(*a*) and to property deriving its value from land including shares in a company or partnership or settled property: I.C.T.A. 1988, s.776(13)(*b*)(i) or any option, consent or embargo affecting the disposition of the land: I.C.T.A. 1988, s.776(13)(*b*)(ii). There is no need for the transaction to be in any way artificial (*Page* v. *Lowther* [1983] S.T.C. 799).

It is necessary for there to be a gain of a capital nature from land, I.C.T.A. 1988, s.776(2), which was acquired with the sole or main object of realising a gain from its disposal, I.C.T.A. 1988, s.776(2)(*a*), or land is held as trading stock, for example, a capital gain on a sale of shares in a land dealing company, I.C.T.A. 1988, s.776(2)(*b*). A direct ownership of land held as trading stock would obviously give rise to a Schedule D, Case I profit. Finally the provisions may apply where land is developed with the sole or main object of realising a gain from its disposal, I.C.T.A. 1988, s.776(2)(*c*), *Winterton* v. *Edwards, Byfield* v. *Edwards* [1980] S.T.C. 206).

There is an exemption for the period of ownership before any intention to develop: I.C.T.A. 1988, s.776(7). On that intention arising it is treated as transferred at market value: I.C.T.A. 1988, s.777(11), C.G.T.A. 1979, s.122).

The gain must be made by the acquirer, holder or developer of the land or any connected person including a company, partnership or trust, I.C.T.A. 1988, ss.776(2)(i), (13), 839, or there must be a scheme or arrangement enabling an indirect gain to be made by a person party to or connected with the scheme, I.C.T.A. 1988, s.776(2)(ii), s.777(2), (5), *Yuill* v. *Wilson* [1980] S.T.C. 460, *Yuill* v. *Fletcher* [1984] S.T.C. 401); *Sugarwhite* v. *Budd* [1988] S.T.C. 533.

There is an exemption for a gain on shares in a land dealing company or a former land dealing company, or parent of a land dealing company in respect of a subsidiary in which it holds at least 90 per cent. of the shares where land is sold by a company as trading stock at full value, I.C.T.A. 1988, s.776(10), *Chilcott* v. *I.R.C.* [1982] S.T.C. 1). There is also a gain where land is effectively disposed: I.C.T.A. 1988, s.776(4). The gain is assessed on the person who realises the profit except where the opportunity to make the gain is transferred, I.C.T.A. 1988, s.776(3)(*b*), in which case the gain is assessed on the transferor with a right for the transferor to recover the tax from the transferee: I.C.T.A. 1988, ss.777(13), 776(5)(*a*).

The Revenue has a right to recover the tax from the transferee if it remains unpaid for six months: I.C.T.A. 1988, ss.776(8), 777(8).

There is no double charge, for example, on the direct transferor nor on the indirect transferor: I.C.T.A. 1988, s.777(12).

17–019 There is deemed to be a single arrangement if there is a common purpose in the transaction: I.C.T.A. 1988, s.776(5)(*b*). The assessment is under Schedule D, Case VI under I.C.T.A. 1988, s.776(3)(*a*) which means that the profit would be taxed as unearned income. It is therefore normally preferable where a charge under I.C.T.A. 1988, s.776 is likely, to arrange matters so that it is taxed as a trading profit under Schedule D, Case I. Although there is no current disadvantage of unearned as opposed to earned income in view of the abolition of the investment income surcharge, and there could in fact be a saving in national insurance contributions, it would nonetheless be impossible to treat the profit as net relevant earnings for pension purposes as it would be if the assessment were under Schedule D, Case I.

The actual computation of the profit is on a just and reasonable basis along Schedule D, Case I lines and taking into account any premiums on leases: I.C.T.A. 1988, ss.776(6), 777(6). This means that although there is a capital profit (or the section would not bite at all) there is no indexation allowance as there is for C.G.T. purposes.

The settlement provisions of I.C.T.A. 1988, ss.660–685 extend to income assessed under these provisions, I.C.T.A. 1988, s.777(10) so that, for example, a capital profit made by a trust in which the settlor had an interest would remain the settlor's income under I.C.T.A. 1988, s.673.

Where a main residence is acquired or developed with a view to realising a gain it is possible to lose the C.G.T. exemption under C.G.T.A. 1979, s.103(3) but there would still be no charge under these provisions: I.C.T.A. 1988, s.776(9).

17–020 It is possible under I.C.T.A. 1988, s.776(11) to apply for clearance from a local inspector by giving him full details of the proposed transactions. The inspector has 30 days within which to respond either by giving or refusing clearance or asking for additional information. There is little incentive for an inspector to give a clearance under this section, because if the taxpayer is asking for a clearance he obviously believes that s.776 might apply in which case the inspector merely has to say that he is not satisfied that the gain would not be chargeable in order to refuse the clearance. He then keeps his options open as to whether or not he will raise an I.C.T.A. 1988, s.776 assessment if the transaction proceeds. This is in marked contrast to the clearance provisions under I.C.T.A. 1988, s.707 on transactions in securities, where clearance applications are dealt with by a specialist unit of the Inland Revenue Technical Division and clearances will normally be granted unless the Revenue would seek to raise an assessment under I.C.T.A. 1988, s.703 on the transaction proceeding. In most cases therefore it will not be advisable to seek clearance under I.C.T.A. 1988, s.776(11).

The provisions apply to capital profits made by non-residents if the land is in the U.K.: I.C.T.A. 1988, s.776(14). In this connection it is worth noting that a capital profit is one which has not been included in any computation of income for U.K. tax purposes under I.C.T.A. 1988, s.777(13).

Where there is a sale by a non-resident the Board has power to direct the purchaser to withhold income tax at the basic rate and account for it to the Revenue under the provisions of I.C.T.A. 1988, ss.777(9) and 349, 350. In practice the Revenue is often unlikely to know of a proposed transaction until it has taken place and it is then likely to be too late to make a direction under I.C.T.A. 1988, s.777(9). The provisions are therefore largely ineffective in practice.

The Revenue has wide powers to obtain information for the purposes of raising assessments under these provisions, although solicitors have a limited privilege, I.C.T.A. 1988, s.778, *Essex and Others* v. *I.R.C. and Another* [1980] S.T.C. 378).

Stamp duty

Stamp duty is payable on a conveyance or transfer on sale of land at a **17–021** rate of 1 per cent. unless the consideration does not exceed £30,000 in which case the stamp duty is normally nil (F.A. 1984, s.109).

Stamp duty is also payable on the grant of a lease at a premium at 1 per cent. on the premium and at rates of up to 24 per cent. on the amount of the average rent reserved under the lease. It is normally better to grant a lease for a shorter period followed by a reversionary lease for a longer period commencing at the end of the first period as if the lease does not exceed 35 years the duty is 2 per cent. but rises to 12 per cent. if the term exceed 35 years but does not exceed 100 years or 24 per cent. of the average rent if the term exceeds 100 years. The first lease cannot be for longer than 21 years under the Law of Property Act 1925, s.149(3). As this would not be a major interest in land for V.A.T. purposes it could cause the loss of zero-rating on the construction of a dwelling which has to be balanced with the stamp duty position.

Stamp duty is not payable on a contract for sale or on the assignment of the benefit of a contract but only on the completion. However a declaration of trust under which the trustee declares he is holding as a bare trustee for a purchaser causes stamp duty to be payable (*Chesterfield Brewery Co.* v. *I.R.C.* [1899] 2 Q.B. 7). An exchange of property on the other hand does not give rise to a stamp duty charge (*I.R.C.* v. *Littlewoods Stores Ltd.* [1962] 2 All E.R. 29, *Portman* v. *I.R.C.* (1956) 35 A.T.C. 349) A sale and lease back however is stampable twice, both on the purchase of the property and on the lease premium and rents (Stamp Act 1891, s.54 and Sched. 1). An option to purchase is stampable on the price paid for the option (*George Wimpey & Co.* v. *I.R.C.* [1975] S.T.C. 248). Where the rent under a lease cannot be ascertained, for example, if it is worded as being by reference to a proportion of the rents arising on sub-letting, no stamp duty is payable.

An agreement for a lease is stampable and an alteration to a lease may amount to a surrender and re-grant which would be taxable. A surrender by operation of law on giving up physical possession of the lease would not be stampable but a surrender by written document would be stampable as a conveyance at 1 per cent.

Stamp duty is a complex subject and only the barest outline of the provisions so far as they affect farmers and landowners are dealt with here.

Stamp duty is not payable on a gift (F.A. 1985, s.82) but if a gift is subject to a mortgage the loan taken over counts as consideration and stamp duty is payable on the amount of the mortgage.

Inheritance Tax

Capital transfers

The impact of C.G.T. and corporation tax on companies' chargeable **18–001** gains in connection with transfers of farms,, farmland or shares in farming companies has been considered in Chapters 15 and 16. This chapter deals with the inheritance tax implications of such transfers. Capital transfer tax became known as inheritance tax (I.H.T.) from July 25, 1986, under F.A. 1986, s.100, and the Capital Transfer Tax Act 1984 became known as the Inheritance Tax Act 1984. The change in title is automatically applied to documents such as wills drafted with reference to capital transfer tax. I.H.T. is in reality three totally distinct taxes: a tax on death which replaces estate duty, a tax on gifts and a tax on trusts. The law governing I.H.T. is largely contained in the I.H.T.A. 1984 and references throughout this chapter are to this Act unless otherwise stated.

Estate tax

As an estate tax I.H.T. is levied as if there had been a transfer equal to **18–002** the entire estate immediately before the death (s.4(1)). If two people die in circumstances where it is not known which one died first they are deemed, under the rule as to commorientes (Law and Property Act 1925, s.184) for all purposes affecting the title to property, to have died in order of seniority so that the younger is deemed to have survived the elder; for I.H.T. purposes they are deemed to have died at the same instant which means that for no part of the estate of either is deemed to pass to the other one (s.4(2)). Estate for this purpose means all property to which the deceased was beneficially entitled other than excluded property (s.5(1) and (2)). Liabilities incurred for full consideration are deductible (s.5(3) to (5)). A person with an interest in possession in a trust is treated for I.H.T. purposes as if he were beneficially entitled to the capital although in this case the I.H.T. on his death will come out of the trust assets, not out of his free estate (s.49(1)).

289

Domicile

18–003   Excluded property includes property outside the U.K. owned by a person not domiciled in the U.K.. Domicile is an important concept for I.H.T. purposes, and the meaning has been considered in Chapter 14.

For I.H.T. purposes a person is deemed to be domiciled in the U.K. for three years after ceasing to have a U.K. domicile for other tax purposes (s.267(1)(a)), which is designed to prevent somebody emigrating during a terminal illness in order to escape I.H.T. and where somebody has been resident in the U.K. for not less than 17 of the 20 years of assessment ending with the year of assessment in which the death occurs or the transfer is made (s.267(1)(b)). Apart from these deemed domicile rules domicile is determined for I.H.T. purposes in the same way as for income tax, except that any dwelling house available in the U.K. is ignored (s.267(4)).

Excluded property

18–004   Other excluded property includes securities free of tax for residents abroad held by a non-domiciled non U.K. resident and various savings certificates, premium bonds, savings bank deposits, etc., held by a person domiciled in the Channel Islands or Isle of Man (s.6(2) and (3)). A reversionary interest in a trust is also normally excluded property (s.48(1)) which is the complement to the person with the interest in possession being treated as being entitled to the entire capital.

It is necessary to aggregate a transfer on death or during life with transfers made within the previous seven years.

Chargeable lifetime transfers are charged at one half of the death scale. However, where the donor dies the tax situation of all gifts made within the previous seven years will be reviewed and the tax charge on each gift computed separately at the death rates at the time of the donor's death but subject to a taper relief on the tax where the gift was made more than three years before the death.

Potentially exempt transfers within seven years of death become chargeable transfers under s.3A. When computing the tax, the value of each gift is the amount of the original transfer (not the current value of property representing it) and each gift is essentially regarded as a "mini-estate" on its own (entirely separate from the estate on death) and is aggregated as the top slice with gifts made in the seven years before (even though some of the earlier gifts may have been made more than seven years before death). The resultant tax is tapered where the death occurs more than three years after the gift (s.7) as follows:

Tax Payable

Gift within 4th year of death—80 per cent. of normal rates
Gift within 5th year of death—60 per cent. of normal rates
Gift within 6th year of death—40 per cent. of normal rates
Gift within 7th year of death—20 per cent. of normal rates

The tax charge on a previously chargeable gift (for example a gift to a discretionary trust) is computed in the same way except (a) where the tax exceeds the tax charge on the original transfer only the excess is payable and (b) where the tax charge is less than the original charge no repayment is given.

In computing I.H.T. on the estate at death the full value of all the gifts **18–005** made within the seven years of death is aggregated with the estate which is regarded as the top slice when applying the table rates.

EXAMPLE

Christopher Alec died on December 31, 1989, leaving an estate of £600,000. He had given Caryll £150,000 on Christmas Day 1986 and Annabelle £50,000 on Christmas Day 1987. He has also used up his annual exemptions.

The gifts were potentially exempt transfers which came into charge on his death.

	£	£
Gift to Caryll 25/12/86	150,000	
I.H.T.		
0–118,000		Nil
118,001 to 150,000 at 40%		12,800
Less: taper relief (20%)		2,560
		———
Tax payable at 80% of normal rates, by Caryll		£10,240
		≡≡≡
Gift to Annabelle 25/12/87	50,000	
I.H.T.		
150,001 to 200,000 at 40%		20,000
		———
No taper relief, tax payable by Annabelle		£20,000
		≡≡≡
Estate at death 31/12/89	600,000	
I.H.T.		
200,001 to 800,000 at 40%		240,000
		———
Tax payable by estate		£240,000
		≡≡≡

Various changes may be made in the distribution of a deceased's estate which are not themselves transfers for I.H.T. purposes (s.17). These include a written disclaimer or variation to a will made within two years after a person's death provided that, for a variation, written notice is given to the Board within six months of the date of the instrument that the variation is to apply for I.H.T. purposes (s.142). A precatory trust created by the will under which, for example, certain personal effects are left to "A" with the request that he divide them among other beneficiaries in accordance with the deceased's wishes, is regarded as a transfer by the deceased if distributed by "A" within two years of the date of death (s.143). A distribution from a discretionary trust created by the will within two years of death is also related back to the date of death (s.144).

Where the distribution left by will or on intestacy is varied to meet a claim under the Inheritance (Provision for Family and Dependants) Act 1975, or by the surviving spouse commuting her life interest under the Administration of Estates Act 1925, s.47A, the I.H.T. is calculated on the revised distribution (ss.146, 145). Similar provisions apply in Scotland to a renunciation of a claim to legitim (s.147).

18–006 One further important exemption for the purposes of I.H.T. on death arises where the deceased's spouse died before November 13, 1974, and left a life interest to the surviving spouse. On the death of the surviving spouse no estate duty would have been chargeable on the capital of the life interest settlement so created in view of F.A. 1894, s.5(2) and this exemption extends to I.H.T. (Sched. 6, para. 2).

Property held in the surviving spouse trust, however, although not itself subject to I.H.T. in view of this exemption, is still regarded as part of the deceased's estate for I.H.T. purposes (s.49(1)) and this can affect the value of similar assets held in the deceased's free estate. For example, if shares in a family farming company were held as to 20 per cent. by the deceased and as to 40 per cent. by the surviving spouse trust set up by the deceased's spouse on death before November 13, 1974, the 40 per cent. interest would be free of I.H.T. (Sched. 6, para. 2) but the 20 per cent. interest would be valued as one-third of a 60 per cent. interest, *i.e.* as a proportion of a controlling interest and not a small minority interest. If this problem is foreseen it might be possible to bring the surviving spouse trust to an end by, for example, renunciation of the life interest before the death or transfer by the life tenant.

Related property

18–007 Another important provision so far as valuation is concerned is that property held by a spouse, or which has within the preceding five years been held by a charity or other exempt body to which it had previously been transferred is related property. When valuing a transferor's estate for I.H.T. purposes it is necessary to include a pro rata proportion of any related property for valuation purposes (s.161). This is a modification of the normal rule that assets are to be valued at the price they might be expected to fetch if sold in the open market at the appropriate time but without any deduction for flooding the market (s.160). If a transferor held 20 per cent. of the shares in a private farming company and his spouse held 40 per cent. of the shares and the husband died and left his shares to his son they would be valued for capital transfer tax purposes as one-third of a 60 per cent. holding and not as a 20 per cent. holding.

If assets valued as related property are sold within three years following death they may be revalued, as at the date of death, as if they were not related property, provided that it was an open market arm's length sale (s.176). In this case the deceased's shares would be valued as a 20 per cent. holding and his spouse's holding of 40 per cent. would be ignored.

Sales after death

There are specific provisions enabling, in effect, the sale proceeds to be **18–008** substituted for the probate value in respect of quoted shares and unit trusts disposed of at less than the probate value in the year following death (ss.178–189). In the case of land sold within three years of death, at market value, the sale proceeds may be substituted for the probate value (s.191(1)) provided that the difference is at least the lower of £1,000 or 5 per cent. of the probate value (s.191(2)). The provisions also introduce a number of rules with regard to changes in the land between the date of death and the date of sale, compulsory acquisition, etc. (ss.190–198).

It may be worthwhile making an election under these provisions where the value of the land has increased between the date of death and the date of sale where, for example, the rate of I.H.T. applicable is lower than the rate of C.G.T. As no C.G.T. is payable on death the excess of the proceeds over the probate would, subject to indexation, be liable to C.G.T., whereas if the probate value is adjusted to the sale proceeds there is clearly no capital gain but I.H.T. would be payable on the excess. Whether or not an election is beneficial therefore depends on the effective marginal rates of I.H.T. and C.G.T. An election could be worthwhile if part of the increase in value was covered by the nil rate band or if there was a reduction in value in the period between death and sale.

There are no provisions enabling the Revenue to adjust the values to the subsequent sale proceeds, *I.R.C.* v. *Marr's Trustees* [1906] S.L.R. 647, although such a sale could cause the probate value to be reconsidered although not necessarily revised.

Gifts tax

I.H.T. is potentially chargeable on lifetime gifts and other transfers of **18–009** value, not necessarily on the value of the asset transferred but on the reduction in the transferor's estate as a result of the transfer. This is a very important concept, particularly when applied to land and shares in unquoted companies. For example, for estate duty purposes it used to be possible to depreciate substantially the value of land by disposing of a perimeter strip which itself was of limited value but which denied access to, and therefore depreciated considerably the value of, the remaining land.

For I.H.T. purposes, however, the reduction in the estate would be the difference between the entire land held prior to the transfer and the land retained following the transfer and the whole of the reduction in the estate as a result of the transfer is potentially chargeable. Similarly, if the transferor held, say, a controlling 55 per cent. of the shares in a farming company and transferred 10 per cent. of the shares to his son the potentially chargeable transfer would be the difference between a 55 per cent. shareholding and a 45 per cent. shareholding which could be considerable as, before the transfer, there was a controlling interest which would have been valued largely on the company earnings, and after the transfer there was a minority interest that would be valued largely on dividends which could be worth proportionately much less. The value of a 10 per cent. interest in isolation, which would normally be the value of the transfer chargeable to C.G.T., would be relatively modest as a small minority interest.

In the case of a series of transfers, say, from father to discretionary trust, it is important to realise that for I.H.T. purposes the reduction in the transferor's estate is charged on a cumulative basis which will normally mean in the case of shares that the proportionate value becomes less and less as the estate before each transfer is reduced, whereas for C.G.T. purposes it is the cumulative transfers within a six-year period which are charged and the relative value of the shares is likely to increase as the cumulative total builds up (F.A. 1985, s.71 and Sched. 21).

Potentially exempt transfers

18–010   The whole basis of I.H.T. lifetime transfers on or after March 18, 1986, was drastically modified by F.A. 1986, s.101 and Sched. 19 which introduced (as s.3A) the potentially exempt transfer and reduced the cumulation period from 10 to seven years (s.7). A potentially exempt transfer means a transfer by an individual (which apart from s.3A would be chargeable to I.H.T.) by way of gift to another individual, an accumulation and maintenance trust (s.71), a disabled persons trust (s.89) or an interest in possession trust (s.49, F(No. 2)A. 1987, s.96). Only actual transfers and not deemed transfers can qualify as potentially exempt transfers and the recipients must be as specified. Gifts by an individual into a discretionary settlement are not potentially exempt. There is a similar exclusion for reversionary interests acquired by a beneficiary (see s.55) and close company transfers under s.94.

In value terms a potentially exempt transfer will only qualify to the extent that the value transferred is attributable to the donor's property becoming comprised in the estate of the donee individual by virtue of the transfer, although this should extend to the diminution in value of the transferor's estate (SP/E13). However, where no property changes hands the potentially exempt transfer is limited to the gain to the donee's estate.

Provided that a donor individual survives for a minimum period of seven years from the date of making the gift, a potentially exempt transfer becomes wholly exempt. However, if the donor dies within the seven-year period it becomes a chargeable transfer and the tax liability is computed in accordance with s.7 at the current table death rates and subject to the appropriate taper relief provided for in that section.

Until a potentially exempt transfer becomes chargeable it is regarded as exempt for all purposes so that it would not be aggregated with a subsequent chargeable transfer (e.g. a transfer to a discretionary trust) for the purpose of establishing the rate of tax appropriate to the chargeable transfer. However, the position is totally changed on the potentially exempt transfer becoming chargeable and this will affect the rate of tax on subsequent transfers. The annual exemption (s.19) is utilised first against chargeable transfers and then against potentially exempt transfers on becoming chargeable.

Gifts with reservation

Property given on or after March 18, 1986, subject to a reservation is **18–011** treated as property to which the donor was still beneficially entitled immediately before his death, if the reservation is still in existence at that time, and is thus liable to I.H.T. as part of his estate (F.A. 1986, s.102). If the reservation ceases before the donor's death, he is treated as having made a potentially exempt transfer, *i.e.* a gift that is exempt from I.H.T. provided that he survives the seven-year gift *inter vivos* period at the time the reservation ceases. F.A. 1986, Sched. 20 introduces rules (similar to the estate duty rules) to cover cases where the property changes hands after the gift or where the donee pre-deceases the donor.

A gift is regarded as a gift with reservation if the property is not enjoyed to the entire exclusion, or virtually to the entire exclusion of the donee. Gifts made under the terms of regular premium insurance policies made before March 18, 1986, and not altered thereafter will be excluded from the definition of gifts with reservation.

The reservation rules do not apply to inter spouse transfers, small gifts, gifts in consideration of marriage, gifts to charities, gifts to political parties, gifts for national purposes, gifts for public benefit, maintenance funds for historic buildings or employee trusts.

A distinction must be made between a reservation of a benefit and the **18–012** extraction of an interest followed by the gift of the remainder subject to that interest. For example, if a donor grants a lease or property to himself and his spouse and gives away the freehold reversion there is no reservation of benefit. If, however, he gives away the freehold subject to a condition of lease back he has reserved a benefit, see *Nichols* v. *I.R.C.* [1975] S.T.C. 278. See also *Chick* v. *Stamp Duties Commissioner* [1958] A.C. 425, *Munroe* v. *Stamp Duties Commissioner* [1934] A.C. 61. Similarly a grant of a lease, even at an undervalue, is not a reservation of benefit if the grantor is not a joint tenant, although it may be a gift of the tenancy.

The material date is defined as the date of the donor's death, where the reservation of interest was retained until death, and in other cases the date on which it ceases to be subject to a reservation.

The particular problems with the reservation of benefit rules which are likely to be met with in a farming context are concerned with a gift of the farmhouse and/or agricultural land by a farmer to his children while he continues to reside on the farm. If he remains in partnership with his children he may have reserved a benefit unless he can show that the land was transferred for full consideration, the consideration being the work done on the farm by the children, *Att.-Gen.* v. *Boden* (1912) 1 K.B. 539 (F.A. 1986, Sched. 20, para. 6(1)(a)) and reflected in the profit shares in that the children's share reflects an increment for the use of the land (Farm Tax Brief Vol. 2, No. 6, p. 46).

If he ceases to be in partnership his continued residence is likely to reserve a benefit at least so far as the farmhouse is concerned. He may be able to grant a lease to a nominee or to himself and his spouse jointly, which gives him a right of occupation in the farmhouse, and then gift the freehold reversion to his children. The lease should be for his reasonable life expectancy but not a lease for life, which would be a settlement, s.43(3). If he outlives the lease he may be protected by F.A. 1986, Sched. 20, para. 6(1)(b) which provides:

(b) in the case of property which is an interest in land, any occupation by the donor of the whole or any part of the land shall be disregarded if:
  (i) it results from a change in the circumstances of the donor since the time of the gift, being a change which was unforeseen at that time and was not brought about by the donor to receive the benefit of this provision; and
  (ii) it occurs at a time when the donor has become unable to maintain himself through old age, infirmity or otherwise; and
  (iii) it represents a reasonable provision by the donee for the care and maintenance of the donor; and
  (iv) the donee is a relative of the donor or his spouse.

Any benefit obtained by an associated operation could be a reservation under F.A. 1986, Sched. 20, para. 6(1)(c).

A gift of shares in a farming company which employs the donor would not be treated as subject to a reservation if the remuneration drawn was reasonable for the work done.

Substitutions and accretions

18–013   The gift with reservation provisions may be traced through a disposal by the donee in exchange for other property unless it becomes settled property or is a disposal of a sum of money in sterling or any other currency. If the donee gives away property subject to a reservation he is deemed still to hold it so that if the reservation ceases it may cease to be treated as a gift with a reservation by the original donor from the date the reservation ceases. Bonus shares or rights issues on shares given to a donee are treated as included in the property of the gift subject to reservation.

Any consideration given for shares, securities, options, etc., by the donee will be allowed as a reduction in the value of the gift subject to the reservation unless the consideration consists of property received by the donor and forming part of his estate on death. Capitalisation of reserves on a bonus issue does not count as additional consideration for this purpose (F.A. 1986, Sched. 20, paras. 2, 3).

Settled gifts

18–014   Where there is a gift into settlement the property comprised in the settlement shall be treated as the property comprised in the gift and if the settlement comes to an end before the material date any property other than property taken absolutely and beneficially by the donor, and any consideration given by him for property so taken, shall be treated as comprised in the gift. If the donee subsequently settles the gifted property, subject to the donor's reservation, the donor is treated as if he were the settlor for these purposes. Where property comprised in the settlement at the material date is directly or indirectly derived from a loan made by the donor to the trustees of the settlement it shall be treated as property originally comprised in the gift. Accumulations of income are not treated as property derived from the gift (F.A. 1986, Sched. 20, para. 5).

Exclusion of benefit

Occupation of land or possession of chattels by the donor is not regarded **18–015** as a reservation of benefit where the asset is rented for full consideration in money or money's worth. An interest in land occupied by the donor as a result of an unforeseen change in circumstances which results in the donor being unable to maintain himself through old age or infirmity or otherwise, and represents a reasonable provision by the donee for the care and maintenance of the donor, who is a relative, will not be treated as a reservation of a benefit. For example, if a farmer retired to Spain and gave the farm and farmhouse to his son, the gift would not be prejudiced if at some later stage the father fell ill and came to live with the son in the farmhouse.

Where there is an insurance scheme consisting of a policy of insurance on the life of the donor or his spouse, or on their joint lives, and the benefits which accrue to the donee are measured by reference to benefits accruing to the donor or his spouse under that or another policy it will be treated as a gift with reservation (F.A. 1986, Sched. 20, para. 7).

Regulations for avoiding double charges, etc.

It will be appreciated that a gift with a reservation can also be a charge- **18–016** able transfer and regulations have been introduced by statutory instrument (S.I. 1987 No. 1130) to avoid such double charges (F.A. 1986, s.104). The regulations provide for the reduction of the value transferred by one transfer by the amount of the other transfer, or by way of credit for the tax payable on one transfer against the tax payable on the other.

Treatment of certain debts and incumbrances

Debts or incumbrances created on or after March 18, 1986, are subject **18–017** to abatement if the loan is received from a person to whom the deceased had transferred property. This consists not only of property derived from the deceased and lent back, but other consideration lent by the donee from property derived from the deceased other than property already included in the consideration, unless it can be shown that the gift was not made to enable or facilitate the loan-back (F.A. 1986, s.103).

The repayment of a loan, which is non-deductible under these provisions, is itself treated as a transfer of value which is a potentially exempt transfer. These provisions were introduced to combat a number of discounted loan I.H.T. avoidance schemes.

A liability in respect of a life assurance policy made after June 30, 1986, can be deducted only if the policy proceeds from part of the estate.

Valuation

Another fundamental aspect of I.H.T. is the inclusion of related prop- **18–018** erty for valuation purposes which applies to chargeable lifetime transfers as well as to transfers on death. This can be of considerable importance in the proper planning of transfers.

For example, if the husband owns say 40 per cent. of the shares in a family farming company, and his wife owns 2 per cent. of the shares a transfer of 2 per cent. by the husband to a discretionary trust would be taxed on the basis of the difference between 49/51sts of a 51 per cent. interest prior to the transfer and 47/49ths of a 49 per cent. interest after the transfer because the wife's shares are included as related property. This could be a substantial transfer as it is a reduction from a controlling interest to a minority interest.

If on the other hand the wife transfers 2 per cent. of the shares to the trust her transfer is the difference between 2/51sts of a 51 per cent. transfer before the transfer and nothing after the transfer. In other words the chargeable transfer is limited to a proportion of the controlling interest which could be very much less than the reduction from a controlling to a minority interest. Any further transfers in both cases would be by references to a 49 per cent. interest, *i.e.* a minority interest remaining after the transfer.

These points are worth bearing in mind where transfers of shares are being contemplated and also apply where other assets such as farm land, where the value of the amount being transferred in isolation can be considerably less than the reduction in the total estate caused by the transfer. It is unlikely to be possible to transfer shares or land to the spouse with a view to a further transfer to the trust as this is likely to be regarded by the Revenue as an associated operation (s.268), or to be regarded as a step introduced for purely tax avoidance reasons and therefore to be ignored following *Ramsay* v. *I.R.C.* [1981] S.T.C. 174, *I.R.C.* v. *Burmah Oil Co. Ltd.* [1982] S.T.C. 30 and *Furniss* v. *Dawson* [1984] S.T.C. 153. In the Revenue view these cases also apply for I.H.T. (Inland Revenue letter dated September 20, 1985, to the Institute of Chartered Accountants in England and Wales published as TR 588).

Sales

18–019 A sale or exchange for full value where there is no gratuitous intent is not a chargeable transfer for I.H.T. purposes provided that it is made at arm's length between unconnected persons or is such as might have been expected to have been made in such a transaction (s.10(1)). In the case of shares in an unquoted company it is also necessary to show that the sale was at a price freely negotiated at the time of the sale, or at a price such as might be expected to have been freely negotiated at the time of sale (s.10(2)).

Rates of tax

18–020 The rate of I.H.T. is 40 per cent. on the excess over the nil rate band which is index linked in accordance with changes in the index of retail prices in the preceding calendar year (s.8). The lifetime rates of I.H.T. are normally one half of those applicable on death (ss.7, 8 and Sched. 1). From 6 April, 1989, I.H.T. does not begin to become chargeable until cumulative transfers total £118,000. This nil rate band is of considerable importance in I.H.T. planning as although I.H.T. is chargeable on a cumulative basis with each successive chargeable transfer added to the preceding transfers in order to fix the rate of I.H.T. payable, this cumulation only applies for a period of seven years and transfers taking place more than seven years before the current transfer are ignored in computing the cumulative total of previous transfers to be included for calculating the tax payable (s.7(1)(a)).

I.H.T. is payable on the assumption that the transferee pays the I.H.T. **18–021** resulting from the transfer. If, therefore, the transferor agrees to pay the I.H.T. that tax is an additional gift which is itself subject to I.H.T. if the gift itself is or becomes chargeable. This requires the amount of the transfer to be grossed up so that the tax on the gross equivalent leaves an amount equal to the value of the assets transferred.

Exemptions

There are a number of important exemptions for I.H.T. purposes, in **18–022** addition to potentially exempt transfers; the most important of which is the complete exemption for transfers between spouses, both of whom are, or both of whom are not, domiciled in the U.K. (s.18). As in a transfer on death where the transferor's spouse is U.K. domiciled but the transferee's spouse is non U.K. domiciled the exemption is limited to £55,000 (s.18(2)) as it would otherwise be possible to avoid I.H.T. by claiming the inter spouse exemption on the transfer to a non-domiciled spouse who would then transfer assets into non U.K. assets which constitute excluded property prior to further transfers to a discretionary trust, for example.

Another important exemption is the annual exemption currently of £3,000 (s.19(1)). This may be carried forward for one year if not fully used, but for no longer, so that the maximum annual exemption is the current year's exemption and any unused proportion of the previous year's exemption. A number of small transfers are therefore likely to be less costly that a single large transfer.

A further useful exemption is that relating to normal expenditure out of income (s.21) under which regular expenditure, such as in particular life assurance policy premiums written in trust, are not treated as chargeable or potentially chargeable transfers, provided that it can be shown that taking one year with another the payments were made out of income and nevertheless the transferor was left with sufficient income to maintain his usual standard of living. It is arguable in the case of a wife maintained by her husband that the whole of her net income, after basic rate tax (as the higher rate tax is the husband's liability in the absence of a claim for separate assessment) is available for normal expenditure out of income.

Full use of these exemptions is the cornerstone of effective I.H.T. planning.

Other exemptions relate to provision for family maintenance, including divorce settlements (s.11), small gifts which do not exceed £250 (s.20), gifts on marriage (s.22), gifts to charities, to political parties, for national purposes and for public benefit (ss.23 to 27), waivers of remuneration (s.14), or dividends (s.15) payments to or from pension funds (ss.12, 151 and 152).

Pension funds

18–023   With regard to company pension funds it is normally possible to write a letter of wishes to the pension fund trustees under which any lump sum payable on death is paid to the desired beneficiaries rather than to the deceased's estate. It would normally be desirable to take advantage of this exemption rather than lose it by, for example, having the pension fund lump sum, which is exempt from I.H.T., paid to the surviving widow.

Suppose, for example, the deceased had an estate of £250,000 and the death cover provided by the pension fund was £150,000, the total amount available would be £400,000 and if he desired to leave this equally to his surviving spouse and his son they would each receive £200,000. This would best be achieved by leaving £50,000 from the estate to the son and sending a letter of wishes to the trustees asking them to pay the death benefit to the son direct. The son would end up with £200,000 and the only chargeable transfer would be £50,000 which would be within the exempt limits and no I.H.T. would be payable. The £200,000 due to the spouse would be exempt. If, however, the pension fund proceeds were paid to the estate and £200,000 was left to the son, and a similar amount to the widow, I.H.T. on the son's inheritance would be payable as it is above the nil rate band.

A similar effect can be achieved with retirement annuity policies and personal pension schemes by writing the policy in trust.

Quick succession and similar benefits

18–024   Quick succession relief is available where the value of a person's estate has been increased by a chargeable transfer within five years before his death or a lifetime transfer of settled property in which the transferor had an interest in possession and where the first transfer was made by reference to the value of the same settled property, either on the making of the settlement or on a subsequent transfer. The relief is calculated as a percentage reduction in I.H.T. payable on the first transfer to the extent that the transferor's estate was increased by that transfer and the amount so calculated is allowed as a deduction against the I.H.T. due on the later transfer. The percentage reduction reduces from 100 per cent. where the transfers occur within one year to 20 per cent. where they fall within the fifth year (s.141).

There are relieving provisions for mutual transfers, that is a gift back by the donee to the donor within 10 years from the date of the donor's transfer but these have been repealed in respect of donee transfers after March 17, 1986 by F.A. 1986, Sched. 19, para. 25.

Close companies

18–025   There are anti-avoidance provisions dealing with close companies so that a gratuitous transfer by a company which is not regarded as a taxable distribution for income tax purposes would be apportioned through to the participators (s.94) with a reduction where the transfer is made to another company in which the participator has an interest (s.95). A transfer is not normally apportioned through to preference shareholders (s.96) and intra group transfers are normally ignored (s.97).

An alteration of share capital is treated as a transfer by the participators (s.98). There are special rules to deal with transfers by companies where the shares are held by trustees and alterations in capital where the shares are similarly held (ss.99 to 101).

Such indirect transfers are not within the potentially exempt transfer provisions.

National heritage

It is possible to obtain a conditional exemption from I.H.T. for transfers **18–026** of national heritage property and maintenance funds set up to fund the upkeep of such property (ss.27, 30–35, 77–79, 207, Sched. 4). Under these provisions the Board of Inland Revenue is empowered (F.A. 1975, s.95) to approve:

  (a) any pictures, prints, books, manuscript, works of art, scientific collections or other things not yielding income which appear to the Treasury to be of national, scientific, historic or artistic interest;
  (b) any land which in the opinion of the Board is of outstanding scenic or historic or scientific interest;
  (c) any building for the preservation of which special steps should, in the opinion of the Board, be taken by reason of its outstanding, historic or architectural interest;
  (d) any any of land which, in the opinion of the Treasury, is essential for the protection of the character and amenities of such a building as is mentioned in paragraph (c) above and
  (e) any object which in the opinion of the Treasury is historically associated with such a building as is mentioned in paragraph (c) above (s.31(1)).

It is possible that farm land surrounding a stately home could itself qualify for the exemption, although it is more likely that the home farm would be held in a maintenance fund and the income used to support the whole estate. The criteria used by the Board have been published in booklet IR67, Capital Taxation and the National Heritage.

It is necessary for the transferee to give the appropriate undertaking that the property will be properly maintained and reasonable access given to the public, and in the case of movable property that it would be kept permanently in the U.K. (s.31(2)). A chargeable event occurs on the breach of an undertaking or on disposal of the property (s.32) which gives rise to a recalculation of the transferor's I.H.T. cumulative total (s.34).

In the context of farming the main interest in these provisions is the ancillary maintenance fund under which farm land may be transferred into such a fund which is basically a discretionary trust. The transfer into the fund is exempt from I.H.T. (s.27) and if property subsequently leaves the fund there will be a charge to tax (Sched. 4, Part II). These provisions are considered further in Chapter 13.

Trusts

18–027   The area of trusts in I.H.T. planning is one of considerable complexity and it is intended here to give only a very brief outline. The basis of a trust or settlement is where property is transferred by the person setting up the trust, the settlor, to persons who hold the legal title to the property transferred, the trustees, to hold not for their own benefit but for the benefit of other persons, the beneficiaries, in accordance with the terms set out in the trust deed. In some trusts one or more of the beneficiaries are entitled to the income of the trust, after meeting the trustees proper expenses, and as such have an interest in possession (*Pearson and Others* v. *I.R.C.* [1980] S.T.C. 318).

Interest in possession trusts

18–028   For I.H.T. purposes a beneficiary with an interest in possession is treated as being entitled to the trust funds which support that interest (s.49). In other words if he is entitled to an interest in the whole of the trust income he is deemed to be entitled to the whole of the capital. If he is entitled to only part of the trust income say, one half, he is deemed to be entitled to the appropriate proportion of the trust capital (s.50). The disposal of an interest in possession is treated as a termination (s.51) and a termination is treated as a transfer of the beneficiary's interest, or a proportionate part thereof (s.52).

As the beneficiary is throughout treated as if he were entitled to the interest in the trust capital, on termination of his life interest if he becomes absolutely entitled to the trust property that is not regarded as a transfer for I.H.T. purposes (s.53(2)). If on the termination of the life interest the trust reverts to the settlor (s.53(3)) or to his spouse or within two years of his death his widow (s.53(4)) there is normally no change to I.H.T. A trust with a revertor to settlor clause is, however, ineffective for income tax purposes in view of I.C.T.A. 1988, s.673, and therefore normally only made over assets which do not have an income, such as insurance policies. The exception in favour of the settlor or his spouse also applies on death (s.54).

A transfer into an interest in possession trust on or after March 17, 1987, is normally a potentially exempt transfer (F.(No. 2)A. 1987, s.96) subject to a number of complex anti-avoidance rules (F.(No. 2)A. 1987, Sched. 7).

Because a beneficiary with an interest in possession is treated as having an absolute interest in the capital of the settled property it follows that a reversionary interest is normally excluded property and may be transferred free of I.H.T.

18–029   It follows from a beneficiary being deemed to be entitled to the capital of the trust that on his death the trust capital is aggregated with his free estate in order to fix the total I.H.T. due. Although the trust bears its own proportion of the total I.H.T. it will nonetheless increase the I.H.T. on the free estate compared with what it would have been if there had been no interest in the trust. This is clearly not a problem if the reversionary interest under the trust passes in the same way as the beneficiary's free estate.

If, however, the beneficiary wishes his free estate to pass to persons other than the reversionary beneficiaries under the trust it may be sensible to give away the free estate before death so that I.H.T. is calculated without aggregation with the trust assets. Obviously if the gift of the free estate is more than seven years before death this would be exempt from I.H.T. if it had been given away as a potentially exempt transfer. If the gift were immediately prior to death there would be no saving in the overall I.H.T. due but the non-aggregation would increase the free estate that would devolve in accordance with the beneficiary's wishes to the detriment of the reversionary beneficiaries of the settled property.

A converse of this treatment is that a reversionary interest, particularly where there is an elderly beneficiary, is a valuable asset which can be transferred free of I.H.T. It may therefore be sensible to skip a generation, so that if, for example, a trust is set up by a husband in favour of his spouse for life with reversion to his children, the children could during the widow's life re-settle their reversionary interests in favour of their own children which would not be a chargeable transfer.

Discretionary trusts

A discretionary trust is one where the interest of the beneficiary in the **18–030**
trust income is not defined but is left to the discretion of the trustees. In such cases it is not possible to treat the beneficiaries as being entitled to any proportion of the trust capital and for I.H.T. purposes the settlement has to be treated as a separate entity.

On setting up a trust in the first instance there is a transfer by the settlor to the trustees which is a chargeable transfer which is not normally potentially exempt.

A discretionary trust is liable to a 10-yearly periodic charge on every tenth anniversary from the creation of the trust and to an exit charge whenever an asset is distributed to a beneficiary.

Periodic charge

The 10-year anniversary of a trust is calculated from the date property **18–031**
was first put into the settlement (s.60) and applies only to anniversaries after March 31, 1983 (s.61). Settlements made by the settlor which commenced on the same day are related settlements (s.62).

The rate of tax to be charged on a 10-year anniversary is 30 per cent. of the effective rate which would be charged at the lifetime rates on the property in the trust fund and any related settlement aggregated with any transfers by the settlor in the 10 years before the commencement of the trust (ss.64 and 66).

In the case of settlements which commenced before March 27, 1974, any prior transfers by the settlor are ignored.

It will be appreciated that if assets are put into a number of different discretionary trusts created on different days within the exempt and nil rate bands the rate of I.H.T. payable on each trust will be confined to the value of assets in the trust and the previous chargeable transfer. Each trust therefore effectively has its own nil rate and reduced rate bands and this method of fragmentation of family assets can result in a worthwhile saving of tax.

Where property has been added to a settlement during the 10-year period I.H.T. on that portion of the trust is calculated only for the appropriate proportion of the 10-year period during which it has been comprised in the trust (s.67).

Exit charge

18–032   Where the trustees make a distribution from the settled property or it otherwise ceases to be comprised in the settlement, tax is charged on the reduction of the estate of the trust as a result of the distribution duly grossed up, if the I.H.T. is to be paid out of the trust assets (s.65).

The rate of tax to be charged is 30 per cent. of the effective rate applicable to the value of the property at commencement plus any previous transfers by the settlor within the 10 years before the settlement, excluding such transfers before March 27, 1974, where the distribution is made before the first 10-year anniversary on which I.H.T. is chargeable. The I.H.T. payable is further reduced to the same number of fortieths of the total I.H.T. applicable to the distribution as there are complete quarters, *i.e.* periods of three months (s.63) in the period from the commencement of the settlement to the date of the distribution (s.68). Adjustments are made for further property to the settlement before the distribution and for the property contained in any related settlement.

In the case of a distribution after a chargeable 10-year anniversary the rate of I.H.T. to be used is that applicable to the 10-year anniversary adjusted for any additions to the trust fund for the appropriate number of fortieths as there are complete quarters (three-month periods) between the 10-year anniversary and the date of distribution (s.69).

EXAMPLE

George who made no previous chargeable transfers, created a settlement on July 4, 1979, having an initial value of £300,000, (£50,000 thereof being on accumulation and maintenance trusts). The 10-year charge arises on July 4, 1989, when the discretionary fund (excluding the accumulation and maintenance element) was valued at £400,000. The settlement was finally wound up on June 20, 1994, when the discretionary element was worth £500,000.

Chargeable transfer:

	£
Relevant property	400,000
Add: Accumulation and maintenance element at initial value	50,000
	£450,000

I.H.T. on £118,000 @ nil	=	nil
£332,000 @ 20%	=	66,400
Effective rate $\dfrac{66,400}{450,000}$	=	14.75%

30% thereof	= 4.425%	
I.H.T. payable at 10-year anniversary, July 4, 1989		
£400,000 × 4.425%	=	£17,700

I.H.T. payable on distribution June 20, 1994

$$\text{I.H.T. payable} = £450,000 \times \frac{19}{40} \times 4.425\%$$ £9,458

19 = completed quarters between July 4, 1989 and June 20, 1994

Favoured trusts

There are a number of special types of discretionary trust which qualify **18–033** for specific exemptions or favourable treatment for I.H.T. purposes. These include charitable trusts (s.58(1)(a)). If property ceases to be held wholly on charitable trusts an I.H.T. charge is levied (s.70). Other favoured trusts are accumulation and maintenance trusts (s.71), pre-1978 protective trusts (s.73), pre-1981 trusts for disabled persons (s.74), employee trusts (s.86), trusts for the maintenance of historic buildings and other national heritage property (s.58(1)(c), Sched. 4, para. 3(1)), superannuation schemes (ss.58(1)(d), 151)), trade or professional compensation funds (s.58(1)(e)) and excluded property (s.58(1)(f)) such as a trust set up outside the U.K. of non U.K. assets by a non U.K. domiciled settlor (s.6(1)).

Protective trusts are those set up with a view to preserving trust assets in the face of the possible bankruptcy or financial irresponsibility of the principal beneficiary under the provisions of the Trustee Act 1925, s.33(1). Although these trusts are in effect discretionary trusts they are nonetheless regarded as trusts in which the principal beneficiary has an interest in possession for I.H.T. purposes (s.88). Trusts for disabled persons are often constituted as discretionary trusts but may be regarded as though a disabled person has an interest in possession for I.H.T. purposes (s.89).

Accumulation and maintenance trusts

Most of these special cases are beyond the scope of this book but **18–034** accumulation and maintenance trusts are of considerable importance in family tax planning. This is a trust where for the time being income or accumulated so far as it is not applied for the maintenance, education or benefit of the beneficiary (s.71(1)(b)) but under which the beneficiaries will become entitled to an interest in possession on or before attaining a specified age not exceeding 25 (s.71(1)(a)). The interest in possession must be obtained within 25 years of commencement of the trust (s.71(2)(a)), or all the beneficiaries must be grandchildren of a common grandparent or children, widows or widowers of such grandchildren who had died before obtaining an interest in possession (s.71(2)(b)).

I.H.T. is only charged on non-qualifying distributions from an accumulation and maintenance settlement (s.71(3)) and the creation of such a trust is a potentially exempt transfer (s.3A). Qualifying distributions are those on a beneficiary becoming beneficially entitled to an interest in possession on or before attaining the specified age or on death before attaining the specified age (s.71(4)). Obviously on obtaining an interest in possession the accumulation and maintenance trust becomes an interest in possession settlement and the beneficiary is regarded as if he had an absolute interest in the trust assets or an appropriate proportion of them in the same way as for any other interest in possession trust (s.49).

This means that it is possible to put assets into such a trust free of I.H.T. unless the settlor died within seven years and there would be no further I.H.T. to pay on the trust assets until these were disposed of by the beneficiary on his death or earlier. It is possible to control the assets by ensuring that the beneficiary has only a life interest in the income from, say, age 25 and to preserve flexibility by giving the trustees power to make advances or appointment out of trust capital to beneficiaries as they deem appropriate. There is no reason why for I.H.T. purposes the settlor should not also be a trustee of the settlement.

Accumulation and maintenance settlements are particularly useful for a grandparent settling funds on a grandchild because any income paid out during the child's minority for his maintenance, education or benefit would be regarded as the child's own income and not that of the grandparent and can therefore be used for example for the payment of school fees or living expenses while at university. A settlement by a parent on his own child is equally effective for I.H.T. purposes although any income distributed during the child's minority actually paid to or for the benefit of the child would remain taxable as the parent's income: I.C.T.A. 1988, s.663(1). The income could be accumulated within the trust during the beneficiary's minority without being regarded as the parent's income: I.C.T.A. 1988, s.664.

In the context of farming it is worth noting that there is nothing to prevent the trustees of a trust from participating directly in the farming and becoming a trading trust. This may be more favourable particularly for C.G.T. purposes than for the trustees to own shares in a farming company, although this would depend on the facts of the case, and there could certainly be merit for such trusts holding minority interests in family farming companies.

Agricultural property relief

18–035   Agricultural property in the U.K., the Channel Islands or the Isle of Man (s.115(5)) may qualify for agricultural property relief, being a deduction of 50 per cent. or 30 per cent. of the agricultural value. Agricultural property is defined as agricultural land or pasture and includes woodlands and any building used in connection with the intensive rearing of livestock or fish, provided that the occupation is ancillary to that of the agricultural land; and it also includes cottages, farm buildings and farm houses, together with the land occupied with them, as are of a character appropriate to the property (s.115(2)). A stud farm for horses counts as agricultural property (s.115(4)).

Agricultural buildings can include those used for growing mushrooms and greenhouses for vegetables.

Although agricultural property may well qualify for business property relief, it is not possible to obtain both reliefs (s.114(1)). The rules relating to business property relief are considered later. In most cases the land and buildings will qualify for agricultural relief and the plant and machinery, animals and stock in trade will qualify for business relief.

Agricultural relief takes the form of a reduction in the agricultural value, that is the value assuming a perpetual covenant prohibiting its use otherwise than as agricultural property, and therefore excluding development value for other purposes (s.116(1), s.115(3)), or hope value. This excess value, if any, will not qualify for either agricultural or business relief.

The agricultural value of agricultural property includes the value of growing crops, cultivations and any unexhausted manurial value, although the instalment basis is not available for the value attributable to growing crops. Sporting rights over the land that are consistent with the agricultural usage are included as part of the agricultural value, but fishing rights are not.

When computing the value of farm cottages, but not the farmhouse, that are occupied solely for agricultural purposes in connection with the farm, no account is to be taken of the fact that the cottage could be used for other residential purposes; this is likely to result in a fairly low value (s.169)). A gamekeeper may not be in agricultural occupation, *Earl of Normantor* v. *Giles* [1980] 1 All E.R. 106.

In cases where the property is occupied with freehold possession, or **18–036** where a right to vacant possession can be obtained within the next 12 months, the relief amounts to 50 per cent. of the agricultural value (s.116(2)(a)). The 50 per cent. relief is also available where "working farmer relief" would have been available had the transfer taken place before March 10, 1981 (s.116(2)(b) and (3)). The sort of case where the old working farmer relief would have applied would have been a family farm where the owner had retired but retained the land and the farming was carried on, for example, by his sons, or another member of his family, between the time of the working farmer's retirement and the time of the transfer of the land on his subsequent death. Working farmer relief would have been available under F.A. 1975, Sched. 8, para. 3(4) even though the family member was in occupation an the owner therefore did not have vacant possession. The 50 per cent. relief is preserved in these circumstances.

Where the working farmer relief still applies the limitations under the old rules of the greater of £250,000 for 1,000 acres are retained. Where only part of the land had been owned for the two years preceding March 10, 1981, the 50 per cent. relief applies only to the proportion so owned and the 30 per cent. relief applies to the remainder (s.116(4)).

In other cases, where the transferor does not have vacant possession within 12 months, the relief is 30 per cent. of the agricultural value. The interests of joint tenants or tenants in common are looked at together in determining whether vacant possession is available (s.116(6)).

Agricultural property does not qualify for relief unless it was occupied by the transferor for the purposes of agriculture throughout the period of two years ending with the date of the transfer (s.117(a)), or was owned through the period of seven years ending with the transfer and was occupied for the purposes of agriculture (s.117(b)). For this purpose occupation by a company controlled by the transferor is treated as the transferor's occupation (s.119(1)).

Thus, to obtain the 50 per cent. relief, it is necessary to have held the land with vacant possession for at least two years and farmed it alone or in partnership. Occupation by a Scottish partnership is deemed to be occupation by the partners despite the fact that a Scottish partnership is a separate legal entity under the Partnership Act 1890, s.4(2) (s.119(2)).

To obtain the 30 per cent. relief for tenanted land it is necessary to have held the property for at least seven years; this is to prevent death bed purchases of investment farm land with a view to obtaining a reduction in the estate for I.H.T. purposes.

Agricultural relief is now given, in the same way as business relief, as a deduction from the value before deducting any available exemptions (s.115(1)(a), s.116(1)). The relief is given before any grossing up for I.H.T. where the donor pays the tax (s.116(7)).

18–037 Agricultural property relief will not be available on a gift unless property comprised in the gift is owned by the transferee throughout the period from the date of transfer to the date of death ("the relevant period") and that throughout the relevant period the property is agricultural property occupied either by the transferee or another for the purposes of agriculture s.124A. Where the original agricultural property comprised shares in a farming company (s.122) the company must have owned the agricultural land throughout the relevant period. Where the transferee predeceases the transferor the conditions must be satisfied at the transferee's death. For transfers after March 16, 1987, the donee's period of ownership of the original property is aggregated with the period of ownership of the shares into which it has been transferred. Where only a part of the transfer can be attributed to the original property, agricultural relief is only available in respect of that part.

Where the transferee has disposed of all or part of the property before the transferor's death, the availability of agricultural relief is retained provided that he applies all the proceeds in the acquisition of replacement agricultural property within 12 months of the disposal and both transactions are on arm's length terms and the replacement property is owned by the transferee at the time of the transferor's death, s.124B.

If agricultural property replaces previously held agricultural property, the combined periods of ownership in the five years before the transfer may be aggregated, and the relief is available if these total two years or more (s.118(1)). In the case of tenanted land, the aggregate period of ownership of land transferred and land which had previously been held must total seven years out of the 10 years before the transfer (s.118(2)). If the replacement land has not itself been held for the requisite period of two or seven years, however, the relief is limited to the agricultural value of the land previously held (s.118(3)). Partnership changes are ignored for determining the period of ownership of land (s.118(4)).

18–038 If there is a binding contract for the sale of the property at the time of the transfer or death, agricultural relief is not available (s.124(1)); this is because the relief is designed to reflect the difficulty of disposing of such property in the short term and this difficulty would not arise where a contract for sale existed. For this purpose a contract for the sale of agricultural property is ignored if it is to a company in exchange for shares as a result of which the vendor would control the company (s.124(1)).

Where the property has been inherited on the death of another person, the transferor is deemed to have owned the land from the date of death; if he subsequently occupies the land he is deemed to have occupied it from that date (s.120(1)(a)). If the land is inherited from a spouse, the spouse's period of occupation or ownership may be aggregated with the transferor's occupation or ownership (s.120(1)(b)). Working farmer relief is carried through where land is inherited from a qualifying spouse on or after March 10, 1981 (s.120(2)).

Where there has been a transfer under which the transferor would have qualified for the relief, the transferee makes a subsequent transfer within the two-year period and one or both of the transfers were on death, the two-year period of ownership for an occupier is waived. The second transferor may have acquired the property from his spouse who inherited it from the first transferor. The property must have been occupied for agricultural purposes either by the second transferor or by the personal representatives of the first transferor (s.121(1)). Replacement property is catered for under these provisions (s.121(2)) and the relief is limited to the agricultural value (s.121(3)).

Payment by instalments

Agricultural property will normally qualify for payment of I.H.T. by 10 **18–039** equal annual instalments on a transfer on death or by a trust or where the donee pays the tax (s.227(1)), either as land (s.227(2)(a)) or as an interest in a business (s.227(2)(c)). Shares in a farming company may also qualify if sufficient shares were held to give control (s.227(2)(b), s.228(1)(a)), or in cases of hardship (s.228(1)(c)), or on death where they account for 20 per cent. of the estate (s.228(1)(b)) or the value exceeds £20,000 and the shares represented 10 per cent. of the nominal share capital or ordinary share capital (s.228(1)(d)).

Companies and trusts

Where part of the value of shares in a company controlled by the trans- **18–040** feror is attributable to the agricultural value of the company's agricultural property (s.122(1)), the relief is given on the appropriate proportion of the value of the shares as if the company's interest in or occupation of the land was the shareholder's interest (s.122(3)). Related property can be included for the purposes of determining control (s.269(2)), unless an election is made within three years of death to have the property valued as if it were not related (s.176, s.122(2)). Relief is not available if there is a binding contract for the sale of the shares except for the purposes of a reconstruction or amalgamation (s.124(2)).

In the case of agricultural land owned by a company, it is necessary not only for it to have been occupied for the two years ending with the transfer of the shares, or to have been owned for a seven-year period ending with the transfer of the shares and occupied for the purposes of agriculture by the tenant, but also the shares must have been held throughout the two- or seven-year period, as appropriate (s.123(1)). There are provisions enabling land to be replaced by further land (s.123(2)). Where the shares have not been held for the requisite period, it is possible to include periods of ownership of other eligible property, such as agricultural property or shares in agricultural land-owning companies where the total period of ownership amounts to at least two years in the five years ending with the transfer, or in the case of tenanted land, for seven years in the 10 years ending with the transfer (s.123(3)). As with other replacements the relief is limited to the relief which would have been available if replacement had not taken place (s.123(4)). Where land is subsequently transferred to a company controlled by the transferor, it is treated as occupied by the company for the period of the transferor's occupation as well as that of the company (s.123(5)).

Agricultural relief extends to agricultural property owned by trusts subject to the same restrictions as apply to an individual (s.115(1)(b)).

Reservation of interest—agricultural relief

18–041   Where an asset disposed of by way of gift is relevant business property, agricultural property or shares representing agricultural property of companies and the property is subject to reservation, business or agricultural relief will be given on the death of the donor or on the property ceasing to be subject to a reservation, as if the property comprised in the gift was a transfer of value by the donee. For this purpose ownership by the donor before the disposal with reservation will be treated as ownership by the donee and occupation by the donor before or after the disposal will be treated as occupation by the donee. In the case of agricultural property of companies the shares must be held by the donee as if it was the agricultural value of agricultural property of the donor.

If the donee dies before the transfer references to the donee include his personal representatives or beneficiaries (F.A. 1986, Sched. 20, para. 8).

Application of business and agricultural relief where transfer partly exempt

18–042   The interaction of the rules for calculating the tax on a partly exempt estate and the agricultural and business property reliefs enabled the chargeable part of an estate to be reduced in full by the agricultural and business reliefs even though the property went partly to an exempt beneficiary. To prevent this exploitation, where part of the partly exempt estate consists of agricultural or business property the value of any specific gifts of such property is taken to be the value as reduced by the business or agricultural relief (s.39A).

Where there is a specific gift of non-business or agricultural property it is abated to a fraction of its principal value. The numerator in the fraction is the value transferred after agricultural and business relief less the net value after such reliefs of any specific gifts of such property. The denominator is the gross value before business or agricultural relief less the gross value of the specific gifts of such assets. The chargeable part of the estate is therefore the net estate after business and agricultural relief less the abated exempt gifts.

Where specific gifts are bequeathed tax-free the grossing up rules in s.38(3) to (5) are applied to the net values after agricultural and business property. The effect of the rules is to encourage specific gifts of property subject to business and agricultural relief to non-exempt beneficiaries.

Grant of tenancy

It is now provided that the grant of a tenancy of agricultural property for **18–043** full consideration in money or money's worth is not a transfer of value by the grantor provided that the land is in the U.K., the Channel Islands or the Isle of Man (s.16(1)).

It is, however, by no means clear precisely what constitutes full consideration. If the rental value of the land in question is, for example, £50 per acre on a rent review to market value of an established tenancy, it could be argued that the grant of a lease at this level would be for full consideration. On the other hand, in a number of cases potential tenants are prepared to pay substantial amounts of key money in order to obtain an agricultural tenancy; it is likely that the Revenue would argue that the grant of a tenancy which did not reflect this key money element, either by way of a premium or its rentalised equivalent, would not be for full consideration and there could therefore be a reduction in the estate as a result of the grant of a tenancy.

The Revenue does accept however, that as an agricultural tenancy is not assignable, it does not have a value on death even when security of tenure passes to other members of the deceased's family. In such circumstances it is not the tenancy that continues, because this is extinguished on death, but the creation of a new tenancy automatically in favour of the deceased's family.

Under the Agricultural Holdings Act 1986, security of tenure is given to the tenant with various rights of succession to members of the tenant's family. This security is not available for a tenancy for a period in excess of one year but for less than two years, A.H.A. 1986, s.36(2)(b), *Gladstone* v. *Bower* [1960] 2 Q.B. 384, or where it is a grazing or mowing licence for a period of less than a year, or if before it was entered into, the County Agricultural Executive Committee approved it as an arrangement not giving rise to security of tenure.

In the case of a Scottish agricultural lease, where the land law is different **18–044** from that of England, any value associated with the prospect of renewal of the lease by tacit relocation is to be ignored (s.177).

A tenancy granted to a company can be effectively assignable by a transfer of the shares in the company and the value of the tenancy would be taken into account in valuing the shares for I.H.T. purposes, see Chapter 16.

So far as the landowner is concerned, there is a considerable difference in value between land with vacant possession and land subject to an agricultural tenancy. The latter may be worth only some 40 per cent. to 60 per cent. of the former. Although it is the general intention that land valued with vacant possession qualifies for the 50 per cent. deduction, and tenanted land qualifies for the 30 per cent. deduction, this does not happen in all cases. Jointly-held land may be subject to a deduction for the joint holding of 10 per cent. to 20 per cent., and yet would qualify for the 50 per cent. relief (s.116(6)).

It should be noted that if the agricultural property is mortgaged, the agricultural property relief is available on the net value only (s.162(4)). It is therefore desirable where possible to secure borrowings on, for example, insurance policies and use the land merely as collateral security.

One way to freeze the current value of land is to grant an option to the intended beneficiary to enable him to acquire the land at a little over the current market value within a period of 21 years. The market value of such an option is unlikely to be high but it is important that the full market value be paid for the option because s.163(1)(a) provides that the existence of an option could only be taken into account to the extent that consideration or money or money's worth was given for it.

A further variation would be a cross-option under which the donor would require the intended donee to buy the land at its market value within the 21-year period of the option, and similarly the donee would be able to require the donor to sell the land during this period. In both cases the price would be the value at the date the option was granted. It is argued that there is no reduction in the donor's estate as a result of the transaction, because he could at any time realise the value at the time of the transfer, but it effectively ensures that any future growth in the value of the land accrues to the donee.

Under the current rules an interest-free loan does not give rise to a chargeable transfer of the interest forgone. An interest-free loan to children or trustees to enable them to purchase part of the farm could be very tax efficient. If the loan is repaid gradually over a period, the effect is to give the donor the equivalent of an income. This arrangement would only be effective, however, if the land had fairly recently been acquired, because the disposal to the children or trustees would crystallise a C.G.T. charge, although this could be held over as explained in Chapter 15.

18–045 Although the granting of a lease for life will normally create a settlement (s.43(3)), this is not so where a lease for life is reserved before a disposal of the freehold reversion (SP/E10). Such a lease reserved by the donor is obviously a depreciating asset, while the freehold reversion, if given away, is an appreciating asset. The value of the freehold reversion at the time of the gift will be subject to I.H.T., but should qualify for agricultural property relief at 50 per cent. The disadvantage is that for C.G.T. purposes the donee's base value will only be the value of the freehold reversion. Clearly the amount of the advantage depends on the life expectancy of the donor. As an alternative, it might be desirable to reserve a lease for a term of years ascertained by reference to the life expectancy of the donor, but not directly related to his death. On death the remaining value of the lease would also have to be valued but is unlikely to be substantial in most cases.

The 50 per cent. agricultural property relief is available where there is vacant possession at the time of transfer, but there is no requirement as to the length of time that vacant possession must remain. It would seem possible, therefore, to enter into an agreement for a lease with, for example, the family farming partnership which is to start in say, one month's time, and in the meantime to dispose of the freehold to a family discretionary trust. In this case the value of the freehold will be reduced by the lease about to be granted to approximately the normal tenanted value, but the 50 per cent. relief would still be available as there was vacant possession at the time of the transfer. It is not thought that the associated operations provisions could apply where the lease and reversion are with different parties.

Partnerships in planning

Subject to any possible charge on the creation of a tenancy, it may well **18–046** be desirable to involve the family in the ownership and running of a farm so that, for example, part of the property owned by a sole trader farmer could be transferred to his spouse; this would normally be a tax exempt transfer. If the farming were then carried on in a partnership between the husband, wife and adult children it would be possible to enter into an agricultural tenancy in favour of the partnership; this would give the benefit of the lower tenanted value for the freehold, although the tenancy, even though non-assignable, would be an asset of the partnership (*O'Brien* v. *Benson's Hosiery (Holdings) Ltd.* [1979] S.T.C. 735) and could have a value in the same way as a company's tenancy, see Chapter 16. Where such an arrangement was entered into before March 10, 1981, it is possible to have both the reduction to tenanted value and the 50 per cent. working farmer relief.

For income tax purposes the rent paid by the partnership to the parents would be a deduction in the firm's accounts and chargeable as investment income of the parents. One disadvantage would be the reduction in relevant earnings for retirement annuity or personal pension purposes. There would also be the loss of retirement relief for C.G.T. purposes. If, however, after a three-year period the parents allowed the firm rent-free occupation of the property, retirement relief for C.G.T. purposes would once more be available for the majority of the period of ownership (F.A. 1985, ss.69 and 70 and Sched. 20) and the associated operation provisions would be avoided for I.H.T. purposes (s.268(2)). If rent is to be waived this should be done before the beginning of the tax year to which the waiver is to relate and it should be done by an appropriate deed of waiver (I.C.T.A. 1970, s.87).

An alternative method of using the partnership might be to treat the **18–047** farm as a partnership asset owned in trust for the partners generally, in which case the value of the land with vacant possession would be credited to the parents' capital account. It would, over a period of time, be reasonable to make transfers from the parents' capital accounts to those of the children active in the partnership as they gradually take over the running of the farm, allowing their parents to reduce the time and effort which they have to devote to the business. The old estate duty case of *Att.-Gen.* v. *Boden* [1912] 1 K.B. 539 should ensure that a transfer of equity commensurate with the change of responsibilities would be for commercial consideration and would not give rise to a chargeable transfer. The fact that the donor remains a partner should not of itself amount to a retention of benefit under F.A. 1986, s.102.

It might also be sensible to accelerate the process of transfers of capital as potentially exempt gifts or by taking advantage of the parents' nil rate bands as well as transfers within the annual exemption limits, currently

£3,000 per donor (s.19). It might be possible to build up funds for any residual I.H.T. ultimately payable by means of a joint whole life last survivor with profits insurance policy written in trust, the premiums for which might qualify as normal expenditure out of income (s.21).

Companies in planning

18–048   Another possibility that might be considered by a farmer or farming family would be to transfer the farming business and land to a company in exchange for shares, rolling over the chargeable gain under C.G.T.A. 1979, s.123. The advantage of such a transfer is that there would be no I.H.T. on the transfer to the company, as there is no chargeable transfer, and any subsequent transfer of shares should be treated as a disposal of shares in a non-quoted company. Owing to the low yields normally achieved by farming companies the valuation of a minority interest is likely to be much lower than the valuation of a similar proportion of the agricultural property itself.

The main problem with a transfer to a company is that if the farm were to be sold, the company would be faced with a capital gain, and if the company were to be liquidated to pass this through to the shareholders, there would again be a further charge on any gain on the value of the shares. This double charge only arises on a disposal of the farm land itself, and where it is probable that a family farm will be kept for future generations, the advantages in being able to deal with minority interest values may well outweight the potential C.G.T. disadvantage.

The Shares Valuation Division of the Inland Revenue is likely to want to value the shares in a farming company on a discounted assets basis, even for a minority interest, but there is no judicial authority for such a proposal. It is clear from *McConnel's Trustees* v. *I.R.C.* 1927 S.L.T. 14 that a controlling interest, in this case 99.8 per cent. of the company's shares, has to be valued by reference to the assets and not the earnings. With a lesser percentage the shares have to be valued on the basis of the rights attached to the shares, (*Borland's Trustee* v. *Steel Brothers & Co. Ltd.* [1901] 1 Ch. 279) largely by reference to dividends and earnings in the case of a minority shareholder (*Holt* v. *I.R.C.* (1953) 32 A.T.C. 402, *Lynall* v. *I.R.C.* (1971) 47 T.C. 375). This is not to say that the underlying asset value would be ignored entirely (*Salvesen's Trustees* v. *I.R.C.* (1930) 9 A.T.C. 43) but it is unlikely that the minority shareholder would benefit very much from the underlying assets except in the case of a sale and liquidation. If this is unlikely there are no grounds for attaching great importance to the asset base.

The Shares Valuation Division is likely to argue that where land is let to the company on an agricultural tenancy, the tenancy has a value of half the difference between the open market value of the land with vacant possession and the value of the tenanted land, even though the tenancy is not assignable. The argument is that the company has perpetual possession and the use of the land can be transferred by selling the shares. This argument has some merit where a controlling shareholding is to be valued, although the difference in value has to be discounted to reflect the inability to realise the tenancy, see Chapter 16.

In the case of a company there is also the opportunity to create shares **18–049** which at the time of creation had little value, for example, deferred shares with no right to participate in a divided or surplus on winding up, or to carry votes for a period of, say, 15 years. If these deferred shares are then transferred the immediate impact on the value of the transferor's estate is relatively modest, but in due course, if the deferred shares eventually rank *pari passu* with the ordinary shares, they will become of considerable value. There could be advantages in transferring such shares into discretionary trusts or accumulation and maintenance settlements for grandchildren. Care is needed in drawing up the rights of the deferred shares to ensure that their rights are engrained from the start and that there is no opportunity of watering them down at a later stage.

A way of freezing the value of shares in a company is to issue further shares to intended donees and providing in the articles that any increase in value goes to the new shares but that the existing value remains with the old shares. The argument is that there is no immediate reduction in the estate, merely a diversion of future growth to the new shares, without there being a reservation of benefit.

Business property relief

Business property relief is similar to agricultural relief in that it is given **18–050** as a percentage reduction in the value transferred before grossing up (s.104(2)) and before deducting annual reliefs (s.104(1)). The rate of relief is 50 per cent. in the case of property consisting of a business or interest in a business (s.105(1)(a)), or of shares which gave the transferor control, or more than 25 per cent. of the votes of an unquoted company (s.105(1)(b)). For this purpose it is possible to include related property (s.161) unless an election has been made to value the property as not related as a result of a sale within three years of death (s.176, s.105(2)).

Control of a company relates basically to voting control on all questions affecting the company as a whole (s.269(1)), including related property shares (s.269(2)). Shares held as trustee are deemed to be held by any person with an interest in possession and not by the trustees unless there is no such individual in which case the first named trustee would have the power to vote the shares, for example, in the case of a discretionary trust, (s.269(3)). Special voting rights in respect of winding up the company or class rights are ignored (s.269(4)).

Business property relief can be available to the trustees in respect of settled land used by a life tenant for his business, *Fetherstonaugh and Others* v. *I.R.C.* [1984] S.T.C. 261, as the life interest is an asset used in the business.

In other cases the rate of business property relief is 30 per cent. (s.104(1)(b)) in the case of unquoted shares which have 25 per cent. or less of the votes (s.105(1)(c)), land or buildings, machinery or plant used for the purpose of a business carried on by a company of which the transferor had control, or by a partnership of which he was a partner (s.105(1)(d)), or similar assets held for the purpose of a business carried on by the transferor which were settled property of a trust in which he had an interest in possession (s.105(1)(e)).

18–051 Where the business is that of, for example, dealing in land or stock or shares, the business property relief is not available (s.108(3)). The presence of a dealing company in a group however would not disqualify the parent of what is primarily a qualifying trading group (s.108(4)(b)). As with agricultural relief if there is a contract to sell shares or land, buildings, plant or machinery (s.113) business property relief is not available. There are exceptions for a transfer of a business or an interest in a business to a company in exchange for shares (s.113(a)) or shares were sold as part of a reconstruction or amalgamation (s.113(b)), or the company is put into liquidation for similar purposes (s.105(5)).

It should be noted that a mere business asset which does not constitute an interest in a business would not qualify for business property relief under s.104(1)(a) at 50 per cent. but nor would it qualify for the 30 per cent. relief unless it was used by a company controlled by the transferor, or by a partnership in which he was a partner under s.105(1)(d).

For C.G.T. it has been held that the sale by a farmer of five acres out of 35 was held to be the sale of an asset and not an interest in the business (*McGregor* v. *Adcock* [1977] S.T.C. 206). A transfer of a business asset, for example a combine harvester, should therefore be accompanied by a transfer of a share in the equity if business property relief is to be obtained. It is also necessary to show that the business is carried on with a view to profit so that a hobby farm would not qualify for relief (s.103(3)).

It is not possible to obtain agricultural property relief and business property relief on the same asset (s.114(1)). A commercial woodland counts as a business and if a deduction is made for I.H.T. on an earlier disposal in respect of which an election for postponement has been made (s.129) the business property relief is applied to the net value (s.114(2)). In other cases the value of business or an interest in a business is the net value of the business itself, *i.e.* the assets, including goodwill, less the liabilities (s.110(a) and (b)). In the case of an interest in the business it is only possible to include assets or liabilities which would be included in the value of the entire business (s.110(c)).

Land or buildings, machinery or plant used in a company in which the transferor had a control or a partnership in which he was a partner, or held by a trust in which the transferor had an interest in possession does not qualify for business property relief unless the business or shares are relevant business property in relation to the transfer (s.105(6)).

18–052 Property qualifying for the business property relief has to have been owned by the transferor throughout the two years immediately preceding the transfer (s.106). This means that the business or interest in the business or shares have to be held for the two-year period, not the underlying assets. If therefore in the case of a farm which had been carried on for more than two years there is a substantial increase in plant and machinery immediately before a transfer or death, the business property relief would apply because the interest in the business has been held for more than two years even though the assets making up the net value have not.

As with agricultural relief there is a provision dealing with replacement property within the two-year qualifying period (s.107(1)(a)). This is limited to the value of the property replaced at the date of the transfer (s.107(1)(b) and (2)) and the total period of ownership must comprise at least two years in total within the five years immediately preceding the transfer (s.107(1)(a)). Partnership changes and the transfer of business to a company controlled by the former owner of the business are disregarded (s.107(3)). The two-year period of ownership requirement is not relaxed in the case of a minority interest in an unquoted company except where there has been a paper for paper reconstruction, or amalgamation within the C.G.T. rules (C.G.T.A. 1979, ss.77 to 86, s.107(4)).

In the case of successions the transferor is deemed to have owned business property from the date of the death under which he inherited the property (s.108(a)). A deceased spouse's period of ownership may be included in determining the minimum period of ownership (s.108(b)). The relief for successive transfers of business property mirrors completely the agricultural relief provisions referred to above (s.109).

Where a company is a member of a group as defined by the Companies Act 1985 (s.103(2)) and it is carrying on a prohibited business of dealing in securities, land or investment holding (s.105(5)(3)) the value of the shares in that company is ignored for business relief purposes unless the business is that of a holding company (s.105(4)) or the investments held are used by other trading companies in the group (s.111).

It is also necessary to exclude excepted assets from business property relief (s.112(1)). An excepted asset is one not used for the purpose of the business throughout the two-year period or acquired for the future use in the business. Use by another group company is acceptable (s.112(2)). In the case of an asset qualifying through being used by a company controlled by the transferor or by a partnership in which he is a partner it must have been used throughout the two years preceding the transfer or have replaced another asset so used so that the total period of ownership of the asset and its replacement constitute at least two years out of the five years preceding the transfer, although the two-year period can be shortened if the successive transfer provisions apply (s.112(3)). A just apportionment may be made where an asset is partly used for the purpose of a qualifying business (s.112(4)) for example because the asset was partly used for the personal benefit of the transferor or a person connected with him (s.112(6)). The relevant period of ownership is the period ending with the transfer (s.112(5)).

Planning

As with agricultural relief the main planning aspect with regard to business property relief is to ensure that so far as possible it is available and, for example, loans are not secured on the business property itself but on other properties such as investment property, insurance policies or stocks and shares so that the net value qualifying for business property relief is not reduced by the amount of the loan (s.162(4)).   **18–053**

Woodlands

As mentioned previously, woodlands can qualify for agricultural property relief (s.115(2)) if part of an agricultural estate. In other cases specific woodland relief applies where land on which trees or underwood are growing is in the U.K. but is not agricultural property (s.125(1)(a)) and the   **18–054** owner has been beneficially entitled to the land throughout the five years ending with his death or for a shorter period where the land was acquired other than for a consideration in money or money's worth, *i.e.* by a gift or inheritance (s.125(1)(b)). For the relief to apply an election must be made within two years of the death, or such longer time as the Board may allow (s.125(3)).

The result of an election is that the value of the trees or underwood is left out of account in determining the I.H.T. payable on death (s.125(2)(a)). However, when the trees or underwood are subsequently disposed of, with or without the land, I.H.T. becomes payable (s.125(2)(b), s.126(1), (3)). Inter spouse transfers are ignored (s.126(2)).

In the case of an arm's length sale the net sale proceeds are charged to I.H.T. (s.127(1)(a)) and in any other case the net value of the trees or underwood at the time of disposal is charged (s.127(1)(b)). If the woodland constitutes commercial woodland so that the business property relief would apply, the value is reduced by 50 per cent. (s.127(2)).

The value charged is subject to I.H.T. by adding the value to the estate at the date of death and calculating the I.H.T. as if it were the highest part of the amended estate value (s.128). The estate value is also increased by the value of any earlier disposals if there has been a part disposal of the woodlands at an earlier date.

If the disposal is a chargeable transfer the value transferred is reduced by the tax payable by reference to the earlier death (s.129). The net proceeds of sale are after deduction of allowable expenses (s.130(1)(b)), including expenses of sale, expenses incurred in replanting within three years of the disposal, or such longer time as the board may allow to replace the trees or underwood disposed of, and expenses incurred in replanting to replace trees or underwood previously disposed of to the extent that this was not allowed on the previous disposal (s.130(2)).

18–055 It will be appreciated that the main advantage of the woodland exemption is where it is likely that the trees will pass on a further death before being sold. The additional tax charged on the disposal is only by reference to the last death on which the trees passed. Ideally if it is not possible to transfer them under a potentially exempt transfer which is likely to remain exempt, a life interest in the trees could be left to an elderly impecunious member of the deceased's family and sold after the death of the life tenant. The I.H.T. payable on the trees would be calculated by reference to aggregation with the life tenant's estate and not with the estate of the original deceased.

The disadvantage of an election is that the trees will continue to increase in value from the date of death to the date of disposal and if there is no intervening death the amount of inheritance tax ultimately payable must be increased over the amount that would have been payable if the election had not been made and inheritance tax paid on the probate value of the trees and underwood.

The tax is payable by the person who would be entitled to the sale proceeds if the disposal were a sale (s.208).

If the disposal is a gift the person paying the tax may elect to pay by 10 equal yearly instalments, the first of which is payable six months after the end of the month in which the transfer is made (s.229).

CHAPTER 19

Value Added Tax

System of taxation

Value added tax (V.A.T.) is chargeable on the supply of goods or ser- **19–001**
vices in the U.K. where it is a taxable supply made by a taxable person in
the course of a business (V.A.T.A. 1983, s.47) carried on by him
(V.A.T.A. 1983, s.2(1)). Taxable supplies are supplies other than those
which are exempt or outside the scope of V.A.T. and which may be taxed
at the standard rate or be zero-rated.

The way in which V.A.T. works is that a registered trader accounts to **19–002**
Customs and Excise for V.A.T. on his taxable outputs or sales and may
reclaim or set off the V.A.T. on his taxable inputs, *i.e.* the V.A.T. suffered
by him on his purchases and expenses incurred for the purpose of making
taxable supplies. Output tax on sales is calculated at the standard rate, currently 15 per cent. (V.A.T.A. 1983, s.9) or may be zero-rated, *i.e.* tax is
charged at 0 per cent. The V.A.T. so charged has to be accounted for to
Customs and Excise on a quarterly basis, or monthly basis if the business
will regularly receive repayments, after taking credit for V.A.T. input tax
on goods and services purchased which have been subjected to V.A.T. by
the supplier (V.A.T.A. 1983, s.14). Only the net amount is payable to, or
recoverable from, Customs and Excise.

In the case of an exempt supply, V.A.T. is not charged by the supplier
and any input tax incurred for making exempt supplies or non-business
activities is not reclaimable by the customer. With a zero-rated supply however the V.A.T. is notionally charged by the supplier at a zero rate which
means that input tax on his related purchases is fully deductible.

A supply for V.A.T. purposes is either a supply of goods or a supply of **19–003**
services; services include anything done by a taxable business for a consideration other than a supply of goods (V.A.T.A. 1983, s.3). A transfer of
the complete title in goods is a supply of goods whereas the hire of goods is
a supply of services (V.A.T.A. 1983, Sched. 2, para. 1). The supply of
goods on hire-purchase is a supply of goods. The supply of a process or
treatment to another person's goods is a supply of goods, (*e.g.* slaughtering

another person's livestock and then preparing the carcasses for sale as meat: V.A.T. leaflet 700/10/84) (V.A.T.A. 1983, Sched. 2, para. 2).

The supply of power, heat, refrigeration or ventilation is a supply of goods (V.A.T.A. 1983, Sched. 2, para. 3); so is the supply of water: S.I. 1989 No. 1114.

The grant, assignment or surrender of a major interest in land is a supply of goods (V.A.T.A. 1983, Sched. 2, para. 4).

19–004   The time of supply of goods is usually when they are removed and of services is when they are performed (V.A.T.A. 1983, s.4). An earlier part payment, or issue of a tax invoice for a standard-rated supply, brings forward the time of supply to that date. Where a tax invoice is issued after a standard-rated supply has occurred its date is usually taken as the time of supply where the tax invoice is issued within 14 days of the supply (or such longer period as has been agreed with Customs and Excise for the business) (V.A.T.A. 1983, s.5).

19–005   Only goods or services supplied in the U.K. are chargeable with V.A.T. (V.A.T.A. 1983, s.6). An export of goods is (or is treated as) a supply in the U.K., although zero-rated (V.A.T.A. 1983, s.16(6)). The importation of goods is liable to V.A.T. at the time of importation (V.A.T.A. 1983, s.2). The V.A.T. charge is not normally included in the quarterly V.A.T. return, but is usually payable at the time of importation. A supply of services is normally made where the supplier belongs: usually the country in which he is resident (V.A.T.A. 1983, ss.6(5), 8). The receipt of certain services is subject to a reverse charge on importation as if the purchaser had himself supplied the services, *i.e.* output tax has to be accounted for, although conversely the supplies may be eligible for an input tax credit as if they had been acquired from a U.K. supplier (V.A.T.A. 1983, s.7 and Sched. 3).

19–006   Where a supplier has issued a sales invoice but the debt becomes irrecoverable he will have accounted for the V.A.T. output tax on the sale. If the customer has become formally insolvent or bankrupt the V.A.T. may be reclaimed under V.A.T.A. 1983, s.22 and the V.A.T. (Bad Debt Relief) Regulations 1986, S.I. 1986 No. 335.

Registration

19–007   Registration is compulsory where a person makes taxable supplies in excess of £8,000 in any quarter or £23,600 in any period of four quarters, or it is reasonable to believe that his taxable supplies in the forthcoming year will exceed £23,600 (V.A.T.A. 1983, Sched. 1, para. 1 as amended from March 15, 1989). If, however, the taxable supplies would all be zero-rated it is possible to request exemption from registration under V.A.T.A. 1983, Sched. 1, para. 11(1). On the other hand a taxpayer who is not liable to be registered, for example because his taxable supplies are below the registration limit, may nonetheless request registration under V.A.T.A. 1983, Sched. 1, para. 5A. Once registered, input V.A.T. can be claimed on pre-registration expenses in certain circumstances (S.I. 1985 No. 886, Reg. 37) and so the timing of registration in this respect needs to be considered carefully quite apart from the prospect of penalties for late registration.

19–008   A group of companies may apply for group registration which treats them as a single entity for V.A.T. purposes (V.A.T.A. 1983, s.29).

Partnerships are registered in the name of the firm and treated as a single business for V.A.T., all partnerships with the same partners being treated as one business (V.A.T.A. 1983, s.30).

So far as a farmer is concerned most of the items which he produces will **19–009** usually be zero-rated as food for human consumption, animal feeding stuffs, seeds or live animals used for producing food for human consumption. On the other hand he may be supplying standard-rated items such as wool, pet food (see *Popes Lane Pet Food Supplies Ltd.* (MAN/86/148), Case 2186), holiday accommodation or services such as hiring out farm equipment or providing services to fellow farmers such as ploughing, seeding or harvesting. These services are all fully taxable supplies and require registration if over the limit in aggregate.

A number of the items acquired by the farmer in the normal course of his **19–010** business suffer V.A.T. including items used in the repair of buildings, services of contractors for the maintenance of buildings, contractors' services for work on the farm, accountants' and lawyers' services, services of land agents and farm advisers, the purchase of fertilisers and other consumable supplies to the farm. It may well be that the input tax on the goods and services purchased exceeds the output tax on those elements of the supplies which are taxable at the standard rate. The farmer will therefore be able to reclaim V.A.T. from Customs and Excise in which case he should consider requesting a monthly accounting period for V.A.T. purposes and submit monthly repayment claims to Customs and Excise.

A cash-flow advantage can also be obtained by ensuring that supplies **19–011** made to the farm business of goods or services eligible for zero-rating are not invoiced with standard rate V.A.T. added due to ignorance on the part of the supplier. For example, builders may overlook zero-rating on the demolition of old farm buildings or civil engineering work, the construction of new farm buildings or civil engineering works, alterations to buildings that are listed and where planning permission had been obtained as required.

Obtaining any zero-rating due on supplies to the farm will be particularly **19–012** important where there is a prospect of Customs and Excise seeking to disallow as a deductible input any standard rate V.A.T. if charged on the supply. This could be where Customs and Excise contend that certain improvements are made with a view to future sale of the farm (an exempt supply) rather than solely for the current running of the farm, or where they contend that works have been carried out for non-business purposes (*e.g.* improvements to the main farmhouse or works to improve the view from the farmhouse; see *Ballacamaish Farms Ltd.* v. *The Treasury* (MAN/86/86); Case 2364). This aspect of V.A.T. law is commented on further below.

Partial exemption

If the exempt outputs, for example from lettings or sales of land not **19–013** qualifying for zero-rating, are material there will be a restriction on input tax. This is normally arrived at on the basis that the input tax on those goods or services that directly relate to the exempt outputs is disallowed in full, apportionment being necessary in some cases (V.A.T.A. 1983, s.15;

S.I. 1985 No. 886, Regs. 29–37; and Customs and Excise Notice No. 706). There are *de minimis* provisions under which no input tax would be disallowed where the input tax wholly or partly related to exempt outputs does not amount to more than £100 per month on average, or less than £250 per month on average and below 50 per cent. of the total input tax, or less than £500 per month on average and below 25 per cent. of the total input tax.

The apportionment is made on a provisional basis on the proportion of the whole use made or to be made of such supplies (Reg. 30(1)(d)(ii)) which is subsequently adjusted (unless Customs and Excise dispense with the adjustment) by reference to the full V.A.T. year (Regs. 33 and 29) and any adjustment is accounted for in the same way as an over-deduction or under-deduction.

Only exceptionally does Customs and Excise permit disallowance of input tax related to exempt supplies to be ascertained by a formula method (Reg. 30).

In practice the record-keeping required to arrive at the correct disallowance for partially exempt supplies can give rise to many problems.

Non-reclaimable input tax

19–014   V.A.T. is not reclaimable on private expenses (including those of directors of a limited company, Customs and Excise Press Release November 12, 1987) as opposed to business expenses, nor on entertaining and staff nor for purchasing motor cars. If goods or services are exchanged, V.A.T. must be charged on the full selling price that would normally have been charged.

In the case of car fuel purchased by a sole proprietor, partner or employee, V.A.T. will be charged on 3/23rds of the car fuel scale benefits which apply to employees. V.A.T. is calculated on one-twelfth of the scale charge per month and reduced by 50 per cent. where the business mileage is over 1,500 miles in a month or averages over 4,500 miles in a three-month period: F.A. 1986, s.9.

V.A.T. on farmhouse running expenses used to be claimed on a flat rate proportion, usually one-third, but Customs and Excise now require a claim to be fully itemised and each item of expense suitably apportioned to the business use. However, V.A.T. Tribunals do not always support this approach (*The Grange Farm* (LON/86/80), Case 2344).

Exempt supplies

19–015   Exempt supplies so far as they normally affect farmers and landowners are set out by V.A.T.A. 1983, s.17 and Sched. 6, Group 1—Land. From April 1, 1989 they are as follows:

Group 1—Land

19–016   The grant (see note (1)) or any interest in or right over land or of any licence to occupy land.

However, the following are standard-rated:

(a) sale of the freehold in a new or partly completed (see note (2)) building or civil engineering work (see notes (4), (5) and (6), apart from any designed as a dwelling or intended for a relevant residential or charitable use (see note (3));

(b) the granting of any interest, right or licence of a right to take game or fish (see note (7);

(c) the provision in an hotel, inn, boarding house or similar establishment (see note (8)) of sleeping accommodation or of accommodation in rooms which are provided in conjunction with sleeping accommodation or for the purpose of a supply of catering;

(d) the provision of holiday accommodation (see note (9)) in a house, flat, caravan, houseboat (see note (10)) or tent;

(e) the provision of seasonal pitches (see note (11)) for caravans, and the grant of facilities at caravan parks to persons for whom such pitches are provided;

(f) the provision of pitches for tents or of camping facilities;

(g) the grant of facilities for parking a vehicle;

(h) the grant of any right to fell and remove standing timber;

(i) the grant of facilities for housing, or storage of, an aircraft or for mooring, (see note (12)) or storage of, a ship, boat or other vessel;

(j) the grant of any right to occupy a box, seat or other accommodation at a sports ground, theatre, concert hall or other place of entertainment; and

(k) the grant of facilities for playing any sport or participating in any physical recreation (see note (13)).

NOTES

(1) "Grant" includes an assignment, other than an assignment of an interest made to the person to whom a surrender of the interest could be made. **19–017**

(2) A building shall be taken to be completed when an architect issues a certificate of practical completion in relation to it or it is first fully occupied, whichever happens first; and a civil engineering work shall be taken to be completed when an engineer issues a certificate of completion in relation to it or it is first fully used, whichever happens first.

(3) Relevant residential or charitable use is as defined for zero-rating purposes (see para. 19–023 below).

(4) A building or civil engineering work is new if it was completed less than three years before the grant.

(5) Subject to Note (6), the grant of the fee simple in a building or work completed before April 1, 1989, is exempt.

(6) However, standard rate applies where the grant is the first grant of the fee simple made on or after April 1, 1989 and the building was not fully occupied, or the work not fully used, before that date.

(7) Where a grant of an interest in, right over or licence to occupy land includes a valuable right to take game or fish, an apportionment shall be made to determine how much of the supply is standard-rated.

(8) "Similar establishment" in (c) above includes premises in which there is provided furnished sleeping accommodation, whether with or without the provision of board or facilities for the preparation of food, which are used by or held out as being suitable for use by visitors or travellers.

(9) "Houseboat" in (d) includes boats or other floating decked structures designed or adapted solely as places of permanent habitation and not having means of self-propulsion (nor being capable of being readily adapted for this).

(10) "Holiday accommodation" in (d) above includes any accommodation advertised or held out as such.

(11) A seasonal pitch in (e) above is a pitch:
(a) which is provided for a period of less than a year; or

    (b) which is provided for a year or a period longer than a year but which the person to whom it is provided is prevented by the terms of any covenant, statutory planning consent or similar permission from occupying by living in a caravan at all times throughout the period for which the pitch is provided.

(12) "Mooring" in (i) above includes anchoring or berthing.

(13) Despite paragraph (k) above, exemption applies where the granting of the facilities is for:

    (a) a continuous period of use exceeding 24 hours; or

    (b) a series of 10 or more periods, whether or not exceeding 24 hours in total, where the following conditions are satisfied:

        (i) each period is in respect of the same activity carried on at the same place;

        (ii) the interval between each period is not less than one day and not more than 14 days;

        (iii) consideration is payable by reference to the whole series and is evidenced by written agreement;

        (iv) the grantee has exclusive use of the facilities; and

        (v) the grantee is a school, a club, an association or an organisation representing affiliated clubs or constituent associations

19–018   An election can be made from August 1, 1989, that supplies which would be exempt by virtue of Schedule 6, Group 1 can be standard-rated instead with the consequence that input V.A.T. incurred by a person making such supplies then becomes recoverable. The detailed provisions are in V.A.T.A. 1983, Sched. 6A, paras. 2–4 and do not apply to dwellings and property for residential or charitable use.

Zero-rated supplies

19–019   The aspects of zero-rating likely to be of interest are contained in V.A.T.A. 1983, s.16 and Sched. 5 Group 1, Group 2, Group 8 and Group 8A, as follows:

Group 1—Food

19–020   The supply of anything comprised in the general items set out below, apart from:

    (a) a supply in the course of catering, which includes any supply for consumption on the premises on which it is supplied and any supply of hot food for consumption off those premises; and for these purposes "hot food" means food which, or any part of which, has been heated for the purpose of enabling it to be consumed at a temperature above the ambient air temperature and is at the time of the supply above that temperature: and

    (b) a supply of anything comprised in any of the exceptions set out further below after the General items (and Notes thereon).

GENERAL ITEMS

Item No.

1. Food of a kind used for human consumption.
2. Animal feeding stuffs.
3. Seeds or other means of propagation of plants comprised in items 1 or 2 above.
4. Live animals of a kind generally used as, or yielding or producing, food for human consumption.

NOTES

(1) "Food" includes drink.
(2) "Animal" includes bird, fish, crustacean and mollusc.
(3) Any supply described in this Group includes the transfer of an undivided share of the whole property in goods (*e.g.* sale of part ownership of an animal) or the transfer of the possession of goods (*e.g.* hire of an animal for its milk).

EXCEPTED ITEMS

Item No.

1. Ice cream, ice lollies, frozen yoghurt, water ices and similar frozen products and prepared mixes and powders for making such products, apart from yoghurt unsuitable for immediate consumption when frozen.
2. Chocolates, sweets and similar confectionery including drained, glacé or crystallised fruits (other than drained cherries and candied peel); and biscuits and other confectionery (not including cakes) wholly or partly covered with chocolate or some product similar in taste and appearance.
3. Beverages chargeable with any duty of excise specifically charged on spirits, beer, wine or made-wine and preparations thereof.
4. Other manufactured beverages, including fruit juices and bottled waters, and syrups, concentrates, essences, powders, crystals or other products for the preparation of beverages; but not including tea, maté, herbal teas and similar products, and preparations and extracts thereof, or cocoa, coffee and chicory and other roasted coffee substitutes, and preparations and extracts thereof, or preparations and extracts of meat, yeast, egg or milk.
5. Any of the following when packaged for human consumption without further preparation, namely, potato crisps, potato sticks, potato puffs, and similar products made from the potato, or from potato flour, or from potato starch, and savoury food products obtained by the swelling of cereals or cereal products; and salted or roasted nuts other than nuts in shell.
6. Pet foods, canned, packaged or prepared; packaged foods (not being pet foods) for birds other than poultry or game; and biscuits and meal for cats and dogs (see *Popes Lane Pet Food Supplies Ltd.* (MAN/86/148, Case 2186).
7. Goods described in items 1, 2 and 3 of the general items which are canned, bottled, packaged or prepared for use:
   (a) in the domestic brewing of any beer;
   (b) in the domestic making of any cider or perry;
   (c) in the domestic production of any wine or made-wine.

Group 2—Sewerage services and water

Item No.   **19–021**

1. Services of:
   (a) reception, disposal or treatment of foul water or sewage in bulk; and

(b) emptying of any cesspools, septic tanks or similar receptacles which are used otherwise than in connection with the carrying on, in the course of a business, of a relevant industrial activity.

2. The supply for use use otherwise than in connection with the carrying on, in the course of a business, of a relevant industrial activity, of water other than:

(a) distilled water, deionised water and water of similar purity, and

(b) water comprised in any of the excepted items set out in Group 1 (Food) above (*e.g.* ice lollies, water ices, etc.).

Group 8—Construction of dwellings, etc.

19–022   Item No.

1. The granting by a person constructing a building designed as a dwelling or number of dwellings or intended for use solely for a relevant residential or charitable purpose of a major interest in, or in any part of, the building or its site.

2. The supply in the course of the construction of:

(a) a building designed as a dwelling or number of dwellings or intended for use solely for a relevant residential or charitable purpose; or

(b) any civil engineering work necessary for the development of a permanent park for residential caravans or any services other than the services of any architect, surveyor or any person acting as consultant or in a supervisory capacity.

3. The supply to a person of:

(a) materials; or

(b) builders' hardware, sanitary ware or other articles of a kind ordinarily installed by builders as fixtures, by a supplier who also makes to the same person supplies within item 2 of this Group, or Group 8A below, of services which include the use of the materials or the installation of the articles.

NOTES

19–023   (1) "Grant" includes assignment.

(2) "Dwelling" includes a garage constructed at the same time as a dwelling for occupation together with it.

(3) Use for a relevant residential purpose means use as:

(a) a home or other institution providing residential accommodation for children;

(b) a home or other institution providing residential accommodation with personal care for persons in need of personal care by reason of old age, disablement, past or present dependence on alcohol or drugs or past or present mental disorder;

(c) a hospice;

(d) residential accommodation for students or school pupils;

(e) residential accommodation for members of any of the armed forces;

(f) a monastery, nunnery or similar establishment; or

(g) an institution which is the sole or main residence of at least 90 per cent. of its residents, except use as a hospital, a prison or similar institution or an hotel, inn or similar establishment.

(4) Use for a relevant charitable purpose means use by a charity otherwise than in the course of furtherance of a business.

(5) Where part of a building is designed as a dwelling or number of dwellings or intended

for use solely for a relevant residential purpose or a relevant charitable purpose (and part is not):

    (a) a grant or other supply relating only to the part so designed or intended for use (or its site) shall be treated as relating to a building so designed or intended for such use;

    (b) a grant or other supply relating only to the part neither so designed nor intended for such use (or its site) shall not be so treated; and

    (c) in the case of any other grant or other supply relating to, or to any part of, the building (or its site), an apportionment shall be made to determine the extent to which it is to be so treated.

(6) Where all or part of a building is intended for use solely for a relevant residential purpose or a relevant charitable purpose:

    (a) a supply relating to the building (or any part of it) shall not be taken for the purposes of item 2 or 3 as relating to a building intended for such use unless it is made to a person who intends to use the building (or part) for such a purpose; and

    (b) a grant or other supply relating to the building (or any part of it) shall not be taken as relating to a building intended for such use unless before it is made the person to whom it is made has given to the person making it a certificate in such form as may be specified in a notice published by the Commissioners stating that the grant or other supply (or a specified part of it) so relates.

(7) The grant of an interest in, or in part of, a building designed as a dwelling or number of dwellings is not within item 1 if:

    (a) the interest granted is such that the grantee will not be entitled to reside in the building, or part, throughout the year; or

    (b) residence there throughout the year will be prevented by the terms of a covenant, statutory planning consent or similar permission.

(8) Where the major interest referred to in item 1 is a tenancy or lease:

    (a) if a premium is payable, the grant falls within that item only to the extent that it is made for consideration in the form of the premium; and

    (b) if a premium is not payable, the grant falls within that item only to the extent that it is made for consideration in the form of the first payment of rent due under the tenancy or lease.

(9) Where the benefit of the consideration for the grant of a major interest as described in item 1 accrues to the person constructing the building but that person is not the grantor, he shall for the purposes of that item be treated as the person making the grant.

(10) References in item 2 above and in the following Notes to the construction of a building or the construction of any civil engineering work do not cover the conversion, reconstruction, alteration or enlargement of any existing building or civil engineering work, or any extension or annexation to an existing building which provides for internal access to the existing building or for which separate use, letting or disposal is prohibited, and the references in item 1 above to a person constructing a building shall be construed accordingly.

(11) A caravan is not a residential caravan if residence in it throughout the year is prevented by the terms of a covenant, statutory planning consent or similar permission.

(12) Item 2 above does not include any supply of services comprising:

    (i) the transfer of an undivided share of the whole property in goods or the transfer of the possession of goods (*e.g.* hire of equipment for construction work), or

    (ii) the use of business goods for private purposes.

(13) In item 3 above, the goods referred to do not include:

    (a) finished or prefabricated furniture, other than furniture designed to be fitted in kitchens; or

    (b) materials for the construction of fitted furniture, other than kitchen furniture; or

    (c) domestic electrical or gas appliances, other than those designed to provide space heating or water heating or both; or

    (d) carpets or carpeting material.

(14) Goods forming part of a description of supply in this Group do not qualify for freedom from import V.A.T. on arrival in the U.K.

Detailed provisions are contained in V.A.T.A. 1983, Sched. 6A to impose charges on a change of use of property within 10 years of its construction where it has been the subject of a supply after March 31, 1989, under the zero-rating provisions governing use for a relevant residential or charitable purpose.

Group 8A—Protected buildings

19–024 Item No.

1. The grant by a person substantially reconstructing a protected building, of a major interest in, or in part of, the building or its site.
2. The supply, in the course of an approved alteration of a protected building, of any services other than the services of an architect, surveyor or any person acting as consultant or in a supervisory capacity.
3. See item 3 of Group 8 above.

NOTES

19–025

(1) "Protected building" means a building designed to remain or become a dwelling or number of dwellings, or intended solely for relevant residential or charitable use, and which is:
   (a) a listed building, within the meaning of:
      (i) the Town and Country Planning Act 1971; or
      (ii) the Town and Country Planning (Scotland) Act 1972; or
      (iii) the Planning (Northern Ireland) Order 1972; or
   (b) a scheduled monument, within the meaning of:
      (i) the Ancient Monuments and Archaeological Areas Act 1979;
         or
      (ii) the Historic Monuments Act (Northern Ireland) 1971.
(1A) Notes (1) to (8) of Group 8 above apply for this Group also.
(2) For the purposes of item 1, a protected building shall not be regarded as substantially reconstructed unless when the reconstruction is completed:
   (a) the reconstructed building incorporates no more of the original building (that is to say, the building as it was before the reconstruction began) than the external walls, together with other external features of architectural or historic interest, or
   (b) of the works carried out to effect the reconstruction, at least three-fifths measured by reference to cost are of such a nature that the supply of services (other than excluded services), materials and other items to carry out the works would, if supplied by a taxable person, be within either item 2 of this Group or item 3 of Group 8 above, as it applies to a supply by a person supplying services within item 2 of this Group; and "excluded services" means the services of an architect, surveyor or other person acting as consultant or in a supervisory capacity.
(3) "Approved alteration" in item 2 above means:
   (a) in the case of a protected building which is an ecclesiastical building which is for the time being used for ecclesiastical purposes (or would be so used but for the works in question), any works of alteration; and
   (b) in the case of a protected building which is a scheduled monument within the meaning of the Historic Monuments Act (Northern Ireland) 1971 and in respect of which a protection order within the meaning of that Act is in force, works of alteration for which consent has been given under section 10 of that Act; and
   (c) in any other case, works of alteration which may not, or but for the existence of a Crown interest or Duchy interest could not, be carried out unless authorised under, or under any provision of,
      (i) Part IV of the Town and Country Planning Act 1971,
      (ii) Part IV of the Town and Country Planning (Scotland) Act 1972,
      (iii) Part V of the Planning (Northern Ireland) Order 1972, or
      (iv) Part I of the Ancient Monuments and Archaeological Areas Act 1979
      and for which, except in the case of a Crown interest or Duchy interest, consent has been obtained under any provision of that Part. "Crown interest" and "Duchy interest" have the same meaning as in section 30 of the said Act of 1979.

(4) For the purposes of paragraph (a) of Note (3) a building used or available for use by a minister of religion wholly or mainly as a residence from which to perform the duties of his office shall be treated as not being an ecclesiastical building.

(5) Where the benefit of the consideration for the grant of a major interest as described in item 1 accrues to the person substantially reconstructing the protected building but that person is not the grantor, he shall be treated for the purposes of that item as the person making the grant.

(6) In item 2 "alteration" does not include repair or maintenance; and where any work consists partly of an approved alteration and partly of other work, an apportionment shall be made to determine the supply which falls within item 2.

(6A) In item 2 construction of a building separate from but within the curtilage of a protected building does not constitute an alteration of the protected building.

(7) Note (12) of Group 8 above applies for this Group also.

Building land developed for resale

From August 1, 1989, provisions detailed in V.A.T.A. 1983, Sched. 6A, **19–026** paras. 5 and 6 take effect to trigger a deemed standard-rated self-supply of land on which construction works are commenced by a person unless:

(a) he holds the freehold and before beginning construction he certifies that he intends to dispose of the land within three years of completing the construction, or

(b) before beginning construction he certifies that he intends to occupy the whole building for at least 10 years from completion and either he intends to be fully taxable throughout that period, or he expects the total value of services supplied to him during construction to be under £100,000, or

(c) he has elected for taxation on or before the beginning of construction.

If a building is being constructed with a view to sale it may be possible to **19–027** enter into partnership with the project managers or property developers rather than paying a fee subject to V.A.T. (*Strathearn Gordon Associates Ltd.* [1985] V.A.T.T.R. 79; (LON/84/294), Case 1884).

In order to dispose of a dwelling as a zero-rated construction it is necess- **19–028** ary to dispose of a major interest in land which under V.A.T.A. 1983, s.48(1) means a sale of the freehold or the grant of a lease for a term exceeding 21 years. The grant of a lease for a lesser period by a constructor of the building will be exempt and not zero-rated, although a lease of over 21 years with break clauses is normally accepted as still being a major interest in land.

In order to qualify for zero-rating the landowner should be actively involved in the qualifying construction, at least to the extent of commissioning the construction of the building, and exercising control over the planning and design rather than merely licensing another developer who builds on the land (*Hulme Educational Foundation* [1978] V.A.T.T.R. 179; (MAN/77/104); Case 625).

It is only construction of any qualifying building and reconstruction of **19–029** protected buildings which qualifies for zero-rating. Other activities such as conversion or alteration, repair or reconstruction are standard-rated so that in some cases it may be preferable to demolish a building and construct it anew rather than reconstruct it. It does not appear necessary for the construction of the building to be completed for any disposal to be zero-rated (*Stapenhill Developments Ltd.* MAN/82/229, Case 1593) but it should have progressed beyond foundation level.

Standard-rated supplies

19–030 Other items subject to V.A.T. at the standard rate of 15 per cent. which are commonly supplied by farms include:

(a) agisting and the care of animals on a contract grazing as opposed to the mere letting of land for grazing purposes which would be zero-rated;

(b) the sale of produce not covered by Schedule 5, Group 1, for example hides and skins of animals, carcasses sold for pet food, packaged foods for birds other than poultry or game, straw sold as animal bedding, manure, fish other than those sold for human or animal feed (see *R.T.L. Tancred* (MAN/86/110), Case 2275), for example coarse-fish farming;

(c) non-exempt rights over land, for example the provision of accommodation or catering facilities, the letting of a house, flat, caravan or houseboat as holiday accommodation, fees for allowing camping in tents or caravans, parking fees, the granting of sporting rights to take fish or game, the granting of any right to fell and remove standing timber, or fees received for letting land for the purposes of exhibitions, point-to-points or other sporting or physical recreation activities: see 19–016.

19–031 Livery charges, stud fees and keep (see *C. & E. Comrs.* v. *Bushbys* [1979] S.T.C. 8) would all be standard-rated as would the provision of drink subject to excise duty such as wine, beer or cider. Farm horses kept for work (see *J. C. & N. C. Ward Ltd.* (LON/83/8), Case 1416) or breeding will be liable to V.A.T. at the standard rate. There is a margin scheme for horses and ponies (S.I. 1983 No. 1099).

19–032 It would appear that a sale of a milk or potato quota is a standard-rated supply unless exempt as part of a sale of land.

19–033 Although a supply is normally thought of as a supply at the net price (100 per cent.) plus V.A.T. at the standard rate (15 per cent.) to give the final consideration, it is in reality a final consideration of which the V.A.T. element, if standard-rated, is 15/115ths or 3/23rds (V.A.T.A. 1983, s.10(2)).

19–034 A cash discount for prompt payment reduces the value on which V.A.T. is calculated, whether or not the discount is taken by the customer (V.A.T.A. 1983, Sched. 4, para. 4).

19–035 There are detailed requirements as to the content of a tax invoice issued by a registered supplier and as to the records to be kept, many of which are dealt with by regulations (V.A.T.A. 1983, Sched. 7; S.I. 1985 No. 886, Reg. 13). Special schemes for retailers could apply to a farm shop which makes standard-rated supplies as well as zero-rated food sales (S.I. 1972 No. 1148, Reg. 2).

19–036 In the case of second-hand cars, motor cycles, boats, caravans, aircraft, firearms and works of art, a supplier of such goods who acquired them second-hand without an invoice showing V.A.T. is liable to standard rate

V.A.T. only on the gross profit margin from his sale. This means that he does not issue a V.A.T. invoice to a customer and no input tax is claimable by the customer (V.A.T.A. 1983, s.18; S.I. 1980 No. 442; S.I. 1981 No. 1741, arts. 4 & 5; S.I. 1983 No. 1099).

A reverse premium paid to a tenant on entering into or taking over a **19–037** lease is subject to V.A.T. (*Neville Russell* (LON/86/708), Case 2484).

Transfer of a business or tenancy

On cessation of a business the disposal of its assets is normally a supply **19–038** of goods and services by the business (V.A.T.A. 1983, s.47(6)) unless it is transferred as a going concern (S.I. 1981 No. 1741, art. 12). If a transferor and transferee apply for the transferee to keep the same registration number as the transferor and the trade is treated as a continuing business for V.A.T. purposes then the transferee is also liable for any past V.A.T. still due from the transferor (V.A.T.A. 1983, s.33 and S.I. 1985 No. 886, Reg. 4(7)).

The V.A.T. position on the transfer of tenant right at the end of a **19–039** tenancy is covered in the following joint memorandum of Customs and Excise and the Central Association of Agricultural Valuers dated July 25, 1984.

The application of value added tax to the transfer of tenant right, live and dead stock, improvements and fixtures and fittings between the parties at the end of a tenancy.

A. Transfers on the grant or termination of an agricultural tenancy giving **19–040** rise to payments between the parties for items such as:
— "tenant's improvements" to the property
— tenant's fixtures, such as machinery and storage bins, affixed to the land, gates and fences, and
— growing crops and cultivations, as well as fertilisers and seeds which have been applied to the land.
1. Such transfers by the landlord to the outgoing tenant, or vice versa, at the end of the tenancy under section 13 of, and Schedules 2, 3 4 and 5 to, the Agricultural Holdings Act 1948 are outside the scope of value added tax.
2. Such transfers which are direct between the incoming and outgoing tenant under the above section and Schedules form part of the consideration for the exempt supply of the right of the incoming tenant to occupy the land, so that no value added tax is chargeable. (Note that, in this situation, the landlord's partial exemption position in relation to other matters may be affected.)
3. Such transfers by the incoming tenant to the landlord which are within the scope of the above sections and Schedules are within the scope of value added tax, but exempt under Sched. 6 (Exemptions) to the Value Added Tax Act 1983.

B. Transfers arising from the grant or termination of an agricultural **19–041** tenancy of specific items giving rise to payments between the parties in the circumstances of A1, 2 and 3 above.

1. Animal feed stuffs, such as home grown hay, straw, silage, cereals and purchased feeding stuffs are within the scope of value added tax, but zero-rated under V.A.T.A. 1983, Sched. 5, Group 1, Item 2.
2. Fuel and oils (such as red diesel oil and TVO) are within the scope of value added tax, but zero-rated under Sched. 5, Group 7, Item 4. (Road fuel is standard rated).
3. (a) Machinery which is not affixed to the land, such as field machinery, vehicles and tools, and
   (b) purchased fertilisers, sprays and other chemicals are within the scope of value added tax and are subject to such tax at the standard rate.

19–042 C. Payment on the transfer of a business.

1. Value added tax is not charged on the transfer of a farm business as a going concern.
   (a) when an outgoing tenant transfers his whole business as a going concern to the landlord who then runs it himself
   (b) where a landlord who has run the business himself transfers the whole as a going concern to an incoming tenant, and
   (c) where an owner occupier of an agricultural property who has run the business himself sells it as a going concern.
2. The exemption also applies in certain circumstances to the transfer of part of the business as a going concern.
3. The above is subject to the conditions set out in Article 12(1) of the Value Added Tax (Special Provisions) Order 1981 (see below).

19–043 D. The grant of a freehold or leasehold interest in commercial land and/or premises used as a business for a single charge which incorporates a payment for fixtures and fittings.

1. In these circumstances the transaction will be treated as a single supply for value added tax purposes and will normally be exempt under V.A.T.A. 1983, Sched. 6, Group 1, Item 1.
2. Items invoiced separately as "fittings" will however normally be liable to value added tax at standard rate.

19–044 The V.A.T. (Special Provisions) Order 1981, S.I. 1981 No. 1741, art. 12(1) provides as follows:
"There shall be treated as neither a supply of goods nor a supply of services:

(1) the following supplies by a person of assets of his business:
   (a) their supply to a person to whom he transfers his business as a going concern;
      (i) where the assets are to be used by the transferee in carrying on the same kind of business, whether or not as a part of any existing business, as that carried on by the transferor, and
      (ii) where, in the case in which the transferor is a taxable person, the transferee is already, or immediately becomes as a result of the transfer, a taxable person or a person defined as such in section 2(2) of the Manx Act;

(b) their supply to a person to whom he transfers part of his business as a going concern:

(i) where that part is capable of separate operation,

(ii) where the assets are to used by the transferee in carrying on the same kind of business, whether or not as part of any existing business, as that carried on by the transferor in relation to that part, and,

(iii) where, in a case in which the transferor is a taxable person, the transferee is already, or immediately becomes as a result of the transfer, a taxable person or a person defined as such in section 2(2) of the Manx Act."

Appeals

Appeals may be made to a V.A.T. tribunal if agreement cannot be reached with Customs and Excise (V.A.T.A. 1983, s.40; Sched. 8; S.I. 1986 No. 590). **19–045**

Penalties

There are substantial penalties for failure to comply with the V.A.T. rules (V.A.T.A. 1983, s.39; F.A. 1985, ss.12–19), which are outside the scope of this book. **19–046**

Leaflets

Customs and Excise have published a number of leaflets which may be of interest to farmers including: **19–047**

701/14 food
701/15 animal feeding stuffs
701/16 sewerage services and water
701/25 pet food
701/37 live animals
701/38 seeds and plant
701/40 abattoirs
708/1 protected buildings (listed buildings and scheduled monuments)
708/2 construction industry
708/3 civil engineering
709/3 hotels and holiday accommodation
709/4 holiday services
726 second-hand horses and ponies
742 land and property
742/1 letting of facilities for sport and physical recreation.

These may all be obtained free of charge from the local V.A.T. office.

CHAPTER 20

Rates

System of tax

20–001   From April 1990 local authorities in England and Wales collect non-domestic rates under the Local Government Finance Act 1988, which repeals the whole of the General Rate Act 1967. Scotland has separate but similar legislation which took effect from April 1989. Rates are levied by reference to units of property known as hereditaments, normally on the basis of occupation, so that a single building in multiple occupation may be divided into a number of different hereditaments for rating purposes (*Allchurch* v. *Hendon Union Assessments Committee* [1891] 2 Q.B. 436). However, properties within a single occupation may nonetheless be split into different hereditaments where they are, for example, capable of being separately let, used for different purposes or are physically separated from one another.

Each hereditament has placed upon it by the valuation officer of the Inland Revenue a rateable value and is included in the local non-domestic rating lists. In England and Wales the list must be compiled on April 1, 1990 and on April 1 in every fifth year afterwards: L.G.F.A. 1988, s.41(2).

The rateable value is the hypothetical rent which could be obtained from a hypothetical letting on the basis that the tenant undertakes to pay all the repairs, insurance and other expenses for the maintenance of the property: L.G.F.A. 1988, Sched. 6, para. 2.

Appeals

20–002   It is possible to appeal against the valuation placed on a hereditament by the valuation officer by making a written proposal under L.G.F.A. 1988, s.55 which specifies the grounds on which the alteration is requested. Any alteration ultimately made may be backdated

Liability

20–003   Rates are payable by the occupier of the hereditament. The occupier must be the actual beneficial occupier and be in permanent exclusive occupation. Where property is unoccupied the owner may be liable for the rates: L.G.F.A. 1988, s.45.

Agricultural property

Agricultural land and buildings are exempt under L.G.F.A. 1988, **20–004**
Sched. 5, paras. 1–8.

Agricultural land is defined by L.G.F.A. 1988, Sched. 5, para. 2 as
including land used as arable, meadow or pasture ground only. Intended
future use for farming is not sufficient (*Fowler* v. *Tavener* (1953) 47 R & IT
39). Purely incidental non-qualifying use does not cause the exemption to
be lost (*Jarvis* v. *Cambridgeshire Rural Assessment Area Assessment Committee* [1938] 4 All E.R. 186) but activities such as putting a caravan on the
land (*Moore* v. *Williamson* [1973] R.A. 172) or turf cutting (*Meriden &
Solihull Rating Authority* v. *Tyacke* [1950] 1 All E.R. 939) would cause an
exemption to be lost.

Land used for a plantation or a wood or for the growth of saleable underwood is exempt; this would obviously include a commercial woodland or a
farm where the sale of wood forms a material element in the farm income.

Certain land is specifically excluded from the definition of agricultural **20–005**
land such as land occupied together with a house as a park, gardens (other
that market gardens), pleasure grounds, land used mainly or exclusively
for purposes of sport or recreation (such as golf courses, games preserves
and fishing rights) and land used as a racecourse: L.G.F.A. 1988, Sched. 5,
para. 2(2).

Agricultural buildings for the purposes of rating exemption are defined
as buildings (other than dwellings) occupied together with agricultural land
or forming part of a market garden and in either case used solely in connection with agricultural operations thereon: L.G.F.A. 1988, Sched. 5,
para. 3.

Agricultural operations are those by way of cultivating the soil or rearing
livestock (*Gilmore* v. *Baker-Carr* [1962] 3 All E.R. 230) which includes
mink farming (*Hallam* v. *James* (1958) 4 R.R.C. 142), cheese making on a
farm (*Covell* v. *Streatfield Hood & Co. Ltd.* (1984) R.A. 193), rearing livestock under controlled conditions (*National Pig Progeny Testing Board* v.
Greenall [1960] 3 All E.R. 556), including embryo transplants (*Northern
Ireland Animal Embryo Transplant Ltd.* v. *Commissioner of Valuation*
(1983) R.A. 183) and mushroom growing (*Beveridge & Co. Ltd.* v. *Perth &
Kinross Assessor* (1967) R.A. 482).

However, agricultural operations exclude the stabling of horses and land
used for gallops (*Thomas* v. *Kenneth Beeston Farms Ltd.* (1958) 4 R.R.C.
1); *Evans* v. *Bailey* [1982] R & V R 136; *Forster and Another* v. *Simpson:
Bailey* v. *Simpson* [1984] R.A. 85), artificial insemination units (*Perth &
Kinross Assessor* v. *Scottish Milk Marketing Board* (1963) S.L.T. 109),
land and buildings used for rearing pheasants and partridges (*Cook* v. *Ross
Poultry Ltd.* (1982) R.A. 187), for an annual point to point (*Hayes* v. *Loyd*
[1985] 2 All E.R. 313), turf growing (*Henshaw* v. *Halton Borough Council
and Watmore* (1984) R & V R 237) and a racehorse stud (*Hemens* v. *Whits-
bury Farm and Stud Ltd.* [1988] 1 All E.R. 72, H.L.).

Buildings in multiple occupation, for example, by a farming co-
operative, are covered by L.G.F.A. 1988, Sched. 5, para. 4.

Buildings used for the intensive rearing of livestock are included as agricultural buildings by Sched. 5, para. 5, as are those used for bee-keeping

(Sched. 5, para. 6); buildings occupied by companies and other corporate bodies are covered by Sched. 5, para. 7 and fish farms by Sched. 5, para. 9. A chicken processing plant did not qualify for exemption in *Prior* v. *Sovereign Chicken Ltd.* [1984] R.A. 73 nor did a mill in *Ipswich Borough Council* v. *Eastern Counties Farmers Ltd. and Tye (V.O.)* [1985] R.A. 111.

Certain rights of fishing are exempt under L.G.F.A. 1988, Sched. 5, para. 10.

Sporting rights are liable to rates which will be collected from either the landlord or the tenant depending on the precise arrangements made, see Chapter 10.

Finance

Market situation

As the business of farming tends to become more and more capital inten-   **21–001** sive the problem of raising the necessary capital becomes more difficult. For a time, with high levels of inflation and increase in land values, the real cost of borrowing money for farm financing was relatively insignificant and farmers tended to find it relatively easy to borrow on the security of their freehold or leasehold interest in the farm. Times have, however, changed and the income of many farmers has dropped in real terms and the rela- tively low levels of inflation and high levels of interest make the real cost of borrowing a substantial burden in many cases. As a result of the drop in farm incomes the price of farm land has also fallen very considerably from its peak which has in turn affected the security of the lenders and has caused them to reappraise loans to farmers and to make the raising of capi- tal much more difficult.

John McQueen, secretary of the Association of Bankrupts, writing in the British Friesian Journal of April 1987, suggested the following rule- of-thumb limits for a financially healthy farm:

(1) Short-term borrowings should not be more than £300 per acre.
(2) Long-term borrowing costs should not be more than £55 per acre.
(3) Net worth of business should not be decreasing annually.
(4) Gross outputs should not be less than £240 per acre.
(5) Gross margin should not be less than £160 per acre.
(6) All other costs should not be over £200 per acre.
(7) Output per £100 of labour costs should not be less than £500.

(8) Leasing and depreciation charges should not be more than £40 per acre.

(9) Tax and drawings should not be more than £50 per acre.

Most lenders would nowadays wish to lend no more than 50 per cent. of the capital value of the farm and would wish to see the interest charges covered at least twice by the net farm profit before interest.

Clearing banks

21–002   The clearing banks are the normal source of short-term finance and often provide working capital by way of overdraft. Interest on overdrafts is normally a fixed percentage over the bank's base rate but unauthorised overdrafts are often charged at a penal rate, often without the borrower being aware of the fact.

Banks also provide long-term loans for the purpose of capital improvements and the purchase of additional farm land. These loans may be in various forms on a fixed or variable rate of interest with repayment periods scheduled to meet the borrower's requirements so far as possible. It is quite common to have a period during which only interest is paid on the loan at, say, quarterly intervals, with capital repayments commencing after perhaps two to five years when the benefit of the capital expenditure is expected to be reflected in the farming profits. Capital repayments may take the form either of direct scheduled repayments or repayment at the end of the loan through endowment life assurance policies on a with- or without-profits basis. Banks might be persuaded to accept repayment out of a commuted pension scheme if the farm is profitable and a long-term annual premium contract is entered into. This is obviously the most tax efficient means of obtaining repayment of the loan because tax relief is obtained for the premiums paid to the pension fund, the investment is within a tax-free fund and commutation used to repay the loan is tax free.

An endowment policy, since the abolition of life assurance relief, has little to recommend it as a means of repayment and in most cases it would be preferable, where a pension fund base scheme is not available, to have a scheduled loan repayment and to provide for the repayment of the loan in the case of death by appropriate term insurance cover.

Insurance companies

21–003   Insurance companies themselves provide a source of funds to farmers on a long-term mortgage basis although these may be expensive if, as would normally be the case, the insurance company would require repayment of the loan to be funded from an endowment policy. If the insurance company insists on a non-profits endowment policy, as many do, the cost becomes even more expensive in the long term.

Agricultural Mortgage Corporation

21–004   The Agricultural Mortgage Corporation is a major source of long-term finance to farmers. In theory the loans may be of up to two-thirds of the value of the land being acquired and for periods of up to 40 years. The rate of interest may be either variable or fixed or a combination of the two, and there is an ability to convert from a variable to a fixed rate of interest at no charge.

Loans for buildings, particularly for the intensive rearing of livestock, tend to be shorter term and the amount lent is more likely to be restricted to 50 per cent. of the total cost. As with all lending institutions these days the ability of the business to repay the loan and to pay the interest are paramount.

Repayment may be by way of a normal repayment mortgage or by an endowment policy or repayment through pension fund commutation. In these last two cases, interest only is paid on the loan to the Agricultural Mortgage Corporation, the repayment coming from the insurance policy or pension fund. An endowment policy is likely to prove an expensive means of repayment.

Lands Improvement Company

Another source of funds, particularly for the acquisition of land, is the **21–005** Lands Improvement Company, which normally lends on terms of between five and 20 years for up to two-thirds of the value of the property. The normal minimum loan is £25,000. Repayment can be on a normal repayment mortgage method or by endowment policy. Arrangement fees are charged for setting up the loan. The Lands Improvement Company is empowered to make loans under the Improvement of Land Acts secured by way of statutory rent charges rather than by way of mortgage which can be helpful in certain cases as the title deeds of the land are not charged. Lands Improvement Company loans are normally at a variable rate of interest although a fixed rate is possible and, in the case of rent charges, is normal.

Pension funds

Other sources of finance include merchant banks and pension funds. **21–006** These tend to be for large loans and the repayment terms would normally be similar to those of a clearing bank. In certain cases it might be possible to arrange with such an institutional investor an equity mortgage whereby the rate of interest is relatively low but the amount to be repaid at the end of the term is related to the then value of the land or to the index of retail prices. Whether or not such loans look attractive depends on the view taken of the likely levels of land prices and inflation over the term of the loan.

Private funding

It is obviously possible in some cases to borrow from private sources **21–007** such as family trusts or direct from members of the family. In such cases the terms of the loan and their repayment would depend on the arrangements made in each case.

From 1981 to 1984 the business start-up scheme and business expansion scheme gave relief for investment in new farming companies. This relief is no longer available as a result of I.C.T.A. 1988, s.297(2)(j).

Sale and leaseback

21–008   Finally it is possible to raise money by selling farm land and leasing it back as an agricultural tenant. This is a part disposal for C.G.T. purposes. The amount realised on the sale of the land is affected by the terms of the tenancy under which it is leased back. As a source of funds a sale and lease-back has the major disadvantage that it does involve a disposal of what is in many instances the prime security of the farm land and this in turn could make short-term borrowing much more difficult in the absence of such security.

Endowment policies

21–009   Although non-profit endowment policies as a means of loan repayment are normally unattractive it might be possible in some cases to persuade the lender to take as a basis of repayment a low-cost endowment policy which is really a combination of a reducing-term policy and a with-profits endowment policy for a smaller amount than the loan if the policy would, on the assumption that profits are accumulated at a reasonably modest level, provide sufficient, on maturity with accumulated profits, to repay the loan and provide a lump sum in addition.

Self-administered pension funds

21–010   Where farming is being carried on by a company it might be possible to set up a small self-administered pension fund for the directors under which premiums would be paid to the pension fund and would be a deduction from the profits at the time of payment. The monies would be invested within the pension fund as a tax-free entity and up to 50 per cent. could be lent back on normal commercial terms to the employing company. Alternatively the pension fund itself could borrow normally a maximum of three times the assets of the fund and use the proceeds to acquire additional farm land which would be let to the company at a normal commercial rent. The rent paid would be deductible by the paying company and would be tax free in the hands of the recipient pension fund.

In order to maximise relief for I.H.T. purposes (Chapter 18) it may be possible to charge non-agricultural assets such as an insurance policy and provide only collateral insurance by way of which a charge on the farm land in order to preserve agricultural relief, in view of I.H.T.A. 1984, s.162(4).

This ploy cannot be used to preserve assets for business relief purposes as it is the purpose of the loan which is important, not the security offered (I.H.T.A. 1984, s.227(7)(b)).

Agricultural Credit Corporation

21–011   Another source of financial assistance is the government sponsored Agricultural Credit Corporation which may provide guarantees for short- and medium-term loans from other financial institutions, in particular clearing banks, for periods of up to 10 years. Loans can be for the purchase of land, buildings, fixed plant and machinery, livestock and working capital. The Corporation normally charges a fee of 2 per cent. of the amount of the guarantee and the scheme is not unlike the small firms loan guarantee scheme operated by the Government and clearing banks.

Agricultural merchants

Many farmers have looked to agricultural merchants for credit, but even **21–012** if available it is often an expensive form of finance either in the form of direct interest charges or higher prices for produce compared with a cash purchase, or commonly a mixture of both. As the merchant, in order to provide credit of this nature, would have to borrow in the first instance, the farmer is effectively paying the merchant's borrowing costs and in addition the merchant's administrative costs and profit in providing the credit facilities. In some cases merchants have arranged with their own bankers a point of sale credit to customers under which the bank effectively provides trade credit, at a price, to the farmer on a minimum enquiry basis. Such finance is likely to prove considerably more expensive than ordinary overdraft facilities. A point of sale finance of this nature has been promoted by Barclays Bank plc as a "master loan facility" and National Westminister Bank plc as "grow cash." These schemes are typically for a maximum period of 18 months at a fixed or variable interest rate with flexible repayment arrangements and for up to 100 per cent. of the cost of the goods supplied through the merchant.

The interest payable would normally be deductible as a trading expense.

Finance costs

Where the cost of raising finance is a capital expense it can nonetheless **21–013** normally be treated as a trading expense as incidental costs of obtaining loan finance under I.C.T.A. 1988, s.77. In the case of a farming company it may, in certain cases, be desirable to raise finance to enable the company to buy its own shares either in order to pay an I.H.T. liability arising on death, or to buy out a retired or dissenting shareholder. This can be done under the provisions of the Companies Act 1985, ss.159–181 but in order to avoid a distribution charge for corporation tax purposes it is necessary to comply with the precise requirements of I.C.T.A. 1988, ss.219–226 in respect of which clearances are available under I.C.T.A. 1988, s.225.

Hire-purchase

Sources of short-term finance include hire-purchase and leasing. Under **21–014** hire-purchase arrangements the property in the equipment does not pass to the purchaser until the final instalment has been paid but capital allowances are available as if the plant were fully owned, as explained in Chapter 5. As the benefit of the capital allowances goes to the hirer it is unlikely that hire-purchase would be cheaper than other forms of short-term credit, for example, a bank overdraft, and in many cases would be substantially more expensive. In a hire-purchase contract the instalments are split between income and capital and the income element, usually known as hire-purchase interest, is claimed as a trading expense for tax purposes over the period of the contract. The interest is often calculated on a straight line basis but any alternative method that is both sensible and consistently applied is likely to be acceptable to the Revenue.

Leasing

21–015   A finance lease contract differs from a hire-purchase contract primarily in that there is no condition in the contract under which the lessee automatically obtains title to the equipment or even has an option to acquire the equipment at the end of the lease. There is often an informal understanding that at the end of the lease the equipment will be made available to an associate of the lessee at a nominal price, and the equipment may then be acquired by the business. Alternatively the lease may be continued at a nominal rent or the hirer may be appointed the lessor's agent to sell the equipment at a commission of, typically, $97\frac{1}{2}$ per cent. of the sale proceeds. Under a leasing contract the benefit of the capital allowances accrues to the lessor and the whole of the lease charges will be claimed as a trading expense by the lessee. For accounting purposes where Statement of Standard Accounting Practice No. 21 is followed equipment used under a finance lease may well be shown in the balance sheet.

An operating lease is one which is defined by S.S.A.P. 21 as any lease other than a finance lease and a finance lease as:

> "a lease that transfers substantially all the risk and rewards of ownership of an asset to the lessee. It should be presumed that such a transfer of risk and rewards occurs if at the inception of a lease the present value of the minimum lease payments, including any initial payment, amounts to substantially all (normally 90 per cent. or more) of the fair value of the leased assets. The present value should be calculated by using the interest rate implicit in the lease (as defined). If the fair value of the asset is not determinable an estimate thereof shall be used." (S.S.A.P. 21, para. 15).

The tax treatment of an operating lease is the same as for a finance lease in that the whole of the lease payments are deductible as a trading expense of the lessee, and capital allowances would be claimed by the lessor.

Credit sale

21–016   Where equipment is acquired on instalment credit, or credit sale, the interest charged would be treated as a trading expense but capital allowances would only be available on the expenditure as it is incurred and not on the whole cost of the asset (F.A. 1985, s.56(2) and F.A. 1971, s.44). This is to be contrasted with an acquisition under hire-purchase. Credit sale arrangements are often made as part of an equipment upgrade under which the old equipment is taken in part exchange and forms the deposit for the new equipment, the balance being paid over an agreed period, often at what appears to be a nominal rate of interest.

In order to compare the advantages of such an arrangement with a hire-purchase or leasing deal it is essential to know the cost of the new equipment in a cash sale which may be substantially less than the price offered in the credit sale, and the amount for which the existing equipment could be sold on the open market which may be less than the trade-in value offered in the credit sale agreement. In some cases the credit sale will provide the cheapest means of financing but it is not always easy to make the appropriate comparisons.

Interest deductibility

Tax law distinguishes between short interest and annual interest. Short **21–017** interest is interest other than annual interest and annual interest is interest on a loan that, at inception, is likely to exceed a period of 12 months (*I.R.C.* v. *Hay* (1924) 8 T.C. 636).

In the case of a farmer carrying on as a sole trader, or in partnership where there is no corporate partner, the distinction between short and annual interest is normally unimportant; interest is deductible as a trading expense under the provisions of I.C.T.A. 1988, s.74, because the prohibition against the deduction of annual payments specifically excludes interest: I.C.T.A. 1988, s.74(*m*).

If, however, the loan is from a non-resident and if it is annual interest it will normally have to be paid under deduction of tax, which will often be unacceptable to the non-resident lender: I.C.T.A. 1988, s.82 and s.349. It is normal in such circumstances to make the loan from the non-resident lender a short-term loan for say, 11 months, so that the interest may be paid gross. The loan should actually be repaid at the end of the 11-month period and should not immediately be renewed as a rolling 11-month credit would probably, on the facts, be regarded as a loan for a period in excess of 12 months and therefore the interest payable would be annual interest and the withholding tax provisions would apply.

In some cases the non-resident resides in a country where there is a suitable double taxation agreement and it is possible with the permission of the Inspector of Foreign Dividends to pay annual interest gross under the Double Taxation Relief (Taxes on Income) (General) Regulations 1970, S.I. 1970 No. 488, para. 2(2). As this requires the appropriate declaration by the lender to the overseas tax authorities this route is often unacceptable in practice.

If interest paid by an individual cannot be claimed as a trading expense it may be claimable as a charge on income under the provisions of I.C.T.A. 1988, s.353 if it is annual interest on a U.K. source loan: I.C.T.A. 1988, s.353(1)(a) or is payable to a bank carrying on business in the U.K. or the Republic of Ireland: I.C.T.A. 1988, s.353(1)(b). The interest must not be on overdraft, I.C.T.A. 1988, s.353(3)(a), and must not be at more than a reasonable commercial rate: I.C.T.A. 1988, s.353(3)(b).

Not unnaturally it is not possible to obtain relief for interest both as a trading expense and as a charge on income: I.C.T.A. 1988, s.368.

The purposes for which interest relief is allowable as a charge on income are strictly limited.

Loans for purchase or improvement of land

Interest is normally eligible for relief if paid on a loan used to acquire an **21–018** interest in land in the U.K. or Republic of Ireland or improving or developing the land or buildings or repaying a loan incurred for this purpose:

I.C.T.A. 1988, s.354(1) and Extra-statutory Concession A28 (1988) for residents of the Republic of Ireland—see Chapter 9.

Unpaid purchase consideration counts as a loan: I.C.T.A. 1988, s.367(3).

Apportionment is permissible for loans only partly qualifying for interest relief: I.C.T.A. 1988, s.367(4).

The relief only applies to a loan of money actually applied for the qualifying purpose, and not previously applied for some other purpose: I.C.T.A. 1988, s.367(2). However the "temporary deposit in a bank or building society, *e.g.* prior to a house completion, is not regarded as an application of the funds" (The Chief Secretary to the Treasury, H.C. Debates, July 16, 1969). Application must be within a reasonable time; the Revenue allows up to six months without further enquiry: I.R.11 (1985), para. 7.

Expenditure on improvements includes maintenance and repair caused by dilapidations before the acquisition of the interest in the land but not otherwise: I.C.T.A. 1988, s.354(2).

Improvements include costs of "street works" servicing the land: I.C.T.A. 1988, s.354(2).

An interest in land includes a large caravan when the caravan together with the land on which it stands is a rateable hereditament for rating purposes, and provided that the occupier or spouse has paid rates covering the period in which the interest is paid: I.C.T.A. 1988, s.354(1), (3).

"Caravan," "large caravan" and "street works" are defined in I.C.T.A. 1988, s.367(1).

A house-boat also qualifies as an interest in land if designed or adapted for use as a place of permanent habitation: I.C.T.A. 1988, ss.354(1), 367(1).

An interest in land does not include a rent charge or the interest of a mortgagee or a creditor's interest as security (H.C. Written Answer, March 21, 1985) I.C.T.A. 1988, s.354(4).

Interest paid by a tenant to a landlord for the purchase of the property is eligible for relief even though title does not pass until a future date: I.C.T.A. 1988, s.354(7).

A sale between spouses does not qualify for interest relief nor does a disposal and purchase of land where it appears that the main purpose was to obtain relief for interest. A sale by a settlor to trustees for the purpose of obtaining interest relief is also disallowed as is a sale to a connected party at an inflated value: I.C.T.A. 1988, s.355(5).

Main residence

21–019 Eligible loans for interest purposes under I.C.T.A. 1988, ss.354, 355, 367 are further restricted to the main residence of the owner, *Frost* v. *Feltham* [1981] S.T.C. 115; I.T.C.A. 1988, s.355(1)–(3) or a separated spouse or a residence let rent free before 6 April 1988 to a dependent relative (as defined in I.C.T.A. 1988, s.367(1)), or to property let at a commercial rent for 26 weeks out of 52 and available for letting or under reconstruction or repair for the remainder of the time. There is a 12-month period of grace for occupying a residence; this period may be extended at the discretion of the Board. For relief for temporary absences see Extra-statutory Concession A27 (1988). "Borrower" means "interest payer" for these purposes.

Interest paid on a loan to purchase a residence where the taxpayer has representative occupation under I.C.T.A. 1988, s.145, is allowable: I.C.T.A. 1988, s.356.

Self-employed occupiers of job-related accommodation can claim mortgage interest relief on another house for their intended future occupation as a main residence, provided that the job-related accommodation is provided at arm's length other than by a company or partnership in which they have an interest. The relief extends to C.G.T. on the intended main residence.

The total amount of all eligible loans for main residences is limited to £30,000 for 1983–84 and subsequent years. There are provisions for the proportionate reduction in the interest if the loan exceeds the eligible amount and loans are dealt with in the order in which they are taken out: I.C.T.A. 1988, s.357.

There are provisions for apportioning the interest on a joint loan. Until **21–020** 1989–90 husband and wife are treated as one for the allowability of loan interest, although where a wife's earnings election is in force interest paid by the wife reduces her earned income only: I.C.T.A. 1988, ss.287, 288. The C.C.A.B. Press Release March 10, 1983, para. 10 explains the position for property held by husband and wife as joint tenants. From 1990–91, however, husband and wife are taxed independently and as regards home loan interest if they are living together they may make an election allocating the deduction and the statutory limit of £30,000 between them as they wish: F.A. 1988, Sched. 3, para. 14.

Up to £1,000 of rolled-up interest may be added to the £30,000 limit with full tax relief available: I.C.T.A. 1988, s.357(6): this covers cases where mortgage repayments are late.

A bridging loan for a 12-month period in respect of a main residence is allowed although the Revenue has discretion to extend this period in appropriate circumstances. But relief was denied where the new house was sold before taking up residence, in *Hughes* v. *Viner* [1985] S.T.C. 235. In essence, interest relief continues on the old loan whether or not the property is being used as the main residence and on the new loan as if there were no old loans. SP 10/80 and Extra-statutory Concession A35 (1988) cover bridging loans in the year of marriage: I.C.T.A. 1988, s.354(5).

Interest paid in respect of property let is available only against rental income on that or any other land and cannot be relieved against income other than from property but can be carried forward against future property income. SP 4/85 sets out current Revenue practice on loans to partners for property used by the partnership.

Interest paid by personal representatives or trustees under a will trust qualifies for relief if, had the deceased survived, he would have been eligible for relief, and provided that the deceased's widow or widower or dependent relative occupies the residence, or where the deceased's widow or widower intends to occupy the property but is presently residing in job-related living accommodation: I.C.T.A. 1988, s.358.

Loans to purchase machinery or plant

21–021   A loan to purchase plant and machinery by a partner entitled to capital allowances qualifies for relief, but only in respect of interest paid within the three years following the year of assessment in which the debt was incurred: I.C.T.A. 1988, s.359(1).

A loan to acquire plant or machinery only partly used for a trade or profession is eligible for interest relief on the appropriate part: I.C.T.A. 1988, s.359(2).

Where an employee borrows money for the purchase of plant or machinery required for the employment and qualifying for capital allowances the loan is eligible for interest relief for a period of up to three years from the end of the year of assessment in which the debt was incurred: I.C.T.A. 1988, s.359(3).

A loan to purchase plant or machinery used partly for the employment qualifies for interest relief on the appropriate part: I.C.T.A. 1988, s.359(4).

Loan applied in acquiring interest in close company

21–022   Interest is allowable if the loan is used to acquire share capital in a close company or in lending money to a close company for the use in its business or repaying an eligible loan: I.C.T.A. 1988, s.360. The close company must be and continue to be a trading company or member of a trading group or must satisfy the income requirements in I.C.T.A. 1988, s.424(4), *i.e.* more than 75 per cent. of its income must be estate or trading income or interest on dividends from 51 per cent. trading subsidiaries.

Interest paid under a guarantee was not interest on a loan in *Hendy* v. *Hadley* [1980] S.T.C. 292.

In order to claim interest relief the taxpayer must either have a material interest in the company or be a shareholder, however small the interest, and work for the greater part of his time, *i.e.* more than 50 per cent., in the management or conduct of the business, and in either case must not have received any capital repayment from the company: I.C.T.A. 1988, s.360(2). The conditions must be satisfied at the time the interest is paid. Where the individual resides in a property owned by the company, where the latter is a property or investment company (albeit satisfying the income requirements of I.C.T.A. 1988, s.424(4)), having a material interest is not sufficient, and the individual is required to have worked for the greater part of his time in the management or conduct of the company or an associated company.

"Material interest" means more than 5 per cent. of the ordinary share capital: I.C.T.A. 1988, s.187(3). After a company ceases to be close, SP 3/78 confirms the continuation of relief as does Extra-statutory Concession A43 (1988) on an exchange of shares on a company reorganisation.

Loan applied in acquiring interest in a co-operative

21–023   There is relief for interest on a loan by an individual taken out to acquire shares in a co-operative (a common ownership enterprise or a co-operative enterprise), or to lend money to a co-operative for business purposes, or for paying off a loan so applied. This applies where the individual works for the greater part of his time as an employee of the body or of a subsidiary of it.

Loan applied in investing in employee-controlled company

Interest relief for employees or their spouses buying ordinary shares in a **21–024** close company is extended to any unquoted U.K. resident trading company or holding company of a trading group where the interest is paid by a full-time employee (or spouse) who has not recovered capital from the company and where at least 50 per cent. of the ordinary share capital and voting power of the company is owned by full-time employees (or their spouses).

Individual employees and their spouses may include a total of 10 per cent. equity interest as an employee holding. If both the individual and his spouse are full-time employees of the company each may count a 10 per cent. shareholding: I.C.T.A. 1988, s.361.

Loan applied in acquiring interest in partnership

A loan to purchase a share in a partnership or to introduce capital into a **21–025** partnership by way of loan is eligible for relief, as is the replacement of an existing eligible loan; I.C.T.A. 1988, s.362(1).

In order to qualify for relief the borrower must have been a partner (other than a limited partner) since the application of the loan and must not have recovered any capital. Relief where a partnership is incorporated into a close company is given by Extra-statutory Concession A43 (1988).

If any capital has been recovered from a close company or partnership or co-operative or employee-controlled company, the eligible loan is deemed to be repaid to that extent and the interest ceases to be allowable for tax purposes: I.C.T.A. 1988, s.363(1).

Capital is deemed to have been recovered if some of the shares are sold or there is a repayment of capital, or if any interest in the partnership is disposed of. A repayment of part of the loan to a close company or capital from a partnership or the assignment of a debt from the close company or partnership is deemed to be a recovery of capital. Non-arm's length disposals are deemed to take place at market value: I.C.T.A. 1988, s.363(2), (3).

Loan to pay inheritance tax

A loan to personal representatives to pay I.H.T. on personalty in order **21–026** to obtain probate, or any replacement loan of such a loan, is eligible for relief for a period of one year from the date of grant of the loan: I.C.T.A. 1988, s.364(1).

Excess interest paid in by personal representatives can be carried back against preceding years' income of the estate, and to the extent unrelieved may be carried forward against future income.

I.H.T. includes interest payable thereon and interest for the one-year period is allowable even if the interest continues to run after that time, although the excess will be disallowed: I.H.T.A. 1988, s.364(4).

Loan to purchase life annuity

21–027 Home re-mortgage schemes are eligible for relief in respect of interest on the loan provided that at least 90 per cent. is used to purchase a single joint life annuity, the annuitants are aged 65 or more, and the loan is secured on land in the U.K. or the Republic of Ireland which is used as the annuitant's main residence. Relief is given on interest on a maximum of £30,000. The loan interest has to be paid by the borrower or one of the other annuitants.

The scheme approved of is to enable elderly taxpayers to improve their income without selling their house. They take out a loan on the security of their home on an interest only mortgage. The loan proceeds are invested in a purchased annuity under I.C.T.A. 1988, s.656 which is taxable only on the income element. The interest paid is allowable for relief and on death the house is sold and the proceeds used to repay the loan: I.H.T.A. 1988, s.365.

Companies

21–028 Interest payable by a company may be treated as a trading expense under I.C.T.A. 1988, s.74(*m*). This is so even if the interest is annual interest: I.C.T.A. 1988, s.337(3), in which case it cannot be treated as a charge on income: I.C.T.A. 1988, s.338(2), *Wilcock* v. *Frigate Investments Ltd.* [1982] S.T.C. 198.

Annual interest paid by a company, however, or by a partnership of which a company is a partner, must be paid under deduction of tax under I.C.T.A. 1988, s.349(2)(a) or (b) unless it is annual interest payable in the U.K. to a bank in the ordinary course of business: I.C.T.A. 1988, s.349(3). Where interest is paid under deduction of tax the interest has to be accounted for to the Inland Revenue by direct payment under the provisions of I.C.T.A. 1988, Sched. 16. Credit may be taken for tax deducted from interest and other annual payments received under deduction of tax.

Annual interest paid by a company other than for trading purposes may be deductible as a charge on income under I.C.T.A. 1988, s.338(3)(a) provided it is treated as a revenue expense and not capitalised (*Chancery Lane Safe Deposit & Offices Co. Ltd.* v. *I.R.C.* (1965) 43 T.C. 83 and *Fitzleet Estates Ltd.* v. *Cherry* [1977] S.T.C. 397). Interest capitalised on the construction of a building may be added to the cost of the building for the purposes of calculating any chargeable gain on disposal if it has not been allowed as a charge on income (I.C.T.A. 1970, s.269). Interest charged to capital is allowed as a charge on income under I.C.T.A. 1988, s.338.

Annual payments between members of a group of companies may be paid gross provided that the appropriate election is in force under I.C.T.A. 1988, s.247.

Anti-avoidance

21–029 Interest is not deductible if paid as part of a scheme or arrangement, the sole or main benefit of which was to produce a deduction for the interest paid: I.C.T.A. 1988, s.787.

Interest paid to a non-resident must be paid under deduction of tax (I.C.T.A. 1988, s.338(4)(a)) or be within I.C.T.A. 1988, s.340 (I.C.T.A. 1988, s.338(4)(c)) or be paid out of income charged to tax under Schedule D, Cases VI or V (I.C.T.A. 1988, s.338(4)(d)).

Under I.C.T.A. 1988, s.340 interest is only allowable if paid to a non-resident gross if it is paid outside the U.K. wholly or mainly for the purposes of a trade carried on outside the U.K. or it is payable in a currency other than sterling and the liability was incurred for the purposes of a trade wherever carried on and the interest is not paid to a connected person. The deduction for interest paid in a foreign currency is restricted to that incurred wholly or mainly for the purposes of a trade carried on outside the U.K. if it is paid to a party under the control of the payer, or both payer and payee are under common control, or the payee has control over the payer.

Bank interest paid by a guarantor called upon to do so was a guarantee payment, not interest (*I.R.C.* v. *Holder & Holder* (1932) 16 T.C. 540). Interest is not paid merely by being debited to an account (*Paton* v. *I.R.C.* (1938) 21 T.C. 626).

Dealing with the Revenue

Tax returns

22–001   It is not within the scope of this book to deal with matters referred to in this chapter in any detail. In the first instance it must be made clear that it is the taxpayer's duty to notify the Revenue of his liability to tax (T.M.A. 1970, s.7) and to submit a return of income (T.M.A. 1970, s.8). In some cases the Revenue may fail to issue the taxpayer with a return of income to complete in which case the taxpayer should ask for the appropriate form unless the whole of his income has been subject to deduction of tax at source and there is no further liability to income tax or C.G.T.

In May 1988 the Revenue issued Statement of Practice SP3/88 which set out its practice concerning delay in submitting tax returns and charging interest under T.M.A. 1970, s.88. This section empowers the Revenue to charge interest from the normal due date for payment in the case of fraud, wilful default or neglect of the taxpayer. The statement of practice relates to new sources of income, continuing sources where inadequate estimated assessments are not appealed against and capital gains. The normal due date for payment of C.G.T. and the higher rates of income tax in December 1, following the tax year in which the gains or income arose. The statement of practice makes it clear that consideration will be given to invoking section 88 where the tax return is not submitted by October 31 following the tax year in question. If it is not possible to submit the tax return by October 31, the Revenue will not raise a charge to interest under section 88 if the inspector is provided by October 31 with sufficient information to enable an adequate estimated assessment to be made.

22–002   The taxpayer under the existing tax law (which is being changed from 1990–91) is the husband, where a man and woman are married and living together, and his return of income should contain full details not only of his own but also of his wife's income: I.C.T.A. 1988, s.279. Many taxpayers end up by paying interest and penalties on undeclared income as a result of bank and building society deposits held by the wife without the knowledge of the husband.

Until 1989–90 either husband or wife can elect for separate assessment under I.C.T.A. 1988, ss.283 and 284 under which husband and wife would independently submit returns of their own income and gains to the Revenue. The Revenue would then add together the amounts shown to arrive at the total income and gains taxable on the husband. It would then compute the tax liability and apportion it in accordance with a predeterminate formula between the husband and wife in accordance with income, gains and allowances. The overall tax liability is not altered by separate assessment and it is normally to be avoided as a time wasting exercise both for the Revenue and the taxpayer unless there are good reasons for each party being unaware of his or her spouse's income and gains. In other cases it is perfectly reasonable for each party to the marriage to bear a fair share of the total tax on the total income by a fair apportionment outside the provisions of a formal election.

On the other hand an election for the separate taxation of wife's earnings under I.C.T.A. 1988, ss.287 and 288 can be valuable and result in a significant saving of taxation. The relief has already been described in Chapter 2.

With effect from 1990–91 husbands and wives will be taxed independently (F.A. 1988, s.32), so each will be responsible only for submitting a tax return of his or her own income and gains.

In the case of partnerships and companies also it is the taxpayer's duty to notify the Revenue of the taxable income and gains (T.M.A. 1970, ss.9 and 10). If income is received by agents the Revenue may obtain information from the agents of the taxable income of the principals under T.M.A. 1970, ss.13 to 17.

P.A.Y.E.

There are material penalties for failing to apply the P.A.Y.E. system **22–003** properly. There is a special book of instructions for farmers on the operation of P.A.Y.E. It is necessary not only to complete the normal weekly or monthly records but also to submit by May 19 the annual form P35 and for each employee earning at a rate in excess of £8,500 and each director, returns of benefits and expenses on form P11D. The Revenue these days is particularly strict on directors who in the past have often drawn money from their own company as an advance of remuneration without operating P.A.Y.E. on the withdrawal. The Revenue argues that this results in a loan in respect of which it can charge an amount equivalent to advance corporation tax under I.C.T.A. 1988, ss.419 and 420 unless it is a drawing on account of remuneration in which case P.A.Y.E. must be operated: see F.A. 1989, s.45.

Assessments

An assessment made by the Revenue may cover a single source of **22–004** income such as rents under Schedule A or several sources, for example, both husband's and wife's earnings under Schedule D, Case I, income for furnished lettings under Schedule D, Case VI and investment income under Schedule D, Case III. A Schedule E assessment covers employment income and benefits in kind, etc.

In each case it is necessary to check the assessment quickly to ensure that the figures are correct and if not to lodge the appropriate notice of appeal within the 30-day period allowed. Although the Revenue has power to accept a late notice of appeal it will not do so unless there is a good reason for the failure to appeal within the time limit. Technically an assessment not appealed against within the appropriate time limit becomes final and conclusive and the tax becomes payable.

Capital gains are normally included on a separate notice of assessment and in the case of companies corporation tax is charged on the total profits of the company including income and chargeable gains.

Discovery

22–005   The Revenue has power to review income that has already been assessed and if it considers that it has been under assessed it may make a discovery under T.M.A. 1970, s.29(3). If the point at issue has already been fully discussed and agreement reached with the Revenue the matter is final and cannot be reopened on discovery (*Olin Energy Systems Ltd.* v. *Scorer* [1985] S.T.C. 218 and *Cenlon Finance Co. Ltd.* v. *Ellwood* (1962) 40 T.C. 176). If, however, the matter has not been determined by the Commissioners, or settled by agreement as if it had been determined by the Commissioners under T.M.A. 1970, s.54, it is possible to raise additional assessments on discovery (*Kidston* v. *Aspinall* (1963) 41 T.C. 371). In order to raise an assessment under the discovery provisions the Revenue does not need to discover any new facts as a mere change of mind is sufficient (*Brodie's Trustees* v. *I.R.C.* (1933) 17 T.C. 432).

Penalties

22–006   The normal time limit within which the Revenue can raise an additional assessment is six years from the end of the year of assessment in which the income is assessable. If, however, there has been fraudulent or negligent conduct, the time-limit is 20 years.

The maximum penalty for failing to make a return is an amount equivalent to the tax lost. Penalties are normally mitigated depending on the co-operation which the taxpayer has shown the Revenue during the negotiations to settle his affairs, see I.R. 73.

The Revenue in appropriate cases has power to seize documents under T.M.A. 1970, s.20 and to request the Commissioners to grant a precept ordering the production of accounts and other records under T.M.A. 1970, s.51.

Appeal hearing

22–007   If agreement cannot be reached with the Revenue on any point, including the suggested alterations as a result of an in-depth investigation, it is open to the taxpayer to appeal against the assessments raised by the Revenue and to go before either the General Commissioners or the Special Commissioners. The General Commissioners are local businessmen to whom it is normal to refer facts in dispute, while the Special Commissioners are lawyers experienced in taxation to whom it is normal to refer technical arguments on the interpretation of tax law. This does not mean however that it is not possible to argue a point of law before the General Commissioners or facts before the Special Commissioners.

The appeal will normally be taken by the taxpayer's professional advisers or by a barrister and the taxpayer will give evidence. Other witnesses may be called by either party to give evidence and documents may be produced as evidence. The Revenue's case is usually put by the inspector or by counsel appointed on behalf of the Revenue.

If either the taxpayer or the Revenue is dissatisfied with the Commissioners' decision an appeal may be taken, but only on a point of law, to the Courts, finally ending up in the House of Lords if necessary. This is done by either party giving immediate notice of dissatisfaction either at the Commissioners' hearing if a decision is given at the hearing or immediately on receipt of the Commissioners' decision if this is given in writing.

An appeal through the courts tends to be extremely expensive.

Interest

If the taxpayer does not pay tax on the due date he is charged interest **22–008** under T.M.A. 1970, s.86, unless there has been tax lost due to fraudulent or negligent conduct in which case he is charged both interest under T.M.A. 1970, s.88 and penalties as mentioned above. If, on the other hand, the taxpayer has overpaid tax he may be entitled to a repayment supplement under I.C.T.A. 1988, ss.824 and 825. Although the rate of interest for the repayment supplement is the same as for overdue tax the starting point for the calculation of the interest is very much in the Revenue's favour and normally only runs from the end of the tax year in which the tax was paid at the very earliest.

If tax is to be paid on a disputed assessment it may well be sensible to pay this right at the end of the tax year so that if the tax has been overpaid the maximum repayment supplement will be available. As an alternative to payment of the tax it is possible to purchase certificates of tax deposit which, if used for the payment of tax, will prevent interest running from the date of the purchase of the certificate. If on the other hand the certificate is not needed it may be repaid in cash, together with interest, although such interest is itself subject to tax. Interest paid on overdue tax is not allowable for tax purposes and repayment supplement received is not taxable.

Collection

The collection of tax is dealt with by a separate branch of the Revenue **22–009** from the Inspectorate which raised the assessment. In theory this prevents fraud within the Revenue, as the people handling the money have no connection with the people raising the assessment. In practice it can give rise to irritating failures of communication where the collector is demanding money with menaces in respect of tax which the inspector has already agreed is not due for payment, or where the assessability is still being argued with the inspector. It is often necessary to request the collector to confirm the collectability or otherwise of the tax ostensibly due before proceeding further. Otherwise the collector is likely to start restraint proceedings or sue for the non payment of tax, and in appropriate cases would commence bankruptcy proceedings or liquidation of a company.

Inheritance tax

22–010   In the case of I.H.T. the taxpayer who makes a gift other than an exempt or potentially exempt transfer is liable to return details of the gift to the Capital Taxes Office. On death it is necessary to pay the I.H.T. due before probate can be obtained and the assets passing on death dealt with.

In-depth investigation

22–011   The Revenue will normally examine the taxpayer's accounts only briefly in most cases and if they appear reasonable they will be accepted. Occasionally, however, accounts will be chosen for an in-depth investigation either because there are consistently low profits compared with the average for similar businesses in the vicinity or compared with the taxpayer's apparent standard of living, or where the Revenue holds information such as particulars of bank interest or information from an auctioneer, etc., which suggests possible irregularities.

The Revenue has admitted that it compares the gross profit percentages of similar businesses, although the percentages applicable to particular businesses are not, in the main, published. Business Economic Notes are, however, beginning to be published by the Revenue as part of a continuing programme to help small firms and the self-employed with tax matters. They provide general guidance on the financial and business background to particular trades and are used by inspectors when they examine business accounts.

The Revenue action in the case of the taxpayer whose accounts are chosen for an in-depth investigation is normally to try and establish a quick settlement on the basis of a business economics review rather than commence a detailed back duty investigation. The procedure will normally begin with a letter probing for possible deficiencies such as details of records maintained, drawings, source of capital introduced and reasons for poor results. If the answer indicates that the records are less than perfect or the drawings or other matters still cause disquiet the taxpayer and his adviser will be invited to an interview. As a result of the interview the inspector will put together a model on business economics lines or prepare capital statements as a result of which he will suggest to the taxpayer what the Revenue estimates his results should have been and call the taxpayer and his professional advisers for a further interview to discuss the additional assessments that the Revenue thinks appropriate.

The taxpayer is often faced with the difficulty of proving a negative, *i.e.* that his income was not as great as that suggested by the inspector, and there is a tendency among some inspectors to assume that any business which produces results below the average is on the fiddle, quite overlooking the fact that the average is itself made up of the results of businesses which are both commercially successful and those which are not.

22–012   Where the results of the farming activities appear poor but loans to the farm are increasing, the inspector may suspect that a different tale is being told to the lender than that being told to the Revenue and mark the case down for further investigation. It must be remembered that inspectors in a particular district deal with a number of local farms and if the farmer's results are against the trend this may require investigation. For example, if his crop is particularly good or bad compared with the majority of farmers in the neighbourhood the inspector will often enquire the reason why. The absence of income from natural by-products would alert the suspicion of the inspector. For example, a dairy farm omitting to record the sale of calves, or a sheep farm selling lambs without accounting for the sale of wool.

When accounts come into the inspector they are marked "A" for accept, "R" for review or "E" for examine. A review is looking for substantial technical adjustments where the accounts are accepted as bona fide. An examination is a fundamental challenge to the whole accounts, see I.R. 72. If in the course of a review it is obvious that, for example, private expenditure has been debited in the profit and loss account, the inspector is likely to reclassify the case as an "E" for an in-depth investigation.

The inspector may be expected to endeavour to apply the business economics approach to farming. A study of the gross margins achieved for particular crops would be recognised by the inspector as a valid means of confirming the accuracy of the accounts and it may be helpful for the farmer or his accountant to prepare such a report if it is thought that the inspector is asking for an in-depth investigation unnecessarily.

It is by no means unknown for disgruntled employees or former employees to allege deficiences in a farmer's accounts which might cause the inspector to review a case more carefully and perhaps begin an in-depth investigation. A well trained inspector in a farming district would be encouraged to listen to the BBC early morning farming broadcasts and study the Farmers Weekly and the local press.

An inspector may, for example, notice on visiting a farm that there is a swimming pool or tennis court and enquire whether agricultural buildings allowances may possibly have been claimed in error on such construction. He will also note whether any fields are let for caravans or camping and whether there are any advertisements of produce for sale, or for bed and breakfast.

Farm records

An inspector considering an in-depth investigation will probably look in **22–013** the first instance for sources of omitted income, very often cash transactions. For example, disposal proceeds for calves and store animals in the case of a dairy farmer, proceeds from the sale of old hens by a poultry farmer, barter transactions whereby a sheep or pig for the deep freeze might be supplied in exchange for private goods or services. Obviously the inspector would be looking for sales of produce at the farm gate for cash and it should be borne in mind that he, or a member of his staff, may well have visited the farm in the guise of a customer. In many farming communities transactions between farmers take place which are traditionally settled in cash, for example, for assistance at ploughing or harvesting, or sales of feed, grain, hay, fodder and livestock.

V.A.T

22–014   Most farmers will be registered for V.A.T. and as Customs and Excise makes regular V.A.T. inspections the V.A.T. records are often better kept and more up-to-date than the normal accounting records. An investigating inspector will normally examine the V.A.T. records and copy returns.

Auction markets

22–015   Farmers will often sell goods at auction markets and it is the practice of many such markets to advance cash; the sale having been recorded, they give a cheque for the balance. The auction market's records are probably correct but the farmer's may not be, as it is not unknown for them to record the cheque receipt and forget the cash advance. Auction market cheques are often endorsed over to other traders in the local community rather than paid into the farmer's bank account. In some cases a bank will provide an immediate cash drawing against an auction market cheque, although the bank statement should show the full amount of the cheque deposited and the cash drawing as a withdrawal.

In some local communities it is still common for "luck money" to be paid by the vendor of an animal to the purchaser. The amounts are individually small, perhaps £1 to £3 per animal for cattle, £1 per pen for sheep, but £5 per animal for in-calf heifers and dairy cattle, although this can be as high as 5 per cent. of the selling price in the latter case. Although the practice of giving luck money is dying out an inspector should be expected to know whether the practice was prevalent in the local community. It should, of course, be regarded as taxable income of the purchaser.

Sundry receipts

22–016   Farms often provide unofficial sites for tents and caravans on a casual basis, for cash, and may also provide accommodation in the farm buildings for casual visitors. Again it is not unknown for the receipts from such activities to be omitted from the farm records.

The income from holiday lettings for farms in tourist areas can be substantial. Some farmers go to the extent of organising family holidays on the farm and these may be advertised locally and nationally. Where farmers keep horses they may well be available for hire and riding lessons may be provided.

Cash receipts can also arise from the casual letting of small areas of land for potato crops, for example.

Other cash crops grown include Christmas trees. Farmers may also sell for cash firewood, timber, straw, hay, manure and surplus feedstuffs. Sales to scrap metal dealers are normally in cash but the dealers are required by the police to keep very detailed records of the time of delivery, description of vehicle the goods were delivered in and the full name and address of the person selling the scrap. Such records can form a fruitful line of enquiry for an energetic inspector.

Cash records

Farmers are normally required by law to keep detailed records of live-stock and other matters related to the farm and an inspector during an investigation would normally call for an example of the statutory statistical records to see how this reconciles with the financial records. Where the farm cash-book is the prime record for the production of accounts the inspector is likely to be interested in the manner in which this is written up, which is likely to indicate that extent to which it fully records all trans-actions. Cheque payments are normally correctly recorded, but incorrect analysis may hide private expenditure, for example, the farmer's private car repairs may be shown as tractor repairs. The extent to which this is detectable by the inspector may well depend on the degree of collusion with the supplier and the extent to which invoices are either vague or delib-erately falsified. In a number of areas accountants will provide a farm book-keeping service which should ensure that the books balance but the inspector will still wish to make investigations to determine whether they tell the whole story. The accountant is, after all, merely processing such information as he is told by the farmer and he can not process what he is not told about. The recording of actual payments in cash may be rather vague and items such as market expenses or haulage, even where recorded to the nearest penny, may, on investigation, prove to be no more than esti-mates or, in some cases, fairly wild guesses.

An investigating inspector will normally look carefully at the year-end cut-off procedure to ensure that expenses have not been brought forward into an accounting period and receipts delayed until the next period.

22–017

Haulage

The inspector is particularly likely to be keen on examining haulage invoices where livestock is sold, as these will normally indicate the destina-tion and very often the number of animals involved and serve as an inde-pendent record of the market sales. The haulage records, for example, could indicate a drop off en route to the market in relation to a private sale or transporting to an out-of-area market, and in both cases the inspector will be keen to ensure that he can trace that the unusual sales have been properly recorded.

22–018

Bank statements

The inspector will normally assume that a reconciliation of the cash-book with the bank has been correctly done by the accountant but may nonetheless call for bank statements to look at the pattern of lodgments which may indicate the necessary retention of unusually large cash amounts to provide for living expenses. Casual receipts from the farm are often regarded as the wife's income and the inspector may use this as the reason for calling for the wife's bank statements, although if the totals are small the income may be covered by the wife's personal allowance in any event. The inspector is unlikely to call for the production of private bank

22–019

statements, deposit accounts, building society passbooks, money market transactions, etc., unless he knows that interest has not been declared or the level of interest in comparison with the declared income make it obvious that the records cannot be correct, or if his preliminary capital reconciliation appears suspect and the amount involved is substantial. The Revenue would like to justify the final settlement both by rough capital reconciliation and business economic review.

Market records

22–020   The inspector is likely to want to reconcile livestock sales with statements from the main local market used by the farmer and he may well request the farmer to authorise the auction market to provide the inspector with a complete list of transactions with the farmer. The suspicion would obviously be that a reconciliation of the market statement for a particular day with the takings records should indicate whether cash advances by the auction market have been properly accounted for as the gross sales figures, less commission, would reconcile whether or not cash advances are recorded by the auctioneers. On the other hand, if a farmer is deliberately under-stating his sales he could be expected to lose the market statement for a day in which he failed to record the sales so that the whole transaction is omitted from the accounts. This would be revealed by reconciliation with the auctioneer's own records.

If the farmer has indulged in such practices he would be well advised to confess very quickly because if he were to refuse to authorise the auctioneer to divulge details of his transactions to the Revenue the inspector is likely to be highly suspicious. The inspector in any event has power to serve a notice under T.M.A. 1970, s.20 on the auctioneer for the production of information relating to the farmer's affairs in view of T.M.A. 1970, s.20(4)(b).

Large purchases of livestock at auction may qualify for a cash discount which itself may be paid in cash and could have escaped being recorded. This will also be shown in the auctioneer's own records.

Dairy farmers

22–021   Most dairy farmers will sell direct to the Milk Marketing Board and be paid monthly. The inspector will check that any expenses or levies deducted from the gross amount receivable on the Milk Marketing Board invoice will not also have been shown as expenses in the accounts if only the net amount received is shown in sales. Where the farmer is a producer/retailer of milk the inspector's problems increase considerably as the amount of cash sales will obviously be very substantial and the inspector will be looking for reconciliation with the Milk Marketing Board levy returns to quantify the production and from that the sales figures. Similarly where the farmer makes butter or cheese on the farm with milk supplied from his own cattle the Milk Marketing Board returns will be used to substantiate raw material usage to ensure that the production is accounted for and part of the production is not sold for cash which is unrecorded.

Arable farmers

A similar use will be made of merchants' accounts for arable farmers **22-022** particularly where a running account is maintained. However, there is normally little opportunity for the farmer to fiddle such records and the inspector's investigation is likely to be rather more cursory.

The inspector will have available to him the yields and seed rates for farm crops and the estimated yields of crops in respect of recent harvests. He will also have the market prices of cereals, the intervention prices, prices of hay, straw and potatoes, gross yields of fruit and vegetables together with average farm gate prices.

Farming notes

The inspector also has available to him sundry important internal infor- **22-023** mation such as average agricultural wages, agricultural contracting scales of charges, grazing and over wintering charges, cost trends for concentrate foodstuffs, egg prices both for packing stations and at farm gate, wool prices and fleece weights, rates of premium under the beef premium scheme, rates of guarantee under the sheep guarantee scheme and other useful information such as grants and premiums available to farmers, sources of loan finance, together with an outline of farming in the U.K. and the marketing of agricultural produce.

An inspector should be expected to have available commercially produced records of prices, yields and income such as

  (i) The Farm Management Pocketbook by John Nix published by the Department of Agricultural Economics, Wye College, near Ashford, Kent TN25 5AH;
 (ii) Farm Incomes in South West England published by the University of Exeter, or the equivalent covering the local area;
(iii) The Agricultural Budgeting & Costing Book published by Agro Business Consultants, The Old Vicarage, Church Lane, Twyford, Melton Mowbray, Leicestershire LE14 2HW;
 (iv) Private Woodlands, a guide to British Timber Prices and Forestry Costings by D.C.E. Hart, Chenies, Coleford, Gloucestershire GL16 8DT; and
  (v) bank publications such as the Midland Bank Farm and Planner.

Livestock

The Movement of Animals (Records) Orders 1960 and 1961 (S.I. 1960 **22-024** No. 105 and 1961 No. 1493) require a farmer to maintain a record of sheep, cattle, pigs and goats moved to and from his farm.

In the case of cattle the required details record the breed, age, sex and earmark or tag number except in the case of animals sent direct to a slaughterhouse or fat stock market. These records are of obvious assistance to the inspector in confirming that all livestock sales have been properly recorded. A farmer whose records are lost or not maintained or not up-to-date is hardly likely to remove the inspector's suspicions that all is not well with his accounts. Most farmers regard the records as important to prevent the spread of disease and a farmer whose records were not up-to-date would not qualify for incentives under the Brucellosis Incentive Scheme.

Census

22–025   The inspector will be familiar with the annual farm Census Return Form under the Agriculture Act 1947 which shows in some detail the area of the holding, the crops and fallow land, grass grown for seed, seasonal use of land, hay and silage stocks, farmers and workers, cattle, pigs, sheep, lambs, poultry and horses on the farm, vegetables grown in the open and under glass, orchards, soft fruit, nursery stock and other horticultural products, etc. Most farmers would be expected to keep a copy of the annual Census Form and it may be required by the inspector, particularly in cases where he is having difficulties reconciling the crop output with the available land or the livestock transactions. If the farmer has not kept a copy he may be requested to ask the Ministry of Agriculture Fisheries and Food for a copy or to authorise the inspector to obtain a copy direct.

Most farmers will keep production records of crops and yields and livestock often in sufficient detail to identify individual animals. This will certainly be true in the case of a pedigree herd as it will be necessary in order to register the animal with the breed society.

Milk Marketing Board

22–026   Many dairy farmers use the Milk Marketing Boards Record Scheme which involves daily record keeping by the farmer and computer print outs of production records to date supplied by the Milk Marketing Board. A farmer receiving aid under the Farm and Horticultural Development Scheme is required as part of the conditions of the scheme to keep detailed records of his farming activities. The inspector may be expected to ask to inspect such records in the course of an investigation where he feels they would assist in confirming the accounts or his suspicions that they fail to reflect the entire financial activities of the farmer.

Dealers

22–027   In many cases farmers will not sell livestock direct to the market but to a dealer who may well be a large scale dealer buying on behalf of multiple butchers, supermarkets, food processors, etc. He may well buy at auction or direct from farmers. The receipt of luck money to such a dealer can be a substantial part of his income where the number of animals handled is large. Small dealers will often buy locally in small numbers, perhaps for cash. Small dealers will often keep the animals on their own land for a period until they have a suitable quantity to take to the market.

Veterinary records

22–028   In some cases the inspector may ask to see the veterinary invoices of a farmer and may examine the Milk Marketing Board invoices for artificial insemination charges. This would be where the farmer appears to have an unusually low number of calves which survive. The inspector would expect most dairy herds to produce 90 per cent. to 95 per cent of the mature herd as calves each year and if it does not do so this is likely to be because the cows are aborting or the calves are not surviving. In either event this will indicate some disease within the herd and a responsible farmer would be expected to seek veterinary advice. The vet's bills may confirm that there has indeed been a disease problem which would account for the poor results under investigation but hopefully remedial action will be possible and results should improve. The artificial insemination charges will confirm that the herd is being properly serviced and in the absence of any confirmation of disease the inspector is likely to suspect that if the number of calves is materially below the average, calves are being born and sold outside the accounting records.

Grants and subsidies

The farmer may well be expected to produce details of grants and subsidies to the inspector during the course of an investigation. The inspector is interested in making sure that the grant has been correctly treated as a capital or revenue receipt and that, for example, capital allowances are only claimed on the net expenditure. However, the inspector may well be able to infer from the grants received the number of animals involved in, for example, the hill livestock compensatory allowance which could help to confirm a reconciliation of livestock.   **22–029**

Contracting receipts

The inspector may be expected to have available the contracting rates recommended by the National Association of Agricultural Contractors and if the amounts allegedly charged for contracting fall far below this level, or the amounts paid materially exceed it, it is likely that the inspector would call for an explanation.   **22–030**

Co-operatives and suppliers

In many cases the merchants with whom the farmer deals are co-operatives in which the farmer has an interest. He will normally be entitled to interest on his money invested and to dividends from the co-operative surpluses. This will be related to his transactions with the co-operative and should be netted off against the cost of supplies. In some cases interest and dividends are paid half-yearly and the receipts may be omitted from the accounts. The inspector is likely to accept the treatment of interest or dividends on such an investment as a deduction from the cost of supplies in the same way as a discount unless the amount is material in which case it would be treated separately as investment income.   **22–031**

In some cases agricultural suppliers also provide a substantial range of goods which the farmer is likely to require for his private purposes and the inspector may well call for the invoices to ensure that private items have been properly so treated in the farmer's accounts. The inspector will be aware in a tightly-knit local community that there is always the possibility of collusion and there was a notorious example not long ago in Scotland where a substantial number of local farmers were claiming capital allowances on the basis of invoices produced for totally fictitious tractors, combine harvesters and other pieces of agricultural equipment.

Private use

22–032   The inspector is unlikely to concentrate on produce consumed by the farmer and his family grown in the kitchen garden as these would not normally be commercially sold unless there is an organised farm shop and therefore no private consumption adjustment is needed. As explained in Chapter 3, under the decision in *Sharkey* v. *Wernher* (1955) 36 T.C. 275, the private consumption of produce otherwise produced for sale should be taken into the accounts at sale price, not at the cost of production. This may be used for the adjustment of animals slaughtered for consumption and produce taken from a farm shop. The inspector may also look at the proportion of the expenditure on the farmhouse to ensure that the amount claimed is reasonable having regard to the size of the accommodation and the business usage and he may also look at the private usage adjustment for motor cars, particularly if more than one car is being run on the business and the non-business use by the family is likely to be substantial.

Domestic heating oil may be misallocated as diesel fuel for farm vehicles. The invoice will normally indicate which is which, as heating oil has a burning rate of 28 seconds and vehicle fuel 35 seconds.

The inspector may ask to see insurance details as the presence of antiques, jewellery, valuable cars, etc., could indicate a standard of living beyond that indicated by the declared profits.

Other items of private expenditure which have been revealed on investigation include the wages of a handyman, gardener, chauffeur or domestic cleaner being shown as agricultural wages.

It is not unknown for substantial improvement and indeed new buildings to be claimed as repairs and this trend is likely to increase with the reduction in the rate of agricultural buildings allowances. An investigating inspector might well ask to visit the site where it is disputed whether an item consists of repair or new building.

Horses

22–033   Normally the Revenue will not argue that horses are kept as part of a trade unless there is a proper riding school or stud, in which case this will normally be regarded as a separate trade from that of farming. However, where the farmer keeps horses for his own use the inspector will wish to make sure that the expenses of upkeep are not lost in the farm purchases. The inspector will know that tack is expensive, that horses need shoeing regularly and that a horse might be expected to consume around £100 of hay and £200 of cereals and other food stocks a year. If the farmer is a member of the local hunt or indulges in point to point or National Hunt racing, and the children belong to pony clubs, etc., the expenses in relation to equine activities could be substantial.

Stock valuation

In an investigation the inspector may well call for the detailed stock valu-   **22–034**
ation, even where animals are valued on the basis of 60 per cent. or 75 per
cent. of the open market value, to ensure that the market value used is
itself realistic and that the 25 per cent. or 40 per cent. deduction has not
been allowed twice.

It should be remembered that the inspector has at his fingertips tables of
market prices. Market prices on a monthly basis are produced by the
Revenue for fat cattle, store cattle, calves, dairy cattle, fat sheep, store
sheep, fat pigs, store pigs and weaners, broiler chickens, capons and hens
and this will be broken down in considerable detail, for example, fat cattle
would be divided between certified and uncertified cattle, steers and
heifers light, medium and heavy. The farmer should therefore be prepared
to justify the market value figures used in his stock calculations. There are
sometimes some inadvertent under-valuations where both the farmer and
his accountant have taken off the 25 per cent. or 40 per cent. deduction
resulting in a double allowance and where the deduction is applied to purchased animals instead of being confined to those that are home grown.

The basis of stock valuation at 60 per cent. or 75 per cent. of the market
value should be made once when the animal reaches maturity and not subsequently increased. It is a common error to re-value mature animals each
year for stock purposes and this would result in an over-valuation and
acceleration in the payment of tax.

The deduction from market value is only available for home-grown animals, not for those purchased as mature animals.

An area which may cause the inspector to consider a particular farmer's   **22–035**
affairs for in-depth investigation is where the stock valuation is suspect. If
the farmer is unable to substantiate the valuation this may lead the inspector to conclude that the rest of the accounts are also suspect.

Stock valuations prepared on the basis of the farmer's costs of production should be supportable from the farmer's records. If the farmer has
already admitted that his production records are not in sufficient detail to
substantiate the detailed cost of production he is unlikely to be able to produce a meaningful valuation at a later stage.

It is important to ensure that the stock valuation includes all stocks
owned by the farmer even if, for example, they are kept away from the
farm, perhaps in a communal grain silo.

The Revenue is not likely to be unduly impressed by a valuation produced by a professional farm stock valuer, because it tends to regard
valuers as being under the influence of the farmer and the valuation is
unlikely, in the Revenue's view, to be meaningful. This is particularly true
if the accounts do not reflect a substantial fee for the valuation. It seems to
be common practice for valuers to produce a valuation on the basis of
information given over the telephone by the farmer rather than from a visit
to the farm. Even where the valuer does visit the farm the valuation may
well be at a different date from the actual visit and even a visit on the day of
the valuation does not necessarily mean that what has been valued has
actually been inspected by the valuer. Valuations have been found in practice to omit recently-born calves or lambs.

Glossary of Farming Terminology*

A–001 AFTERMATH — The grass which remains after a crop of hay or silage is cut or (more commonly) the new growth after such a cut is taken. Also called LATTER-MATH, EADISH and EDDISH.

AGIST — To take in livestock for a period or (more commonly) to provide grazing for another person's livestock for a payment (called an agistment).

ARTIFICIALS — Inorganic fertilisers.

AT FOOT — Suckling; eg a ewe with suckling lambs is said to have lambs at foot.

AWAY GOING CROP — A crop grown by a quitting tenant which he is allowed to harvest after he has quit the farm. Also called FOLLOWING CROP or OFF-GOING CROP.

BABY BEEF — Meat from cattle fed intensively for slaughter at or before the age of 12 months.

BACONER — A pig kept to be turned into bacon as distinct from a Porker which is slaughtered at a younger age.

BAG — The udder of a cow.

BAG-UP — To have an enlarging udder due to growth of the milk vessels shortly before the birth of a calf.

BARE FALLOW — Land left uncropped.

BAR PIG)
BARROW PIG) — See 'Hog'.

BARREN —
a. Unable to bear offspring.

b. (of cows). A barren cow is one that is neither in calf nor in-milk; it may not in fact be infertile.

* by kind permission of the Inland Revenue

BARREN LAND	Land incapable of growing useful crops. **A–001**
BELL-WETHER	The leading sheep of a flock.
BILLY OR BILLY GOAT	A male goat over two years of age.
BOAR	A male pig kept for breeding.
BOBBY-CALF	A young calf fit for manufacturing purposes such as meat extracts but not good enough to be sold as veal.
BOTTLE	A truss of straw.
BRAN	a. The larger bits of the outside of the grain separated from the flour and other parts when wheat is milled.
	b. A boar.
BRASH	A soil containing a large amount of stone chippings especially limestone.
BRAWN	Another word for a boar.
BREWER'S GRAINS	A feeding stuff made of the remains of barley which has been brewed to make beer.
BRIM	Used of pigs to mean to mate or to be on heat.
BRING OFF	To hatch out eggs.
BRITCH	Coarse inferior wool round the thighs of a sheep.
BROCK	a. Swill.
	b. Potatoes not fit for sale.
BROILER	Fowl fattened for killing.
BROKEN-MOUTHED	Refers to an old sheep that has lost some of its teeth and is unable to cope with mountain grazing.
BROOD	A hatch of young birds.
BROODY	The state of a motherly hen wishing to sit on her eggs to hatch them.
BROWSE	To feed upon twigs, leaves, tussocks etc.
BUCKLING	A male goat between one and two years of age.
BULL	An adult entire male bovine animal.
BULLING	Of the right size and age to be mated with a bull; ie in, or soon to be in, oestrus.

365

A–001 BULLOCK

A neutered bull (occasionally to mean young beef cattle whether male or female).

BURY

A clamp in which potatoes or other vegetables are stored.

BRUSSEN)
BUST   )

A ruptured pig.

BUTTERFAT

A type of fatty substance found in milk from which comes cream and butter.

BUTTERMILK

The liquid left behind after cream has been churned into butter.

CAD PIG

See Runt.

CADE

A young animal brought up as a pet.

CAKE

Cattle food compressed into slabs.

CALF

The offspring of a cow from birth until one year old.

CAPON

A neutered male chicken, fattened for killing at 12–14 weeks old.

CARR

An area of wet boggy ground or land reclaimed from a bog by draining it.

CARSE

A Scottish term for a stretch of alluvial flat land.

CASE

(Of eggs) means 360.

CASH CROP

One that yields produce for sale as distinct from one eaten by animals on the farm.

CATCH CROP

One that occupies the land for a short time between two main crops.

CEREALS

The cultivated members of the grass family such as wheat, barley, oats, rye.

CHILVER

A ewe lamb.

CLAMP

See Bury.

CLEARING CROP

One that is grown in widely spread rows between which hoes and cultivators can work to keep the land clear and free from weeds.

CLIP

To take the wool off the sheep, or shear it. The wool so removed whether from one sheep of a whole flock is called the clip.

CLOSED GILT

Pregnant gilt.

CLOVER HAY	Hay made from a crop of clover: also used to mean hay from temporary grass land as distinct from meadow hay from permanent grass land.	**A–001**
COARSE GRAINS	Cereal grains used for livestock feeding.	
COCK	An entire male chicken over 18 months of age.	
COCKEREL	An entire male chicken under 18 months of age but too old to be called a chicken.	
COLE) COLESEED)	Another name for rape.	
COLESTRUM	An especially rich and nourishing milk-like liquid produced by cows and other mammals for the first two or three days after giving birth.	
COOLER	A piece of dairy equipment used for cooling milk as it comes from the cow.	
COTT	A fleece that has become matted, generally as a result of ill health in a sheep.	
CORN	A grain crop—wheat, barley or oats.	
COUPLE	a. Ewe in lamb. b. Any dam with young at foot.	
COW	Normally a cow is an animal that has calved but in some areas animals are not termed cows until they have calved twice.	
COW-HEIFER	(Sometimes) a cow that has only had one calf.	
CRET	Seen Runt a.	
CRONE	An old worn out broken mouthed ewe.	
CULL	Take out animals that are too old or of poor quality.	
DEADSTOCK	A farmer's stock other than livestock, eg corn and hay, sometimes used to include tools and machinery.	
DIAMOND RAM) DINMONT RAM)	A young entire male sheep from its first shearing to its second.	
DILLING	See Runt a.	

A–001 DOWN-CALVER — A cow or heifer near calving time.

DRAFT — (–ewes) used by sheep farmers to denote ewes which are to be sold for breeding on easier farms.

DRIED GRASS — A feeding stuff from grass which has been dried artificially.
It is not the same thing as hay.

DRY FLOCK — A flying flock of store sheep.

DUTCH BARN — A large open shelter used chiefly for storing hay and straw.

EADISH)
EDDISH) — See Aftermath.

EARLING — A young lamb.

EASEMENT — A right to use, or to restrict the use of, land belonging to someone else, eg a right to draw water from a neighbouring farm, or to erect an electricity pylon.

EATAGE — Grass used for grazing especially the aftermath of a hay crop.

EILD — Dry—not giving milk.

ENSILE — To pack material into a clamp or silo so as to turn it into silage or haylage.

ENTIRE — Whole—not neutered.

EWE — A female sheep.

EWE LAMB — A female lamb from birth until the autumn of its first year when it becomes a ewe hoggett (or teg).

FALL — The birth of an animal: a fall of lambs is the number born in a flock during the season.

FALLOW — Land left without a crop during the growing season.

FALLOW CROP — A crop which is grown for ploughing in.

FARROW — a. To give birth to piglets.

b. A heifer or cow not in calf—barren.

FAT — Of the right weight and condition for market.

FELL	a. A hilly or mountainous moorland. **A–001**
	b. The hide of an animal especially with wool or hair still on it.
FEN	A flat low lying area of land very rich in humus and usually very fertile when drained.
FLUSH	To feed concentrates to a female animal shortly before service, with the object of increasing fertility.
FLYING FLOCK	A non-permanent flock of sheep—sometimes ewes past their best, sometimes stores brought on to a farm for only a season before being sold fat. A flying flock of store sheep is also called a dry flock.
FLY WOOL	The loose locks of wool that come away from the fleece when a sheep is shorn.
FODDER	Conserved food for cattle and horses especially dry food such as hay or straw.
FOG	Grass left standing throughout the winter.
FOGGAGE	Late grass—autumn grazing.
FOLD	a. An enclosure for grazing.
	b. In poultry a small combined house and run that can be easily moved about.
FOLDING	A method of feeding where animals (usually sheep) eat roots etc in the field, their progress being regulated by the movement of temporary "folds".
FOLLOWERS	a. Young stock in a dairy herd intended as replacements in the production herd.
	b. Cattle turned on to a pasture to clear it up after the dairy cows have had first pick.
FORAGE	Food for ruminants especially hay and straw.
FORAGE CROP	A crop for animal feeding and not human consumption.
FREE-MARTIN	Heifer which is sterile and abnormal in structure (often the twin of a bull calf).

A–001	FRIDD	See Inbye.
	GELD	a. Dry or barren.
		b. To neuter.
	GELT	a. Gelded (ie neutered).
		b. Another word for gilt.
	GILD	Alternative spelling of geld.
	GILT	A female pig up to the time of farrowing her first litter.
	GIMMER	A North Country term for a shearling ewe. (South Country equivalent is a theave.)
	GOATING	A young goat between one and two years of age.
	GOBBLER	An adult male turkey.
	GRATTEN	Stubble.
	GREEN CROP	One which provides fresh fodder for stock such as lucerne or maize.
	GREEN MANURE	A crop grown for ploughing back into the soil.
	GROATS	Oats or other grain husked and split.
	GROWER	A chick that has reached the stage of being able to do without artificial heat—say about 6 weeks old.
	GUMMER) GUNNER)	Weaned pig.
	HALF-CREASE	Owner of land allows it to be grazed by another's sheep in return for half the lambs born.
	HANDLING	The getting together of flocks etc for dipping and clipping sheep. May refer to the meeting of shepherds who undertake the rounding up of several flocks on a co-operative basis.
	HARD FRUIT	Apples and pears as distinct from stone fruit and soft fruit. (But note that "*top* fruit" *does* include cherries and plums.)
	HAY	Cut and dried plants usually grass or clover used for fodder.
	HEEDER	A ram lamb.

HEFT	A group of sheep grazing a particular range of mountain pasture. Hefting is the propensity of mountain sheep to remain on "their" part of unfenced mountain grazings.	**A–001**

HEFTED · A flock of sheep going with the farm taken over by the newcomer is "hefted" with the farm.

HEIFER A young cow until she has given birth to her first (or sometimes her second) calf.

HEN A female chicken after completing her first laying year; (before then it is a pullet).

HIGGLER A wandering dealer who buys poultry, eggs etc at the farmer's back door.

HILT See Gilt.

HIRSEL The area of land over which a shepherd has charge; or the number of sheep in his charge.

HOG A neutered pig.

HOGG   ) A young ewe or wether from the time
HOGGETT) of the summer lamb sales until first shorn in the following spring. Also called Teg.

HORNED STOCK A general term for cattle as distinct from other farm animals.

HOVEL An open or closed shed used for keeping cattle in.

HULL The outer covering of certain vegetable seeds especially pods of peas and beans.

HUMUS Partly decayed matter, leaves, stalks etc.

ILT See Gilt.

IN– Following by the name of an animal means pregnant, eg in calf or in pig.

INBYE In hill farms the enclosed fields round the homestead. (In Wales the term is Fridd.)

KEEP Grass or green food on which animals are grazed.

A–001

KID	A young goat less than 12 months old.
KINE	Cows.
LACTATION	The period during which an animal produces milk.
LATTERMATH	See Aftermath.
LEA	General term for open ground whether grass or arable but also another word for ley.
LEY	Temporary grass especially grass meant to be ploughed up a few years after sowing.
LEY FARMING	A system in which fields in turn grow such crops as corn and roots, then are laid down to temporary grass for a few seasons.
LOAM	Type of soil: somewhere between clays and sands and containing some of each.
LUCERNE	A perennial plant grown for fodder (Alfalfa).
MAIDEN HEIFER	A heifer not yet served.
MAIZE	Cereal plant grown in Britain chiefly as a fodder crop. Sweetcorn.
MEADOW	Originally permanent grassland used for the production of hay as distinct from pasture used for grazing. In practice the terms used are often based on historical rather than current use.
MEADOW HAY	Hay taken from permanent pasture. (Hay from a ley is called "seeds hay".)
MULTIPLE SUCKLER	System of rearing beef calves whereby one cow may be feeding up to three calves before they are weaned.
NEAT	An ox, bullock, or heifer.
NEATSFOOT	Calvesfoot or Cowheel.
NISGULL	See Runt a.
NITRATE	A compound of nitrogen, oxygen and a third element.
NIGGLEDRICH	See Runt a.
NORFOLK ROTATION	The four course rotation of crops—turnips, barley, clover, wheat.

NURSE-COW	A foster cow—one who suckles and looks after the calf of another. **A–001**
OIL CAKE	A cattle food made from various oily seeds such as linseed or cotton seed.
OPEN GILT	A gilt not yet served.
OUTGOING	Tenant right.
OUTLIER	A cow or other beast staying out of doors in winter.
OUT RUN	In Hill farms the wide open spaces away and usually separate from the homestead.
OX	a. A bovine animal. b. A bull especially when used as a beast of burden or draught.
PAN	A hard layer of tightly packed soil below the depth reached by the plough.
PIGHTLE	A small enclosure or field.
PLOUGH SICK	A term used for land which has grown arable crops for so long it shows signs of exhaustion and needs laying down to grass.
POLL (TO)	To dishorn.
POLLED	Having no horns.
PORKER	A pig kept for turning into pork as distinct from a baconer.
PRODUCTION RATION	That part of the ration over and above what is needed to maintain an animal, given to make it produce milk or reach fat condition.
PULLET	Young fowl in their first year of laying.
PUR LAMB	Male ram lamb.
QUEY	Scottish for heifer.
RACK UP	To fill up a hay rack with hay or straw.
RAM	An entire male sheep. Age sometimes given by number of times shorn, eg three shear ram.
RAPE	A plant which produces seeds rich in oil. May be grown as a cash crop or for feeding to animals.

373

A–001 RATLING PIG — See Runt a.

RISING — In giving the age of a beast means "nearly".

RIT)
RITLING) — See Runt a.

ROWEN — The aftermath or second growth of grass which is then left ungrazed till winter.

RUNNER — Weaned pig.

RUNT — a. A poor below standard creature especially the smallest pig in a litter.

b. A store beast.

SCREW — A very inferior animal.

SCRUB — a. A wilderness of brushwood, stunted trees and shaggy vegetation.

b. Of animals—poor, stunted.

SCUR — A horny knob that sometimes appears on polled cattle in place of a horn.

SECOND CUT — The second crop of conserved grass off a field in the same season.

SEEDS — A mixture of grasses and clovers. Another name for temporary grassland.

SEEDS HAY — Hay from temporary grassland.

SEEDS LEY — Another term for temporary grassland.

SEG — See Stag.

SHEAR — a. To cut or clip with shears.

b. The word is also used in giving the age of a sheep. A two shear ewe is one who has twice been shorn and may be between two and three years old.

SHEARLING — A young sheep from its first shearing until its second.

SHEEP AT HALVES — See Half-Crease.

SHEEP WALK — A stretch of grassland, usually rough, on which sheep are pastured.

SHIELING — In Scotland a bit of rough pasture land or a rough hut or shelter put up on such land.

SHIPPEN) SHIPPON)	A cow house.	**A–001**
SILAGE	Grass and other greenstuff preserved in a moist state generally with molasses or other additives, either in a clamp, or in a silotower.	
SILO	An airtight container into which fresh cut grass and other greenstuff is packed to make silage.	
SINGLE SUCKLER	System of rearing beef calves whereby a cow will bear and rear one calf until it is weaned.	
SINGLETON	A single lamb born to a ewe.	
SITTER	A broody hen.	
SLINK CALF	A calf born before its time.	
SLIPS	West Country term for newly weaned pigs sold at market.	
SNIVELDRAFT	See Runt a.	
SOILAGE	Green crops such as kale cut and fed in their fresh state to animals.	
SOW	A female pig kept for breeding is called a sow after she has had one litter, before that she is called a gilt.	
SPRINGER	An in calf heifer.	
SQUEAKER	A piglet.	
STAG	A bull or boar neutered later in life than usual.	
STEADING	The farm homestead: sometimes used to mean the buildings without the farmhouse.	
STEAM UP	To give a pregnant animal an increasingly rich diet with a view to an extra good lactation.	
STEER) STOT)	A young male ox, usually a neutered one, intended for beef.	
STEG	See Stag.	
STELL	A roofless stone shelter for cattle and sheep in open moorland places.	
STIRK	A heifer under two years of age. Sometimes used to describe a bullock.	
STORE	a. An animal sold or bought for fattening.	

A–001

	b. Being kept for fattening later.
	Generally any cattle intended for the beef market.
SUCKLER HERD	See "Single Suckler" and "Multiple Suckler".
SUCKER PIG) SUCKING) SUCKLING)	An unweaned piglet.
TACK	a. Pasture rented for grazing.
	b. Sheep sent away to be wintered elsewhere are said to be on tack.
TEG	A young sheep from the time of the summer lamb sales until it is shorn in the following spring. Also called Hogg.
TENANT RIGHT	The right of a tenant farmer to receive compensation for hay, straw, cultivations etc which he must leave behind him when he quits the farm he is renting.
TERRIER	A book in which records are kept of the farm's fields etc on an estate.
THEAVE	A shearling ewe kept for breeding purposes generally so called from the time it is first shorn until it is two years old. Also called Gimmer.
TILLAGE	Cultivation, manuring and seeding of the land. Arable land classed as tillage means land under crops or lying in fallow to distinguish it from arable land under permanent grass.
TOP FRUIT	Apples, pears, plums, cherries and nuts; the complementary classification to "soft fruit".
TUP	A ram—an entire male sheep.
TWO SHEAR	Of sheep between second and third shearing.
UNEXHAUSTED MANURIAL VALUE	The amount and worth of fertiliser or manure which remains in the land after one or more crops have been grown.
UNTHRIFTY	Animals which do not respond to normal feeding and care.

| VETCH | A plant grown for fodder. | **A–001** |

VETCH | A plant grown for fodder.

WATER MEADOW | A meadow adjoining a river, by which it is flooded from time to time.

WEANER | a. A piglet from the time of weaning to the time it becomes a fattener (or feeding pig).

b. A weaned calf.

WEDDER | See Wether.

WETHER | A neutered male sheep.

WETHER LAMB | A neutered male lamb so called until the time of the store sales.

WETHER HOGGET)
WETHER HOGG   )
WETHER TEG   ) | A neutered lamb from the time of the store sales in the late summer of the year of birth until it is shorn in the following spring or summer.

WETHER SHEARLING | A neutered male sheep after being shorn for the first time.

WICK | A dairy farm.

WOOL CLIP | All the wool off a sheep; or all the season's wool from the whole flock.

YELD | Dry—not producing milk.

YELT | See Gilt.

Metric/Imperial Conversion Tables

Metric	Rough Approximation (i)	Example	Imperial Equivalent
Hectare (ha)	$2\frac{1}{2}$ acres	football pitch	2.471 acres
Metres (m)	$3\frac{1}{4}$ feet		3.279 feet
Kilometre (km)	$\frac{1}{2}$ mile		0.621 mile
Kilogramme (kg)	$2\frac{1}{4}$ lbs		2.205 lbs
1 kg of milk (0.971)	$1\frac{3}{4}$ pints		1.712 pints
50 kg	1 cwt	sackful/"bag"	0.984 cwt
90 kg (pigs)	10 score	a "baconer"	9.9 score
500 kg	10 cwt ($\frac{1}{2}$ ton)	a dairy cow	9.84 cwt
1000 kg (a tonne)	1 ton		0.984 ton
1 litre (l)	$1\frac{3}{4}$ pints		1.76 pints
1 litre of milk (1.03 kg)			
10 litres	$2\frac{1}{4}$ gals	a bucketful	2.2 gals
45 litres	10 gals	a milk churn	9.9 gals
1p/kg	£0.5/cwt		£0.508/cwt
100p/kg	£9 a score		£9.1 a score
1 kWatt	$1\frac{1}{3}$ HP		1.342 HP
1 Megajoule (MJ)	250 calories		239 Kilocalories

A–001	Imperial	Rough Approximation (i)	Example	Metric Equivalent
	Acre	$\frac{1}{2}$ hectare	$\frac{1}{2}$ a football pitch	0.405 ha
	Yard	0.9 m		0.914 m
	Mile	1.6 km		1.609 km
	1 lb	$\frac{1}{2}$ kg		0.454 kg
	1 cwt	50 kg	sackful/"bag"	50.8 kg
	10 cwt	500 kg	a dairy cow	508 kg
	ton	tonne (1000 kg)		1.016 tonnes
	10 score	90kg	a "baconer"	90.7 kg
	1 pint	$\frac{1}{2}$ litre		0.568 litres
	1 gallon	$4\frac{1}{2}$ litres		4.546 (l)
	1 gallon of milk	$4\frac{3}{4}$ kg		4.681 kg
	10 gals	45 (l)	milk churn	45.46 (l)
	£1/cwt	2p/kg		1.967p/kg
	£1/score	11p/kg		11.02p/kg
	HP	$\frac{3}{4}$ kW		0.746 kW
	100 Kilocalorie (Calorie)	$\frac{1}{2}$ Megajoule (MJ)		0.418 MJ

(i) These rough approximations are often used.

Wildlife and Countryside Act 1981

Circular 4/83 **A–002**
(Department of the Environment)
(Ministry of Agriculture, Fisheries and Food)

Circular 6/83
(Welsh Office)

Joint Circular from the
Department of the Environment
2 Marsham Street, London SW1P 3EB

Ministry of Agriculture, Fisheries and Food
Whitehall Place
London SW1A 2HH

Welsh Office
Cathays Park, Cardiff CF1 3NQ

Sir   31 *January* 1983

Financial Guidelines for Management Agreements

1. A commencement order has been made to bring into effect on 28 February 1983 the provisions of Sections 32 and 41 of the Wildlife and Countryside Act 1981. These impose new duties upon Agriculture Ministers with regard to Sites of Special Scientific Interest and in certain other areas of the countryside, notably National Parks.

2. These Sections also impose a duty upon the Nature Conservancy Council and relevant authorities, as defined in Section 41(5), in certain circumstances to offer a management agreement (in the case of the latter under Section 39 of the Act, and in the case of the Nature Conservancy Council under Section 16 of the National Parks and Access to the Countryside Act 1949 or Section 15 of the Countryside Act 1968.) These circumstances are where an Agriculture Minister has refused an application for a grant under a scheme made under Section 29 of the Agriculture Act 1970 (farm capital grants) in consequence of an objection made by the Nature Conservancy Council or the relevant authority as appropriate.

3. Section 50 of the 1981 Act further provides that payments made under agreements offered in these circumstances (and under all agreements offered by the Nature Conservancy Council, following notification to it of proposed operations which would be damaging to Sites of Special Scientific

Interest) shall be of such amounts as may be determined by the offeror in accordance with guidance given by the Ministers.

4. This guidance has been prepared in consultation with a wide range of interested parties, and is set out in the Appendix to this Circular. While great care has been taken in preparation of the present guidance, the Ministers will monitor its application closely and will review its content as necessary in the light of experience gained in practical use.

A–003 5. The Ministers wish to draw attention to the following comments and additional information:

(i) *EC and other capital grants:* Although Sections 32 and 41 of the Act impose an obligation on the authority concerned to offer an agreement only in respect of the refusal of grant proposals under a scheme made under Section 29 of the Agriculture Act 1970, there are other grant schemes made under other legislation—principally the Agriculture and Horticulture Development Scheme—and grants for forestry operations are also available. The Nature Conservancy Council has stated that where it makes objection to proposed operations it will voluntarily apply its obligation under Section 32(2) to all types of farm capital grant, and, normally, to grants or felling permissions under forestry legislation. The Ministers have welcomed this commitment to the spirit of the legislation and express the hope that National Park Authorities and relevant authorities for any other areas which may be specified under Section 41(3) will consider whether to make similar arrangements, where this is seen to be appropriate.

(ii) *Scope of management agreements:* Management agreements may often be principally concerned with the negotiation of restrictions on potentially damaging farming operations and with maintaining existing farming practices, but in some cases there may also be benefit in providing for additional measures to be undertaken, to help improve the general amenity of the land. The guidance is, however, confined to determination of payment for agreed restrictions on proposed activities; where positive requirements are included the manner in which these objectives are achieved will be a matter for the parties concerned.

(iii) *Wider application of guidance:* By statute, the guidance must have effect in the circumstances described above, defined by Section 50(1) of the Act. However relevant authorities may wish to consider the extent to which its provisions may be applicable to other management agreements designed to protect the conservation value of land under the power conferred upon them in Section 39 of the Act. Grants may be available from the Countryside Commission to assist authorities to finance such agreements.

(iv) *Alternative means of protection:* In some special cases voluntary purchase by agreement or lease of the area concerned—with possible lease-back or resale with appropriate restrictions—may be preferred by either party as an alternative to entering into a management agreement. Again, grants may be available to local authorities from the

Countryside Commission to assist towards the acquisition or the net costs of the land transaction as appropriate.

(v) *Obligation to offer a management agreement:* The Nature Conservancy Council or relevant authority is obliged to offer a draft management agreement only where the Agriculture Minister's refusal of grant is in respect of expenditure to be incurred. If in any circumstances an operation is begun without prior notification or before the Agriculture Minister has given his decision following a formal objection, a draft agreement need not be offered.

(vi) *Withdrawal of offer:* The Nature Conservancy Council or relevant authority may withdraw a formal objection made under Section 32 or 41 of the Act at any time prior to the making of the relevant decision on grant by the Agriculture Minister; and in such circumstances may withdraw any undertaking given previously to enter into a management agreement in respect of the operation upon which grant was claimed.

(vii) *Distribution of guidance and forms:* It should be the responsibility of the Nature Conservancy Council or the relevant authority, as appropriate, to make available a copy of the guidance for the offeree's inspection and to provide him with the necessary forms in accordance with Annex A, B or C as appropriate. Worked examples will be made available to assist the offeree in their completion.

(viii) *Adjusting payments:* Paragraph 41 of the guidance refers to the **A–004** requirement that annual payments, provided under the terms of a management agreement, should be adjusted to reflect changes in farm productivity and profitability. It is the intention in due course to produce standard indices covering a range of types of agreement, to assist bodies making payments to calculate such adjustments without need for individual review (although, by agreement, the parties concerned may adopt different arrangements, if they wish). Preparation of these indices will inevitably require analysis of a wide cross-section of agreements entered into by the Nature Conservancy Council and relevant authorities. Accordingly some time must elapse before such indices can become available for general use. As an interim measure, therefore, a single national index will be produced for the purpose of revising payments due. The first such index will be made available to the Nature Conservancy Council and all relevant authorities in time for use when the first agreement reached under this guidance falls due for adjustment. The index may be revised from time to time thereafter and payment for each agreement should be updated on the first and subsequent anniversaries of its effective date, using, where appropriate, the index in force in the month in which that anniversary falls. When the more comprehensive range of indices is available, retrospective adjustments should be made, during its first year of application, to equalise any discrepancies between payments due under those arrangements and those actually paid using the interim index; and offerees should be informed of this arrangement.

(ix) *Confidentiality:* Information provided by the offeree for the drawing up of an agreement under this guidance is confidential. However, to assist in preparing the indices described above, the offeror should supply the Ministry of Agriculture, Fisheries and Food (Economics Division 1, Room 602, 3 Whitehall Place, London SW1A 2HH) with copies of information provided, in accordance with Annex C of the guidance, adjusted if appropriate to reflect the terms agreed during subsequent negotiations between the parties, or a decision on arbitration.

(x) *Arbitration:* Section 50(3) of the Act provides for the offeree to require arbitration in the event of dispute over the determination of payment. However, the Ministers have expressed the hope that resort to arbitration will be infrequent, and that all parties will make every effort to secure agreement by mutual consent.

(xi) *Implications for rent reviews:* Paragraph 8 of the guidance prescribes the basis of annual payments made to a landlord under the terms of a complementary management agreement. The landlord should be entitled on rent reviews as and when appropriate to recover a proportion of the payment made to the tenant as though the management agreement did not exist and the tenant was in no way restricted from farming in accordance with the Agricultural Holdings Act 1948.

(xii) *Criminal liability:* Nothing in the guidance detracts from the liability to criminal proceedings of a person who contravenes Section 28(5) or 29(3) of the 1981 Act.

We are, Sir, your obedient Servants

A. FLEXMAN, *Assistant Secretary, Department of the Environment*
P. W. MURPHY, *Assistant Secretary, Ministry of Agriculture, Fisheries and Food*
H. K. TRIMNELL, *Assistant Secretary, Welsh Office*

The Chief Executive
  County Councils ⎫
  District Councils ⎬ In England and Wales
  London Borough Councils ⎭
The Town Clerk, City of London
The Director General, Greater London Council
The National Park Officer
  Peak Park Joint Planning Board
  Lake District Special Planning Board
The Director General, Nature Conservancy Council
The Director, Countryside Commission

[DOE DRA1/30/12/1]
[MAFF LU/9557]
[WO CL/88/21/01]

APPENDIX

WILDLIFE AND COUNTRYSIDE ACT 1981— FINANCIAL GUIDELINES FOR MANAGEMENT AGREEMENTS

Scope of Guidance

1. Section 50 of the Wildlife and Countryside Act 1981 provides that where   **A–005** management agreements are offered, in specified circumstances, by either the Nature Conservancy Council or a "relevant authority"[1] for the purposes of conservation or amenity, payments under such agreements shall be determined by the body concerned (the offeror) in accordance with guidance given by Ministers.

2. These provisions have statutory application in the following circumstances:—

(i) *Sites of Special Scientific Interest* (SSSI)

— If the *Nature Conservancy Council* (NCC) has been notified[2] by an owner or occupier of a proposal to undertake an operation specified as likely to damage an SSSI and,

— in consequence offers to enter into a management agreement with him.

(ii) *Sites of Special Scientific Interest, National Parks and other specified areas*

— if, in respect of land within an SSSI, National Park, or other specified area,[3] an owner or occupier proposes to undertake an operation on which *farm capital grant* will be claimed under Section 29 of the Agriculture Act 1970 and,

— either the *NCC* or the *relevant authority* objects,[4] in consequence of which the grant application is refused with effect that the objecting body *is obliged* to offer a management agreement.

3. The financial terms of these agreements must be proposed by the offeror in accordance with guidance by Ministers, and such guidance is given in these notes. However, an individual owner or occupier (the offeree) may wish to make his personal contribution to conservation of the land concerned within the terms of the agreement by offering to accept a lesser payment, or alternative arrangements (including payments in kind).

4. The Nature Conservancy Council has further stated that it will voluntarily apply the provisions of Section 32(2) of the Act to applications made for *all* types of farm capital grant (including grants under the Crofting Coun-

---

[1] In *England and Wales* the "*relevant authority*" means:
    (a) as respects land in a National Park, the county planning authority;
    (b) as respects land in Greater London, the Greater London Council or the London borough council; and
    (c) as respects any other land, the local planning authority;
    in *Scotland* it means the authority exercising district planning functions.
[2] Under Section 28(5) or 29(4) of the Wildlife and Countryside Act 1981.
[3] Specified by Ministers under section 41(3) of the 1981 Act.
[4] Under Section 32(1)(b) or 41(3)(b) of the 1981 Act.

ties Agricultural Grants Scheme) and, normally, for grants or felling permission under *forestry* legislation; and that agreements so offered will accord with the terms of the guidelines (subject, as above, to any abatement proposed by the owner or occupier himself).

Wider aspects of agreements

5. The immediate purpose of a management agreement offered under these provisions is to conserve the countryside by providing for agreed *restrictions* on operations which would damage the conservation value of the land and to ensure that all necessary steps are taken to *maintain* the land in its present state. However, the parties may also agree to include other requirements, whereby an owner or occupier undertakes to implement measures aimed, as appropriate, at *enhancing* the natural beauty or amenity of the land: eg creating or improving public access where appropriate, or specific land management works. Any payment in respect of such further requirements should in general be determined in accordance with the merits of the individual case.[5]

"Dual responsibility"

6. Where a Site of Special Scientific Interest falls within a National Park or other specified area, proposed agricultural operations on that site should first be considered by the Nature Conservancy Council, who will consult the relevant authority and agree with it which body or bodies should lodge objection, if any, to the payment of farm capital grant and which of them should subsequently take the lead role in negotiations with the owner or occupier. Subsequently the bodies concerned should agree between themselves an appropriate apportionment of payments due under any agreements thus concluded.

Landlords and Tenants

A–006 7. Where land is let and a tenant proposes to accept an offer of a management agreement, the terms of the agreement may have implications for his tenancy agreement[6] and for his landlord. The offeror should therefore require a tenant's written assurance that his landlord has been informed:

— of the proposed operation where this is required;
— that the offer of a management agreement is likely to be accepted by the tenant;
— of the area and the location to be covered by the agreement and the parties to that agreement.

8. *Complementary agreements:* A tenant's interest will be limited and any management agreement with that tenant cannot bind a future occupier. In view of this, it may best serve to ensure the long-term protection of a site if the offeror seeks to include the landlord's interests in the principal agreement, or otherwise seeks a complementary agreement with him. Normally

[5] Statutory provisions govern compensation for *access*.
[6] Or in Scotland, lease.

there should be no need to make more than a nominal payment, to encourage the making of such agreements, as the landlord's interest will be secured by the Agricultural Holdings legislation. As part of an agreement of this type the landlord should normally undertake not to serve a notice to remedy contrary to the intentions of the management agreement.

9. *End of tenancy:* The agreement with the landlord should provide that, where he plans to take the land in hand at the termination of the tenancy,[6] he should give *6 months'* advance notice to the NCC or relevant authority of his wish to terminate the agreement. Where this happens or where the landlord leases the land to a new tenant,[7] a management agreement should be offered on terms similar to those enjoyed by the previous occupier.

Mortgagees and Other Interested Parties

10. Mortgagees, and other parties having interests in the property (eg fishing rights), should also be consulted, as appropriate, by the offeree before he enters into a management agreement. **A–007**

First Steps Towards Agreement

Minimising loss

11. From the date of initial notification of his proposed operation, an owner or occupier should not take any action which may increase the sum eventually payable to him under these guidelines, eg entering into any contract or commitment with third parties relating to the proposed work. *If he enters into such a contract or commitment during that period he will have no claim on the offeror if it is inconsistent with restrictions which are contained in a management agreement subsequently completed, and involves him in financial loss.*

Short-term agreement

12. Detailed negotiations to conclude the terms of a proposed management agreement may extend over a period of some months, and the negotiating parties are recommended to take early action (ie within the period of 3 months following notification) to enter into a short-term agreement. Such an agreement would be for a fixed term (between 6 and 12 months, as appropriate) and would include a commitment by the owner or occupier not to undertake the proposed operation while discussions on the subsequent long-term agreement continue. A nominal sum should be payable to the offeree on entering into a short-term agreement of this type and the right of the owner or occupier to arbitration on the payment proposed under the subsequent agreement (see paragraph 33 below) would be retained.

Long-term agreement; methods of payments[8]

13. For the purposes of the long-term agreement, owners or owner-occupiers may choose *either* lump sum payment *or* annual payments, as

---

[7] In Scotland, where a lease is acquired by a successor.
[8] Different considerations apply in forestry cases—see Annex A.

described below, and should be encouraged to state their preference as early as possible. Only annual payments are available to *tenants* and, in general, *landlords* would be expected only to enter into an agreement of the type described in paragraph 8 above.

(a) *Lump sum payment: (owners or owner-occupiers)*

14. Payment, at the commencement of the agreement, of a single *lump sum* for a management agreement over a 20-year period (or such other period, possibly operating in perpetuity, as may be agreed between the parties). *The amount should be equal to the difference between the restricted and unrestricted value of the owner or owner-occupier's interest*, calculated having regard to the rules for assessment in respect of the compulsory acquisition of an interest in land, as set out in Section 5 of the Land Compensation Act 1961[9] so far as applicable and subject to any necessary modifications; and in so far as there is no statutory eligibility for compensation in this respect under Section 30(2) of the 1981 Act.

15. *Offerees choosing this method should be requested to provide the information listed at Annex B.*

(b) *Annual payments: (owners or owner-occupiers)*

16. Payment of *annual sums* for an agreement over a 20-year period, or such other period as may be agreed between the parties. *The payments should reflect net profits forgone because of the agreement* (or, in the case of payments to landlords, be nominal—see paragraph 8), the last payment to fall due 12 months before expiry.

(c) *Annual payments: (tenants/occupiers)*

17. Annual payments as above while the offeree is in continued occupancy. Lump sum payment is *not* available.

Assessment of annual payments

A–008   18. *Individual assessment* will be appropriate in most cases to calculate the sums payable. *The offeree should provide the information listed at Annex C* and indicate the amount he seeks. However, *by agreement of both parties*, provision of more limited information may be sufficient in certain cases. In some cases, eg involving non-agricultural operations, different information will be required; this should be as agreed between the parties.

19. *"Standard" payments:* Where well-defined categories of land and restrictions on farming operation are involved the offeror may wish to determine[10] and periodically revise,[11] *standard annual payments* as an alternative option. Such payments would be offered as a tariff, with

[9] In Scotland, Section 12 of the Land Compensation (Scotland) Act 1963.
[10] In consultation with representative bodies of farmers and landowners.
[11] Not less frequently than each five years, in consultation as above.

uniform payments per hectare for particular categories of agreement. The use of standard sums in this way may be found convenient by both parties, since the need for detailed assessment of a proposed operation is eliminated, and agreement can therefore be reached more quickly.

20. If standard payments are available, the offeror should ask the offeree whether he wishes to choose that option. Where the offeree opts for individual assessment he will be unable subsequently to change to a standard payment.

The draft agreement[12]

21. Where the Minister has refused an application for farm capital grant,[13] in consequence of an objection by the Nature Conservancy Council or relevant authority, the Act requires that body to submit a *draft agreement* to the applicant within *3 months* of its receiving notice of the Minister's decision.

22. Such an agreement should be as complete as possible, drawn up in the light of progress in the negotiations between the parties at that stage. In particular the draft agreement should record:

— the main heads of agreement including the proposed restrictions and the period;

— the method of payment where this has been decided;

— if appropriate, the relevant standard payment (where this option is available);

— the sum or sums payable or, if not yet determined, that payment will be determined in accordance with this guidance by further negotiation, or following arbitration (see paragraph 33 below);

— if agreement is not complete, a date (not more than 6 months ahead) by which a formal offer will be made (see paragraph 32 below).

Other Factors Determining Payment

(i) *Deemed eligibility for grant*

23. For the purpose of calculating payment under a management agreement it should be assumed that, but for the conservation considerations, farm capital grant *would* have been payable on a proposed agricultural operation which the offeree undertakes in the agreement not to carry out.   **A–009**

24. *Exceptions:* However, if:

(a) the proposed operation is ineligible[14] for such grant; *or*

(b) the business is not an agricultural business, or is an agricultural business which does not satisfy[14] the sufficient employment test, where that applies in respect of the proposed investment; *or*

---

[12] Under Section 32(2) or 41(4) of the 1981 Act.
[13] Grant under Section 29 of the Agriculture Act 1970.
[14] Under the published criteria of the Agriculture Departments.

(c) if the operation is begun before the appropriate Agriculture Minister determines the grant application;

neither the Nature Conservancy Council nor relevant authority is under a statutory obligation to offer an agreement; and agreements offered voluntarily[15] by the Nature Conservancy Council in such circumstances (which by virtue of Section 50(1)(a)(i) of the Act are also subject to this guidance) should assume that farm capital grant would not have been paid. (Advice to the offeror on items (a) and (b) above will be given on request by the local office of MAFF.[16] As regards (c), the Headquarters of the Agriculture Department concerned will inform the offeror and offeree of the date of the Minister's decision.)

25. *Investment limits:* The farm capital grant schemes prescribe limits on the overall level of investments on which a farmer may claim grant. If expenditure on a proposal would in whole or in part exceed the relevant limit, the excess would not be eligible for grant until such time as it could be accommodated within that limit. The offeror should invite the offeree to consent that MAFF[16] notify the former whether the proposed expenditure is within the offeree's investment limit and, if not, when (in the absence of other claims) it would become eligible for grant. *If the offeree refuses consent to such disclosure, payment under the agreement should be calculated without allowance for grant.*

A–010 26. If the offeree consents, the local office of MAFF[16] will notify the offeror of the amount of investment eligible for grant on the effective date of the agreement (see paragraph 38 below). Where the cost of the proposed operation exceeds the amount so notified, the offeree should be asked whether he would still be prepared to accept from the effective date an agreement with abatement in respect of that part of the cost which would not be eligible for grant until a later date. If the offeree declines, he should be invited to withdraw his notification on the understanding that he may represent it at a time when the whole cost of the operation would be within his eligible investment limit. Once an agreement has become effective it should not be amended to take account of investments claimed subsequent to the effective date and during the life of the agreement.

27. *Rate of grant:* The local office of MAFF[17] will also advise the offeror on the *rate of grant* for which a proposed operation would have qualified on the effective date of the agreement (see paragraph 38 below) and under which grant scheme the application was made.

(ii) *Phasing in of proposed operations*

28. The offeror should have regard to the practical limit on the amount of work which can be undertaken by the offeree in a given period of time and that a development may take a number of years to implement or give a full return. In such cases it will be appropriate for lump sum payments to be

---

[15] Following notification under Section 28(5) or 29(4) of the 1981 Act.
[16] In Scotland DAFS; in Wales WOAD.
[17] In Scotland DAFS; in Wales WOAD.

correspondingly smaller, or for payments on an annual basis to be phased in during the early years of the agreement.

(iii) *Professional fees and other expenditure incurred*

29. *Fees:* On completion of a management agreement the offeror should pay the reasonable costs of the offeree incurred in retaining professional advisers to assist him in connection with the agreement. (VAT on such fees should be met only where the offeree is not registered for VAT purposes.)

30. *Other costs:* A management agreement may also provide for payment:—

(a) for expenditure reasonably incurred within the previous 12 months of the date of notification which has been rendered abortive or in undertaking work rendered abortive by the agreement (subject to minimising loss—see paragraph 11 above);

(b) for any loss or damage directly attributable to the agreement;

in so far as there is no statutory eligibility for compensation in this respect under Section 30(3) of the 1981 Act, and no relevant payment is available under other provisions of these guidelines.

31. *Abortive negotiations:* Where the offeror withdraws from negotiations it should defray costs incurred by the offeree, unless the latter has acted in an unreasonable manner: eg if the offeree carries out operations which would render the agreement abortive.

The Formal Offer

32. The formal offer should record full details of the financial and other terms of agreement proposed, *the offeror appending thereto the following statement:*   **A–011**

"These terms constitute the formal offer of a management agreement. Under Section 50(3) of the Wildlife and Countryside Act 1981 you have the right to dispute the offer, within one month of receipt, and to require that determination of the payment offered shall be referred to arbitration."

Arbitration

33. Where an offeree disputes the financial terms of the agreement he may, within *one month*[18] of receiving the formal offer, require determination by an arbitrator[19] of the amounts payable.

34. In default of agreement on the appointment of an arbitrator[19] he will be appointed by the appropriate Secretary of State[18] on the recommendation of relevant professional bodies.

---

[18] Under Section 50(3) of the 1981 Act.
[19] In Scotland, arbiter.

35. If the making of the offer was *mandatory*—ie made in the circumstances described at paragraph 2(ii) above—and the amounts payable determined by the arbitrator[19] exceed those determined by the offeror, the latter must amend its offer accordingly.[18]

36. If the making of the offer was *non-mandatory*—ie made in the circumstances described at paragraph 2(i) above—the offeror may choose *either* to amend the offer to give effect to the arbitrator's[19] determination, *or* to withdraw that offer.[18]

37. *Costs of arbitration:* Costs will be awarded by the arbitrator.[19]

Commencement of Agreement

A–012 38. For the purpose of this guidance the *effective date* of the management agreement should be the date of the agreement or *3 months* after the date of receipt by the authority of the owner's or occupier's notification of details of the proposed operations, whichever is the earlier.

39. Where a *lump sum payment* has been chosen the date of valuation should be as at the effective date. Payment should fall due on the completion of the agreement and interest[20] should be payable in respect of the period between the effective date and completion. Where *annual payments* have been chosen the first payment should be backdated to the effective date and paid with interest[20] on completion; thereafter payments should be on regular dates specified in the agreement.

40. The lump sum payment or the first annual payment under the management agreement should be abated by the nominal sum paid to the offeree under the short-term management agreement where applicable (see paragraph 12 above).

Adjusting Payments (Annual Payments Only)

41. Agreements providing for annual payments should include provision for making adjustments in such payments to reflect *annual* changes in farm productivity and profitability, using *indices* which will be provided for this purpose (except that, by the consent of both parties, other bases of adjustment may be employed, eg: where the agreement is non-agricultural).

42. In addition, where the management agreement is based upon individual assessment, it should be open to either party at intervals of not less than *five years* to require full reassessment of further annual payments due, and to refer any consequent dispute to arbitration.

Breach, or Voiding, of Agreement

A–013 43. If an agreement is breached by deliberate action of the owner or occupier, the offeror should, by the terms of that agreement, be entitled:—

---

[20] *Interest* should be assessed at a rate prescribed by regulations under Section 32 of the Land Compensation Act 1961 or, in Scotland, under Section 40 of the Land Compensation (Scotland) Act 1963.

(a) in the case of *lump sums*: to require *proportionate repayment* by the offeree, ie of a sum which *at the time of breach* would represent the value of a notional agreement offered for the unexpired period (subject in the case of a 20 year agreement to a minimum recoupment of 1/20 of the original payment for each year remaining of the terminated agreement or, in the case of an agreement made for a different period,[21] a corresponding fraction). Repayment should be net of any liability to capital gains tax[22] attracted by the original payment;

(b) in the case of *annual payments*: to cease further payments, *but there should be no provision for recoupment of those paid over before the breach of the agreement.*

44. An agreement may be terminated in the event that it is no longer possible to achieve its purposes by virtue of the action of a third party or by natural or accidental causes. In such circumstances:

(a) in the case of *lump sums*: the offeror should be entitled similarly to require proportionate repayment as above but may at its discretion abandon its claim or seek a reduced sum. If a claim is made it should be open to the offeree to require an arbitrator[23] to determine whether repayment should be waived or reduced after consideration of the full circumstances of the case including any hardship likely to arise from enforcement of a requirement to repay;

(b) in the case of *annual payments*: the offeror should be required to honour the next payment due under the agreement, after which further payments should cease.

Partial damage

45. The agreement should also provide for the possibility that partial damage to the land may occur. In such cases it should be open to the offeror to terminate the agreement on terms as in paragraphs 43 and 44 above suitably modified; and to offer to enter into a new management agreement in respect of the changed situation.

Renewal of Agreement

46. *Two years* before an agreement with a fixed termination date is due to expire, the offeror should initiate negotiations about possible renewal, *and should in all cases inform the offeree at least one year before expiry whether the offeror wishes, in principle, to renew the agreement.*   **A–014**

Capital Taxation

47. Various fiscal reliefs are available to owners of heritage land. These are explained in detail in the explanatory memorandum "Capital Taxation and

---

[21] An agreement operating in perpetuity should be treated as if it were for a period of 60 years for this purpose.

[22] Or corporation tax in respect of a capital gain.

[23] In Scotland, arbiter.

the National Heritage" (issued by the Treasury in December 1980 and soon to be revised). In particular conditional exemption from capital transfer tax may be given in respect of land of outstanding scenic, historic, or scientific interest; and such land may be designated if a maintenance fund is set up to provide for its upkeep and the provision of public access. If land is conditionally exempted or designated the owner is statutorily required to give undertakings that reasonable steps will be taken for the maintenance of the land and the preservation of its character and for securing reasonable public access. These undertakings will normally impose broadly the same restrictions on land use as a management agreement because both are designed to secure the same general heritage objectives.

48. Accordingly, management agreements should require the offeree to notify the authority with which the agreement was concluded of the granting of conditional exemption or designation in respect of land which is subject to a management agreement, and such agreements (whether made with an owner, a tenant or both) should provide for their termination on the date from which conditional exemption or designation is granted: *annual payment* under the agreement should thereby cease from that date; in the case of an agreement based on *lump sum* payment partial recoupment of the sum expended should be effected in accordance with the provisions of paragraph 43 above. If conditional exemption or designation relates to only part of an area of land which is the subject of a management agreement, or if the conditions of the management agreement were more onerous than the statutory undertakings, fresh agreements should be negotiated accordingly, with recourse to arbitration in the event of dispute. Agreements should also provide for renegotiation in the case of conditional exemption or designation being withdrawn since in that event the statutory undertakings will no longer apply.

49. *Position of tenants:* Where a landlord gains conditional exemption or designation in respect of his land it will clearly be necessary for him to secure the undertakings he is required to give in return by an agreement or arrangement with his tenant. Such an agreement or arrangement will no doubt take account of the financial implications of the restrictions imposed on the tenant's use of the land by the undertakings given for conditional exemption or designation. This applies equally whether the land has been subject to a management agreement or not.

ANNEX A

FORESTRY OPERATIONS

A–015   1. Conditions which may be sought on forestry operations are likely to fall into the following main categories:

> (a) *Outright prohibition*
> — Planting of bare land
> — Felling of woodland
> The prohibition should be for a stated number of years.

(b) *Modification of management practices*
— New planting and restocking (eg choice of species)
— Maintenance operations (eg application of fertiliser)
— Clearfelling (eg design of coups)

Modifications may apply throughout the whole crop rotation or only to part of it.

2. The Forestry Commission will advise the offeror whether forestry grant would have been payable in the circumstances of the application and, if so, the rate of grant for which the proposed operation would have qualified. This information should be taken into account in assessing the financial terms of the management agreement.

3. *Lump sums* should be determined by individual assessments of net revenue forgone based generally on a comparison of *discounted streams of expenditure and income over the appropriate period* and calculated (a) with, and (b) without, the constraints imposed by the management agreement. The rate of discount should be set in real terms and be that generally used by woodland owners in calculating present worth as agreed between the representative body for private forestry and the Nature Conservancy Council.

4. Alternatively the offeree may elect to receive payment based on the *depreciation in value* of the land or woodlands concerned. Payment should be calculated having regard to the rules for assessment in respect of the compulsory acquisition of an interest in land, as set out in Section 5 of the Land Compensation Act 1961[24] so far as applicable and subject to any necessary modifications.

5. Where *annual payments* are required, these should be derived from the **A–016** lump sum (calculated as in paragraph 3 above) rentalised (amortised) to produce a flow of annual payments based on an estimate of current market rates of interest over the period of the management agreement.

6. Because of the wide variability in circumstances, payment based on a *standard sum* is not appropriate for conditions placed on forestry operations.

7. Forms will be available from the Nature Conservancy Council.

8. The general provisions of the guidance in relation to wider agreements, landlords and tenants, mortgagees, minimising loss, short-term agreements, phasing in of proposed operations, professional fees and other costs, abortive negotiations, the formal offer, arbitration, commencement of agreement, breach or voiding of agreement, renewal of agreement and capital taxation, apply equally to forestry as they do to agricultural operations.

[24] In Scotland, Section 12 of the Land Compensation (Scotland) Act 1963.

ANNEX B

LUMP SUM PAYMENT

A–017 The owner should provide full details of the land over which restrictions are proposed (including its area) and should state the sum which he wishes to claim as payment for the agreement. Information should, if possible, be supplied to the offeror within one month of being requested and should include the following:—

(a) Interest in the land—whether freehold[25] or leased; the terms of any tenancy agreement; rent passing; and date of the last rent review;

(b) details of restrictions, easements[26] or rights of way affecting the land; and of any land charges and mineral rights;

(c) Names and addresses of professional advisers acting for the offeree.

Brief details should also be given of any adjoining land which the offeree holds (including a plan of the total holding). The offeror may, in addition, request such other relevant information as considered necessary.

ANNEX C

ANNUAL PAYMENTS FOR NET PROFITS FORGONE

A–018 1. If the owner or occupier chooses individual assessment, then he should provide the offeror, if possible within one month of being requested, with the required information as set out in the following forms. In addition, he should supply:

— A plan showing the total area farmed by him, including the area proposed for improvement;
— If requested by the offeror, confirmation of his financial ability to carry out the proposed operation: eg a bank manager's letter;
— Any other relevant information that the offeror may request.

2. The offeror will either accept the sum claimed or enter into negotiations with the offeree.

3. The information given in the forms should be as complete as possible because it will form the basis for negotiations. Nevertheless, other factors may emerge which will need to be included.

---

[25] In Scotland, dominium utile.
[26] In Scotland, servitudes.

IN CONFIDENCE

Information submitted in support of a claim for annual payments based on **A–019** an individual assessment of net profits forgone.

I claim the sum of £................. as detailed in Part 5 of this form

Signed .... Date ....

Information on the claimant

Full name ....

Address: .... Postcode ....

If the address of your farm is different, please enter it here ....

.... Postcode ....

Please give the name(s) and address(es) of your professional adviser(s) ....

....

.... Postcode(s) ....

Is your interest in the land freehold* ☐ or leasehold ☐ ?
(*Please tick the appropriate box*).

If you lease the land, state the name and address of your landlord, the terms of tenancy, rent passing and date of last review; and confirm that the landlord has been notified as required by paragraph 7 of the guidance ......

....

....

....

Please detail any restrictions, easements† or rights of way affecting the land

....

Please detail any land charges or mineral rights ....

....

In Scotland: *=dominium utile. †=servitudes.

A–019   *County (in Scotland, Region)* ....

1—Information on current crops, livestock and labour

Please complete these summaries of the cropping area, number of livestock and labour engaged for the whole farm.

Crops, etc	Hectares*		Stock	Number
Wheat			Dairy cows	
Barley			Beef cows	
Oats			Beef heifers	
Other cash crops (please specify)			Fattening/rearing cattle	
			Ewes	
			Sows	
			Fattening pigs (over 20kg)	
Permanent grass				
Temporary grass			Laying hens	
Fodder crops			Broilers	
Total crops and grass			Other stock (please specify)	
Rough grazing				
Buildings, Roads, Woodlands etc.				
Total farm area				

Labour	Number
Farmers and partners	
Farmers' and partners' spouses doing farmwork	
Regular full-time workers	
Regular part-time workers	
Seasonal and casual workers	

1 acre=0.40 hectares

2—Description of the current situation and the proposed improvement

What is the extent of the area you propose to improve? ⬚ hectares*

*1 acre=0.40 hectares

Please describe:

—The current situation (*eg "The area is mainly rough grazing (mostly heather), supporting about 50 ewes and lambs throughout the summer. Lambs are sold as stores"*)

—The proposed improvement (*eg "The area is to be drained, limed, fertilised and reseeded"*)

—The proposed practice after improvement (*eg "the stocking rate would be raised to 200 ewes during the summer. Lambs would still be sold as stores"*)

A–019

3—Annual financial effect of the proposed improvement.

Include all revenue and cost changes (variable and fixed) which would be expected as a result of the improvement. Use current values and costs.

If there are substantial year by year differences (eg because of phasing of improvements) use a fresh sheet for each year. The final sheet should reflect the full effects.

Budget for 19..........

Extra Variable and Operating Costs

Number and type of Units eg 2 tonnes compound fertilizer	Cost/Unit £	Total Cost £
Total		(a)

Extra Revenue

Number and type of Units eg 10 heifers	Value/Unit £	Total Revenue £
Total		(c)

Revenue Forgone

Number and type of Units eg 10 lambs	Value/Unit £	Total Revenue £
Total		(b)

Variable and Operating Costs Saved

Number and type of Units eg 200 kg N	Value/Unit £	Total Cost £
Total		(d)

Annual Benefit from Proposed Improvements (before fixed cost adjustment)

$$(c + d) - (a + b) = £ _____(A)$$

4—Capital requirements of the improvement (*if any*)

Please give details of any additional capital expenditure for buildings, drainage, fencing, machinery etc.

Items			Total Cost £		*Annual Charge Factor	Annual Charge Net of Grant £
No	Description	Gross unit Cost (£)	Gross	Net of Grant		
		Total				(B)

5—Amount claimed

Estimated Net Annual Profits Forgone | A − B = £_____(C) |

*The multiplying factor used should reflect the economic life of the item and the appropriate interest rate: eg the factor is 0.163 for an item with a 10 year life at 10% interest.

Index

Property—*cont.*
investment company, held through, 9–027,
9–028
loan interest, 9–012
one estate, land managed as, 9–013
plant and machinery used in maintenance
of, 9–015
premium on letting, 9–020
rent. *See* **Rent.**
right to repurchase, sale with, 9–023
sale of, treatment of proceeds, 9–017
tenancies, categories of, 9–008
tenant's repairing lease, 9–008
unfurnished lettings, 9–005—9–007

Race Losses,
income from letting, 3–005
Rates,
agricultural property, on, 20–004, 20–005
generally, 1–036
occupier, paid by, 20–003
replacement of, 20–001
sporting rights, on, 10–004, 10–005, 20–005
tax system of, 20–001
valuation,
appeal against, 20–002
net annual value, 20–001
Redundancy Payments,
trading expenses, as, 3–025
Rent,
allowable deductions, 9–009, 9–011,
9–014
contribution to repairs, including, 9–008
deduction for tax purposes, as, 3–037
definition, 9–005
electric line wayleaves, in respect of, 9–026
expenses, 9–009—9–011
furnished holiday lettings, from, 9–002,
9–003
furnished property lettings, from, 9–004
mines, etc., payable in connection with,
9–025
non-residents, and, 9–019
not paid, relief for, 9–018
unfurnished property lettings, from,
9–005—9–007
waiver of, 9–018
Repayment Supplement,
entitlement to, 4–049
Residence,
company, 14–012, 14–013
determination of, 14–004
dual, 14–008
husband and wife, of, 14–009
meaning, 14–005
ordinary, 14–005, 14–007
Revenue practice, 14–006
Retention of Title,
stock valuation, and, 6–012
Retirement Annuity Relief,
net relevant earnings, 4–048

Schedule B,
abolition, 12–008
former taxation under, 12–005, 12–006
Schedule D,
woodlands, election for taxation of,
12–007
Scientific Research,
capital allowances, 5–052
Sea Wall,
capital expenditure on, 9–014
Settled Property,
capital gains tax, 15–081
gifts with reservations, 18–014
hold-over relief, 15–044
meaning, 15–080
trustees of, 15–081
Share Farming,
herd basis, application of, 7–015
Shares,
calls on, capital gains tax, 15–030
conversion of securities, 16–021
hold-over relief, 15–046
pooling election by partners, 16–013
reorganisation, 16–020
unquoted,
company reorganisation, capital gains
tax on, 15–019, 15–020
valuation of, 15–006, 16–046
valuation of,
asset backing, 16–041, 16–042
dividend yield, 16–045
factors involved, 16–041, 16–042
percentage shareholdings, 16–043
price earnings ratio, 16–044
rights attaching, effect of, 16–042
unquoted company, in, 15–006, 16–046
value shifting, 16–015
Sheep,
farming methods, 1–010, 1–011
hill, 1–012
intensive rearing, 1–011
intervention scheme, 1–013
sale of, 1–013
specialist breeders, 1–012
wood production, 1–014
Sites of Special Scientific Interest,
compensation for, 13–028
designation of, 13–027, 13–028
management agreements. *See* **Management
Agreements.**
objection to designation, 13–028
Sole Trader,
accounting date, change of, 2–013
accounts, optimum date for, 2–008
advantages of, 2–001
assessment,
cessation of trade, on, 2–005—2–007
optimum date for accounts, 2–008
penultimate and ante-penultimate years,
2–009, 2–010
preceding year, basis on, 2–002—2–012